"*Geographies of Global Change* is a remarkable achievement. It provides the first comprehensive and broadly accessible treatment of global change and is one of those rare books that should command a broad intellectual readership. The essays are very clear and well written, and highlight the central geographical issues and debates, covering the political, social, cultural and environmental aspects of geographical dimensions of global change. There is no better text for helping to grasp the breadth of issues implied by 'global change', and for getting a sense of what needs to be done."

Neil Smith
Rutgers University

"The book captures the complexities of important issues but presents them in an easy, straight-forward manner. It confronts globalization in a critical but constructive way and powerfully unveils contemporary trends and processes that challenge traditional static views of world processes. The chapters are superbly written and amply illustrated."

David Wilson
University of Illinois

"The book is a remarkably coherent collection and altogether a significant accomplishment. It is notable for the high standards achieved by the individual contributions and also for the contemporary relevance of the arguments marshaled. Accessible and informative, it should be indispensable reading for *every* geography major."

Kevin R. Cox
Ohio State University

"At last, a comprehensive text that brings together the very best ideas and literatures of the 1990s to address contemporary processes of social and spatial change. *Geographies of Global Change* is a *tour de force* – wonderfully rich and invigorating mapping of late modern geographies; essential reading for anyone striving to understand the complexity and diversity of the contemporary world at the end of the twentieth century.

Geographies of Global Change is clearly written, rigorously argued, and gripping reading. It redefines what we mean by a 'textbook' and sets new standards for teachers and students alike. Few other textbooks deal with complex issues and ideas in such depth and clarity, and few others have been able to provide such a rich array of interesting interpretations of socio-spatial change and the dizzying experience of hypermodernity. *Geographies of Global Change* will be a boon to teachers and lecturers and a goldmine for undergraduate and graduate students seeking a clearer understanding of the diversity and richness that typifies much of contemporary geography."

John Pickles
The University of Kentucky

Geographies of Global Change

Remapping the World in the Late Twentieth Century

Edited by

R. J. Johnston,
Peter J. Taylor and
Michael J. Watts

BLACKWELL
Oxford UK & Cambridge USA

First published 1995

Blackwell Publishers Ltd
108 Cowley Road
Oxford OX4 1JF

Blackwell Publishers Inc.
238 Main Street
Cambridge, Massachusetts 02142
USA

British Library Cataloguing in Publication Data

A CIP catalogue record for this book is available from the British Library.

Library of Congress Cataloging-in-Publication Data

Geographies of global change: remapping the world in the late twentieth century / edited by R. J. Johnston, Peter J. Taylor, and Michael J. Watts.
p. cm.
Includes bibliographical references and index.
ISBN 0-631-19326-X (acid-free) — ISBN 0-631-19327-8 (pbk. : acid-free)
1. Geography. I. Johnston, R. J. (Ronald John) II. Taylor, Peter J. (Peter James), 1944– . III. Watts, Michael.
G128. G474 1995
910—dc20 94-26900
CIP

Typeset in 11 on 13 pt Sabon
by Graphicraft Ltd, Hong Kong
Printed in Great Britain by Hartnolls Ltd, Bodmin, Cornwall
This book is printed on acid-free paper.

Contents

List of Plates

List of Figures

List of Tables

Preface

We have designed a book that engages with globalization from a specifically geographical perspective. The latter entails a broad overview of contemporary social practices and we have attempted to produce a reasonably comprehensive guide to global change. To this end we commissioned twenty relatively short chapters that cover the five major areas of change that we identify: economic, political, social, cultural and environmental. We are grateful to our contributors, who were given big tasks in relatively little space and responded splendidly to the challenge. However good the concept of the book, ultimately its success rests upon the quality of the content and on this count our authors have done us proud. We think the end product is an informative and lively text for students which will enable them to understand better the world they live in.

Acknowledgments

The editors and publisher gratefully acknowledge the following for permission to reproduce copyright material:

Table 15.1, from the report commissioned by the ITV Network Association, from the study by Booz, Allen·Hamilton; figure 3.2, reproduced by permission of Oxford University Press, from N. Thrift and A. Amin, *Globalisation – Institutions and Economic Prospects*; table 4.1, reproduced by permission of Guilford Press, from P. Dicken, *Global Shift: the Internationalization of Economic Activity*; table 14.2, reproduced by permission of Cambridge University Press, from Paul Knox and Peter J. Taylor (eds), *World Cities in a World-System*; figure 18.3 reproduced by permission of Routledge, originally published in Mary Douglas, *Risk and Blame: Essays in Cultural Theory* (Routledge 1992); figures 18.1 and 18.2, tables 18.1–18.3, reproduced by permission of Cambridge University Press, from B. L. Turner II et al., *The Earth Transformed by Human Action*; material from various chapters reproduced from *1993 World Investment Report*, by permission of the United Nations, Publications Board.

The publisher apologizes for any errors or omissions in the above list and would be grateful to be notified of any corrections that should be incorporated in the next edition or reprint of this book.

1

Global Change at the End of the Twentieth Century

Peter J. Taylor, Michael J. Watts and R. J. Johnston

Introduction

What sort of a world do we live in? For most of the readers of this book their world is a comfortable one. Consumers of textbooks are typically drawn from that minority of the world's population who are able to live relatively affluent lives. But what of the rest, the majority of humanity? Here affluence is replaced by struggles for survival of various forms where textbooks and many other "normal" commodities of the comfortable world are an unnecessary luxury. But these are not two separate worlds, as they are sometimes conveniently portrayed. "Worlds of comfort" and "worlds of struggle" are interweaved in complicated geographies as they penetrate one another's spaces in ever-increasing ways. Hence, as we come to the end of the twentieth century the sense that we all live in "one world" has probably never been stronger: "their" struggle is related to "our" comfort.

To say we live in an interconnected "one world" is only a very partial answer to the question we began with. What is "in" this one world, what is it made up of? One way to begin is to look at the mass media, perhaps the most overt expression of the "globalization" of our world. What is deemed newsworthy by media élites gives a particularly relevant description of our world because it provides the day-to-day window through which we look beyond our own direct experiences. In what follows, a selection of news stories which appeared in the days between Christmas Day and New Year's Eve 1993 are briefly listed. The world we find has a mixture of hope, concern and despair. The "big stories" concerned hopes for peace in Israel/Palestine, South Africa and Northern Ireland. But in the "run of the mill" stories there was much less hope, with many regions seemingly vying with one another to take the place of the above as "intractable" political problems. In Liberia gunmen were attacking aid centers, in the Congo there was "ethnic" violence, we were reminded of Somalia by a bomb blast in Mogadishu, while all the time Sarajevo was being shelled during the failed Bosnian Christmas truce. But violence was not limited to these civil wars, it appeared in "religious" stories: eight tourists were wounded in a gun attack on a bus by Muslims in Egypt, and after a bomb attack on a Roman Catholic cathedral in the Philippines two mosques suffered grenade attacks. In addition there was a row between the Anglican Church and the Muslim government of the Sudan, while on a more peaceful note a giant statue of Buddha was unveiled on a hill overlooking Hong Kong. There were also stories about language: the Serbs, Croats and Bosnian Muslims are trying to prove that they speak different languages rather than a single Serbo-Croat language as previously assumed, while the French Assembly has passed a new law forcing French radio to play at least 40 percent of pop songs in the French language. And the *Wall Street Journal* has decided to bring out a Polish-language edition. These cultural concerns are matched by political stories relating to international travel: British immigration officials target a flight from Jamaica; Fidel Castro's daughter seeks asylum in the USA; and the recent Russian election winner, the nationalist Zhirinovsky, is expelled from Bulgaria and refused entry into Germany.

Environmental stories are also popular: we learn about a threat to rhinos in Africa; of evacuation of French and Spanish beaches because detonators are being washed ashore; that 15 percent of Russian territory is an "environmental disaster zone" with 100,000 people living where radiation levels are too high; and that the US government admits having carried out secret radiation tests on its citizens. Perhaps most disturbingly we hear from the head of the

China National Nuclear Corporation that "Nuclear industry and technology are a yard stick for measuring the overall strength of a country" as part of his promotion to export nuclear power stations. Finally, for light relief, Euro Disney is thinking of changing its name to Paris Disney and there is a report of claims by World Bank economists that economic success in a country is largely a matter of "luck."

What a *pot pourri* of stories – what can we make of this diverse content of newsworthy items over just a few days? The first point is that these are not presented as exotic tales, like travellers' stories in another era. The world no longer consists of "faraway places of which we know little" but is most definitely constructed for us as one world. There are interconnections implied simply by the selection of stories for our consumption. Muslim gunmen, Russian nationalists, and Chinese technocrats, in their quite different ways, are seen as potential threats to our comfortable lives, for example. But the connections are by no means clear between, say, some of the cultural items and the environmental issues that were reported. It is the prime purpose of this collection of essays to investigate such connections so as to help readers make sense of their world as a global whole. After reading this volume a student should be able first to relate her or his life experiences to the wider world and secondly to see that world as a diverse unity. The aim of this introduction is to set the scene for this task by broaching some key questions for a systematic study of our world that cannot be gleaned from cumulative intake of media stories, however thorough. We focus on our title, first introducing questions concerning the nature of contemporary global change; and then we enquire into the nature of the geographies which are both a result of, and contribute to, this dynamic world. In the twenty subsequent systematic essays these questions will be tackled in a wide variety of ways and contexts to fulfill the book's purpose. We do not claim to find answers for all the important questions that confront us as members of one world – to claim so would be foolhardy in the extreme – but the remainder of the book does provide materials out of which a good understanding can be obtained so that the questions will begin to look that much less daunting.

All Change!

As we come to the end of the twentieth century there does seem to be a global ambience of pervasive change. People in all walks of life in all regions are confronting an uncertainty about present conditions and where they are leading. In other words recent past experiences are becoming less and less a sound basis upon which to plan our actions. But we must be careful in interpreting this contemporary claim to be particularly changeable. Social change in all its many manifestations is a normal feature of our social formation and has been so for many centuries: in Marx's memorable phrase, for societies built upon capitalism "all that is solid melts into air." So what is so different about today? The claim is that contemporary change is both quantitatively and qualitatively distinctive. It is argued that cumulative "normal" change has come to a head to produce an unprecedented degree of change in our times. In addition, this "speeding up" of our world has created a global scope as never before, making contemporary globalization a qualitatively new phenomenon. Both of these positions are contentious but must be treated seriously.

The most impressive feature of the quantitative claim to our living in "new times" is the sheer range of acute changes that are being experienced. The whole gamut of social activities is being turned upside down leaving no nook untouched. In economic processes the certainty of the Fordist regime of accumulation based upon large vertically-integrated companies in loose alliance with their labor force has given way to alternative organization, although the nature of "post-Fordism" is not yet fully clear. Certainly it will include the antithesis of Fordism in a more "flexible" regime with a more polarized workforce. In the realm of politics the changes have been equally dramatic, with various forms of dismantling of welfare states in the richer countries and the virtual abolition of state responsibility for its citizens in the rest of the world. This has been particularly painful in what were the communist states before the end of the Cold War. The demise of the latter has totally transformed international relations, but hopes for a "new world order" under US tutelage were short-lived as what has been ironically dubbed a "new world disorder" emerged instead. People have responded to the uncertainty and chaos in different ways across the world. Where rural economies have collapsed, "mega-cities" are forming on an unprecedented demographic scale while others are forced to become border refuges, again to a degree hitherto unknown. More conventional international migration ranges from economic élites moving between world cities, the control centers of capitalism, to "economic" refugees attempting to cross the "poverty curtain" separating the rich countries from the rest. But there are also cultural resistances to these changes which in fine dialectical fashion add their own impetus to the new dynamism. Emerging from what had been considered to be a secularizing world we have witnessed a remarkable revival of religious fervor in all regions of the world. Hindus in South Asia, Muslims in the Middle East and Christians in the Americas are the three largest regional movements toward their respective "fundamentalisms." But at the same time the supposed reasons for the "success" of the "West" – markets and democracy – have never had such widespread support outside their core regions. And through all this social turmoil the Earth as a life support system has been discovered, or rather our threats to it in that role have become recognized. Culminating with the "Earth Summit" in Rio de Janeiro in 1992, our uncertainty now extends to whether we will even have a world that we can live in in the future. It is this massive economic, political, social, cultural, and environmental change simultaneously burdening our lives that has persuaded so many that we do indeed live in "new times."

All the above changes have one prominent feature in common: they are global in scope. At the beginning of the twentieth century the political geographer Halford Mackinder (1904) argued that we were entering a qualitatively different era since the whole world had come under the European orbit: henceforth international politics would have to operate in a "closed world". As we come to the end of the century this notion of one world has come to impinge upon all aspects of social activities and not just relations between states. New flexible accumulation in economic activities will often mean transfer of production across world regions, the political conflicts of the new world disorder can be watched nightly in our living rooms on the American news channel CNN, world cities lead the way in expressing new social polarization almost as microcosms of the whole world, "terrorism" reaches the parts ordinary military activity cannot reach and everybody's pollution contributes to making our world less liveable. This is the justification for the terminology employing the prefix "geo," meaning world, used below for ordering the chapters into five sections: as well as the familiar geopolitical and the increasingly popular geoeconomic, we have geosocial, geocultural and geoenvironmental sections. The sum of all these processes has come to be called globalization, implying a qualitative change from the state-centered human activities of previous eras: the new times are global times.

It is in this common global scale of operations that we can begin to see some of the connections between the various changes identified above. If we discount the possibility that it is a mere coincidence that these processes have become global at the same time – the unlikely model of parallel globalizations – then we move toward the idea that they constitute a single holistic movement. The big question is, of course, how do they fit together? Here we must be very careful. In fact it is easier to eliminate possibilities than to come up with generally acceptable answers. For instance we can be sure that globalization is not the result of a simple cause-and-effect model where we reduce the complexity to one prime explanatory factor. Such reductionism usually highlights the economic sphere as the source of change and since we have chosen to begin with the geoeconomic, below, this warrants some explication here. We adhere broadly to a materialist position in which the geoeconomic has a crucial role in creating globalization. But we do not believe this to be the "economic base" above which other human activities are mere "superstructure." This famous Marxist architectural metaphor is sorely wanting as a world model. Our geoeconomics has an important enabling role not least because of its expression as "Western" material success to be emulated elsewhere. But we do not subscribe to discredited diffusion models of "modernization" (Blaut 1994). Political, social and cultural activities are not merely responses to material differences; they are integral to the constitution of those differences since economics never takes place in a non-economic world. Hence as well as economics being implicated in, say, cultural movements of resistance, culture is equally implicated in the nature of economic complexes. One metaphor that captures some of this being and becoming is the notion of a great vortex purposively mixing up all these activities as they are sucked into a holistic globalization (Wallerstein 1984). The destructive implications of this metaphor bring us starkly to the geoenvironmental question of whether the Earth's fragile ecology can be sustained. In fact, rather than arguing an economic reductionist position, we would suggest that it is this first ever secular glimpse of the end of the world which makes our new global times qualitatively different from the past.

Our organization of this volume into five sections could be interpreted as developing a global perspective on five "adjectival" geographies: economic geography, political geography, social geography, cultural geography and the geography of the environment. No doubt some people will use the essays that way, which is fine, but it is not how we intend them. These divisions go against the spirit of holism which is central to our argument and, in our view, is geography's only legitimate *raison d'être* in any case. Our organization is a pragmatic one, putting together similar topics for pedagogic reasons, but with no fundamental disciplinary divisions implied. Nonetheless this is a book of multiple geographies, albeit not in the usual adjectival way, and they constitute our way of unravelling the globalization vortex.

Multiple Geographies

One of the major intellectual developments in social science's attempts to grapple with the new global times has been a better understanding of the role of space and time in constituting social formations. Space and time had for long been treated in a taken-for-granted manner as "containers" of human activities in a totally inert way: space as a mere stage upon which events occurred, time as a mere medium we experience consecutively. The problematizing of both space and time in social analyses repositioned a previously marginal geography toward the center of the realm of social sciences (Gregory and Urry 1985). Good geographical analysis should be particularly sensitive to space and time. We add time to space here because it does not make sense to consider one without the other despite some traditional geographical claims to the contrary. Hence for us, "all geography is historical geography." Focusing on the contemporary can never mean ignoring the past; for our arguments here the whole notion of "new times" has presupposed comparison with the past as "non-new," for example. But bringing space and time together into our discussion does not mean pursuing the familiar analogy between the two, such as equating time periodization with space regionalization. Rather we need to consider both, in their many social guises as multiple spaces and multiple times. It is these concepts that generate the multiple geographies for tackling the globalization vortex.

Social spaces and social times have a variety of meanings; here we focus on their respective use as models of time spans and spatial scales. An extreme interpretation of contemporary globalization would treat contemporary human activities as representing a new epoch for humanity. This is to focus upon a particular space–time analysis that privileges the global scale and the long-term perspective – Braudel's (1980) *longue durée* – in understanding where we are today. A more sensitive analysis, as promoted in this volume, does not neglect other scales and spans that continue to exist however powerful globalization might be or even become. As we begin to consider other spans and scales we move along a route toward doubting the uniqueness of our new global times. In other words we locate the present into a context where globalization is more a continuation than something new, more increasing in scale than specially global. For this initial discussion we identify three time spans and three spatial scales.

For Braudel (1980), the opposite to *longue durée* is the eventism with which we began this introduction, using newspaper stories. This is a short-term view of social time where the world consists of multiple events. Some events are more important than others, of course, and some aspire to be crucial, the turning points of history. Obviously, earlier this century, the stock market crash of 1929 was an event where before and after could not be the same. For our times the revolutions of 1989 and their aftermath represent another "world shock." As well as marking the end of the Cold War, these events ushered in a new world in the sense that the demise of communism touched much more than international rivalry. Despite all the things that go to make up globalization having been readily apparent before 1989, it is only after the revolutions that a notion of new times becomes generally accepted as normal. But in its turn the shock of 1989 can be located within medium-term time spans, with cyclical analyses of social change. Long economic waves of approximately half a century, Kondratieff cycles, fit the post-World War II period quite neatly as a growth phase – the "postwar boom" to *c.* 1970 – followed by a stagnation phase to the present. In this context our new times are an example of restructuring which occurs regularly in the wave-like progression of capitalism. This relates also to even longer "hegemonic cycles" when one country dominates the world-economy as the USA did in mid-century. Our new times, therefore, also represent decline from such hegemonic heights, again a process not unique to the present. But the point is not to contrast the different social spans; rather we try and understand how they interact: introducing the short and medium-term spans does not disprove the existence of singular new global times – logically all times have all spans – but it does sharpen our awareness of how we should consider the question.

Returning to the spatial, the opposite to global is local, which is again where events take place. As with spans, so with scales a sensitive analysis does not argue for one against another but focuses upon their relations. Local communities may be buffeted by global forces but they are not helpless victims with no coping strategies. However, neither can they be autonomous of the world they inhabit, so that their strategies will invariably involve consequences beyond their direct control. In this case geographers deal with a local–global dialectic, where local events constitute global structures which then impinge on local events in an iterative continuum (Watts 1992a). In addition, in terms of scales, the national scale of the states represents the medium level. This is an important locus of formal power wherein exists the capacity to affect both local communities and the wider international scene. Globalization is hypothesized as a transfer of real power "upwards" from states to the global, but a proper treatment of scales can modify such a simple notion. Despite many predictions concerning their demise, states are still around, acting as dangerously as ever, and any reasonable interpretation of our present must make room for some degree of continued salience for states (Taylor

1994). Again this is not to undermine the idea of new global times – different spatial scales will exist in every society above small isolated communities – but it does sharpen our perspective on the question once again.

The really difficult matter is trying to combine scales and spans within a single analysis. This is where using globalization as our starting point, in its fusion of *longue durée* with the global, may help. There are multiple geographies – historical geographies – to be constructed at different spans and scales which together will help understand the vortex of globalization, which often seems to blind us by its power. The essays in this volume, in their very many different ways, provide perspectives on the question of the nature of our present predicaments through multifarious geographies. They will provide some answers, but not always consistently, to the questions we have raised in this opening chapter but they will also take the arguments much further. In the best tradition of geography, this collective effort aspires to make sense of our complex and ever-changing world.

PART I

Geoeconomic Change

Introduction to Part I: The Reconfiguration of Late Twentieth-Century Capitalism

> *Is it any wonder that buckets of foreign money – around $40 billion, or one quarter of all cross border equity investment – continues to slosh into the emerging stockmarkets of the world's poorer countries?* The Economist, *January 8, 1994, p. 16*

When the history of the late twentieth century is written, there seems little doubt that mobility – of capital, labor and meaning – will be one of its touchstones. The ceaseless search for profitability within the interstices of a world market has propelled a radical restructuring of national economies around the world. The global corporation dominates the economic iconography of the post-1945 period. Its leitmotifs are speed, innovation, progress, and what geographers have come to call space–time compression (Harvey 1990b). Joseph Schumpeter's (1952) invocation of creative destruction as the "essential fact" of capitalism provides us with a powerful image to understand the geo-economic restructuring wrought by capitalist impulses amidst a world-wide rhetoric of monetarism, laissez-faire, deregulation and market triumphalism.

By the early 1990s there were 37,000 transnational corporations (TNCs) in the world with over 170,000 foreign affiliates (UN 1993). The total stock of foreign direct investment (FDI) accounted for by this universe of corporations was in excess of $2 trillion in 1992, and totalled over $5.5 trillion in worldwide sales. Transnational capital is, of course, highly concentrated geographically, sectorally and in terms of the share of foreign assets controlled by the largest firms; 90 percent of TNCs are headquartered in the advanced capitalist states (five major home countries account for over half of the developed country total), and the triad of Japan, North America and Western Europe produced 70 percent of global foreign investment inflows and 96 percent of global outflows in 1991. Roughly 1 percent of parent TNCs own half of the total FDI stock and the largest 100 TNCs account for 14 percent of the total world stock of outward investment.

Since the mid-1970s, flows of foreign investment have followed an upward trend averaging 13 percent per annum. Commonly understood in terms of the seemingly unstoppable globalization or "transnationalization" of the world economy, there have been in fact two striking foreign investment surges, one from 1978 to 1981 (growth averaged 15 percent per year) and another between 1986 and 1990 (averaging an astonishing 28 percent per annum). The unprecedented increase after 1985 was unquestionably facilitated by a period of economic growth in the wake of the recession of the early 1980s and by a period of intense mergers and acquisitions, but there were also long-term structural forces at work. In particular the accumulation of FDI stock since the first oil boom, the proliferation of integrated international production ("the world car"), and the radical changes in the macro-ideological environment which, under the auspices of monetarism in the core and the growing hegemony of the International Monetary Fund and the World Bank at the periphery, precipitated extensive trade liberation, exchange rate reform and privatization (particularly in Western Europe, Latin America and increasingly in the former COMECON countries). The virtual disappearance of nationalization since 1974 and the explosion of privatization – in essence the selling off of state properties and enterprises – has been especially striking. Notwithstanding the recent GATT (General Agreement on Tariffs and Trade) agreement, 79 new legislative measures adopted during 1992 in 43 countries liberalized the rules on FDI (ibid.).

The rapid growth of FDI has been accompanied by important shifts in its sectoral composition. During the 1950s foreign investment was concentrated in primary and resource-based manufacturing. By the 1990s services accounted for close to 50 percent of all FDI and they absorbed almost two-thirds of annual flows. While integration and capital mobility is proceeding at different rates across different industries and functions, financial services is probably the most global of corporation activities, stimulated by electronic transfers and 24-hour trading. The rise of an integrated global agrifood system, propelled by recent regulatory changes imposed by NAFTA (North American Free Trade Agreement) and GATT, confirms, however, that all sectors have felt the press of global capital flows. Indeed, transnational capital in the food preparation and processing sector has emerged as one of the most dynamic fronts along which the market is opening up the former Soviet sphere. Nonetheless, hypermobility of capital and frictionless space is hardly the norm, and a truly global research and development and manufacturing system remains restricted to a relatively small number of firms in limited sectors and branches.

Worldwide FDI flows have, however, slowed since 1991 – for the first time in fact since 1982 – largely as a result of contractions by Japanese and West European business operations. Nonetheless, there is still substantial scope for further growth and, as the quotation from *The Economist* cited at the outset of this discussion suggests, the inflow of foreign investment into the developing world has been especially dynamic. Developing countries received 25 percent of all inflows in 1991 (equal to their share in the first half of the 1980s!); the figure grew substantially in 1992. Inflows to East and Southeast Asia, and increasingly in Latin America, have been especially dramatic. Investment flows into Asia and the Pacific rose by almost 10 percent in 1992 and investment in East and Central Europe, while spatially uneven, has been singularly impressive. The transnationalization of capital has been, in this respect, a propulsive force in the rapid industrialization of some newly industrializing states (most recently in Thailand, China and Malaysia, which represent a sort of second wave of newly industrialized countries (NICs) following on the heels of Taiwan, Singapore, South Korea and Hong Kong). During their industrial revolutions of the nineteenth century, Britain and the US took a half century to double their incomes; the Asian Tigers and their new companions are achieving this within a decade. Whether this growing heterogeneity within the less developed world, and the rapidity of industrialization in particular, warrants the sort of optimism expressed by *The Economist* (8 January 1994, p. 16) however – "rapid changes in the balance of the world economy" and the NICs having "consigned to the rubbish bin the old notion that the rich world . . . towers over the whole world economy" – is another matter entirely.

Within the broad parameters of geoeconomic restructuring, capital flows and their effects are nonetheless highly uneven. Some parts of the South – sub-Saharan Africa in particular – almost slid off the economic map during the 1980s. Indeed, the impact of structural adjustment and liberalization during the 1980s produced horrifying contraction and austerity in large parts of the Third World, with little evidence of subsequent economic recovery and growth. In sub-Saharan Africa, economic output and exports tumbled by 30 percent between 1980 and 1986; private net transfers declined from a positive inflow of $2.5 billion in 1980–2 to a net *outflow* of $7 billion in 1985–7. The standard of living of the "average African" is, according to the World Bank, lower now than at independence. Latin America did not fare much better (Watts 1991a). So-called "shock therapy" in Russia and Eastern Europe has also taken its toll, reflected in the terrifying rates of unemployment, plant closures and economic insecurity. Life expectancy in Russia dropped from 62 to 59 years over twelve months, the largest single year drop ever recorded in a developed country, and the death rate soared by 20 percent in 1992. Plagued by inflation and economic chaos, Russia looks increasingly like an archetypical Third World state. More than anything, the demise of the former socialisms reveals that markets cannot simply be wished into existence but have to be *constructed*. Often they are not, and what emerges from the ashes of socialism is not a free market wonderland but what has been referred to as market Stalinism. Market liberalization as often as not produces illiberal politics, and not infrequently wildly distorted markets.

In short, amidst the euphoria of liberalization and global economic restructuring driven by new trade agreements and international capital mobility, the end of the Cold War looks as though it has been replaced by new forms of North–South dependency. According to the United Nations Development Program, global polarization of income and wealth has increased substantially; between 1960 and 1990 country differentials between the wealthiest and the poorest 20 percent increased from 30 to more than 60, while real disparities between people, as opposed to country averages, became even more stark (UNDP 1993). Since the Third World was in one sense the creation of the Cold War, the New World Order implies, as some have suggested, the end of the Third World. However, it has been replaced, after a decade of severe discipline meted out by the global regulatory agencies, by a map of North–South dependency which conveys an extremely bleak prospect for large parts of the global economy (Arnold 1993; Bello 1994).

The costs of the economic restructuring during the 1980s are no less in evidence among the advanced capitalist states, of course. Unemployment in the OECD (Organization for Economic Cooperation and Development) states stands at a historic high, while in different ways and at different velocities the postwar Fordist arrangements – as corporatist systems of production, consumption, and regulation – have been dismantled and reconfigured (Sayer and Walker 1992). High monetarism and neoliberal orthodoxy perpetrated by Kohl, Thatcher, and Reagan have eroded the political economic landscapes of both the EU and North America. Whether this represents a shift to a distinctively new system of post-Fordist or flexible accumulation is a matter of intense debate (Scott and Storper 1993). What is less arguable is that the shape and character of national economies, and the scope and nature of public intervention, have been radically altered since the mid-1970s. What new sorts of social contracts and private–public networks will emerge in the twenty-first century is anyone's guess. What seems clear, however, in debates over economic modernization in the North and South alike, is that the uncritical adoption of simple market or state models no longer carries much cachet. And to this extent the role of civil society – the third pole of economic development – in reviving flagging economies is drawing considerable attention (Putnam 1993a). None of this should imply that new forms of regional trade integration or deregulation (NAFTA, EEC, GATT) or integrated international production are of no consequence. But it would also be foolhardy to assume that the current trend toward further deregulation, free trade, regional integration and hypermobile capital are historical inevitabilities. Fin de siècle capitalism cannot be so readily captured and contained.

2

A Hyperactive World

Nigel Thrift

Introduction

That light overhead, the one gliding slowly through the night sky, is a telecommunications satellite. In miniature, it contains the three main themes of this chapter. Through it, millions of messages are being passed back and forth. Because of it, money capital seems to have become an elemental force, blowing backwards and forwards across the globe. As a result of innovations like it, the world is shrinking – many places seem closer together than they once did.

The satellite is itself a sign of a world whose economies, societies and cultures are becoming ever more closely intertwined – a process which usually goes under the name of *globalization* (Giddens 1991). But what sense can we make of this process of globalization? Again, the satellite provides some clues. Those millions of messages signify a fundamental problem of *representation*. Simply put, the world is becoming so complexly interconnected that some have begun to doubt its very *legibility*. The swash of money capital registering in the circuits of the satellite comes to signify the "hypermobility" of a new *space of flows*. In this space of flows, money capital has become like a hyperactive child, unable to keep still even for a second. Finally, the shrinking world that innovations like the satellite have helped to bring about is signified by *time–space compression*. Places are moving closer together in electronic space and, because of transport innovations, in physical space too. These three simple themes of legibility, the space of flows, and time–space compression can therefore be seen as "barometers of modernity" (Descombes 1993), big ideas about what makes our modern world "modern."

In the first part of this chapter, I will examine these three barometers carefully because they inform so much current discussion about modern global geographies. My purpose is simple. It is to show that they are partial accounts made into a whole. In the second part of this chapter I hope to show, through a double-take on the world of international money, just how partial these accounts are, even in that sphere of the world's economy which might be expected most closely to approximate to them. Then in the last part of the chapter I want to propose the beginnings of a more moderate account of the signs of the times.

Preamble

I shall begin by taking a reading of the barometers of modernity. I will do this by examining illegibility, hypermobility and time–space compression through the writings of the three authors who have most often deployed them; Fredric Jameson, Manuel Castells and David Harvey.

Each of these authors paints a picture of the world as having come under the sway of a new form of capitalism – whether it is called late or multinational or informational or global capitalism. This form of capitalism usually involves a combination of ingredients, and most especially: the accelerated internationalization of economic processes; a frenetic international financial system; the use of new information technologies; new kinds of production; different modes of state intervention; and the increasing involvement of culture as a factor in and of production. The three authors disagree on how swiftly the new form of capitalism has taken hold of the world. For Harvey, for example, it has been a comparatively rapid process. For Jameson, in contrast, the new form has crept up on us; "we have gone through a transformation of the life world which is somehow decisive but incomparable with the older convulsions of modernisation and industrialisation, less perceptive and dramatic, somehow, but more permanent precisely because more thoroughgoing and all-pervasive" (Jameson 1991, p. xxi). But each of the three authors agrees that at the heart of this change has been *space*. The new system of capitalism attacks and suppresses distance, *and* our notions of distance, producing a new global economic space in which global capitalism can play.

These three cartographers of global capitalism map its presence in different ways. They use different means of locating its presence. For Fredric Jameson, one of the defining characteristics of global capitalism is simply that it is hard to locate. Global capitalism's labyrinthine complexity makes everything less and less legible, less and less easily read. This is a world where electronically generated images prevail. As a result it is a world which is increasingly difficult to touch or tie down; a world of confused senses which can only sense confusion. This brave new world of global capitalism has nefarious consequences. First, the *effects* of power are all too obvious – poverty, famine, war, disease – but the exact causes of oppression are more and more difficult to discern. The Four Horsemen of the Apocalypse still stalk the world but all we can see are the hoof-prints. Secondly, the modern city becomes suspended in a global space "in which people are unable to map (in their minds) either their own position or the urban totality in which they find themselves" (Jameson 1991, p. 51). Thus the observer experiences a kind of vertigo, a sense of an unseen abyss over which humanity teeters. Jameson's answer is to call for "new cognitive maps" that will help us to retreat from the edge.

The barometer of hypermobility is best expressed in the work of Manuel Castells, who identifies a new type of economic space – a mobile space of flows – which is the precondition for the coming into existence of a worldwide informational economy: "the enhancement of telecommunications has created the material infrastructure needed for the formation of a global economy, in a movement similar to that which lay behind the construction of the railways and the formation of national markets during the nineteenth century" (Castells 1993, p. 20). This space of flows "dominates the historically constructed space of places, as the logic of dominant organisations detaches itself from the social constraints of cultural identities and local societies through the powerful medium of information technologies" (Castells 1989, p. 6). Increasingly, in other words, electronic trade winds blow across the globe, creating a new economic atmosphere.

Finally, the barometer of time–space compression is nowadays usually associated with the work of David Harvey, who uses the idea in two main ways: first, to express a marked increase in the pace of life brought about by innovations like modern telecommunications and the effects that this has on the topology of human communication – a seeming collapse of space and time – and, secondly, to signal the subsequent upheaval in our experience and representation of space and time that this speed-up brings about; "time–space compression always exacts its toll on our capacity to grapple with the realities unfolding around us" (Harvey 1989, p. 306). Just as in Jameson, there is a call for cognitive maps that can be used to navigate "through a period of excessive ephemerality in the political and private as well as the social realm" (Harvey 1989, p. 306).

These three accounts share much in common. Each is concerned with drawing out the lineaments of fundamental transformations in economies, societies and cultures – and in the nature of time and space. Each displays a considerable degree of apprehension about our ability to comprehend these transformations. Each places much of the blame for this state of affairs on the roadrunner pace of modern life, which blurs our understanding. But each of them also believes that it is possible to find a theoretical space from which it is possible to look out and explain what is going on.

What ought we to make of these accounts? I want to make a critique in two stages. First of all, I want to suggest that these accounts are simply the latest manifestation of a tradition of thinking that goes back a long, long way into history. They need to be taken with "a large pinch of déjà vu" (Porter 1993, p. 16). Secondly, I want to suggest that these accounts are in danger of constructing global capitalism as a more abstract system than it actually is. I will want to show this by reference to the world of international money and finance.

Critique 1: Antique Barometers?

Each of the three accounts briefly outlined above has a long history. Indeed they go back so far in time that they may even have reached their historical sell-by date. To illustrate this point I will take each account in turn.

The debate over the idea that the world, as it becomes increasingly globalized, has become increasingly illegible has exact resonances in nineteenth-century reactions to the expanding metropolis as a disconcerting mixture of multiplicity, movement and decenteredness (Prendergast 1992). These reactions were threefold. There was nostalgia for the old days, a nostalgia which was not much more than a demand for a return to the more secure and hierarchical social taxonomies of the past where everyone could be located in their "proper place." There was the idea that the city could be controlled through new forms of visualization which systematically refused to see significant forms of difference, division, and conflict – as in many Impressionist paintings. Then, lastly, there was a simple flight from the city, back into a "rural" world of psychic peace and soothed subjectivities. In each case, these reactions still exist. Further, they still exact their charms – look only at the content of some modern television advertisements.

Again, the account of a space of flows is growing hoary with age. It dates from at least the eighteenth century, and ideas of "circulation" – of desires and letters circulating in the body of the nation state. But it comes into its own in the nineteenth century with the spread of the railway and then the telegraph (Thrift 1990). In France, for example, writers used it as a convention to describe the new spaces of continuous movement and circulation that were springing up as a result of these innovations. Modern life is drawn in terms of speed and flow – everything moves too fast. In the twentieth century, the innovations may change but the phenomenology of speed and flow remains much the same.

This same phenomenology can be found in the account of time–space compression. One side of Harvey's account, the annihilation of space by time, was a favorite meditation of the early Victorian writer: "it was the topos which the early nineteenth century used to describe the new situation into which the railroad placed natural space after depriving it of its hitherto absolute powers. Motion was no longer dependent on the conditions of natural space, but on a mechanical power that created its own new spatiality" (Schivelsbuch 1986, p. 10). An article published in the *Quarterly News* in 1839 exactly captures the sense of struggle that resulted:

> For instance, supposing that railroads, even at our present simmering rate of travelling, were to be suddenly established all over England, the whole population of the country would, speaking metaphorically, at once advance en masse, and place their chairs nearer to the fireside of their metropolis by two thirds of the time which now separates them from it: they would also sit nearer to one another by two thirds of the time which now respectively alienates them. If the rate were to be sufficiently accelerated, this process would be repeated: our harbours, our dockyards, our towns, the whole of our rural population, would again not only draw nearer to each other by two thirds, all would proportionally approach the national hearth. As distances were thus annihilated the surface of our country would, as it were, shrivel in size until it became not much bigger than one immense city. (cited in Schivelsbuch 1986, p. 34)

The idea of the annihilation of space by time was recycled by writers like Marx later in the nineteenth century, surfaced again in geography textbooks of the 1920s and 1930s, and was then resurrected once more in the geography of the 1960s and 1970s as the phenomenon of "time–space convergence." The other side of Harvey's account, the effects that the upheaval of our experience of space and time have on our powers of representation and, by implication, identity, has clear resonances with ideas that date from at least the eighteenth century that the increased pace of life would lead to a kind of general hysteria in society (Porter 1993); time–space compression leads to time–space depression. It is the kind of depiction of volatile, fragmented subjects for volatile, fragmented times which Virginia Woolf captured so brilliantly in the 1920s in her discussion of the "atomism of the city" which is staged not only as "a problem of perception" but also as one "which raised problems of identity" (Williams 1975, pp. 241–2);

> After twenty minutes the body and mind were like scraps of torn paper tumbling from a sack and, indeed, the process of motoring fast out of London so much resembles the chopping up small of

> identity which precedes unconsciousness and perhaps death itself that it is an open question in what sense Orlando can be said to have existed at the present moment. (Woolf 1926, cited in Prendergast 1992, p. 193)

Of course, just because these three accounts are starting to show their age does not mean that they are without a certain kind of narrative power. They make for a wonderful modernist detective story, full of mystery (illegibility), spirit (the space of flows), and pace (time–space compression). But do these three barometers of modernity convincingly represent the modern world? This is not a question that we can answer directly. What we can say instead is that they produce a partial representation which does not recognise its own partiality. Clearly the world has become more difficult to understand as it has become more and more complex. Certainly, there is an electronic space of flows. Of course, the world has speeded up. But because the world is more difficult to understand doesn't mean that nothing is understandable. The space of flows doesn't reach everyone. Speed isn't everything. The story is as much in what is missing from these accounts as in what is there.

These points can be made better by considering a real example. This example, which forms the second part of this chapter, is the world of international money. This is a world that is often regarded as the most telling example of a brave new world of flows: abstract, complex, instantaneous. But what do we find? Nothing of the kind.

Critique 2: Hooked on Speed

Take 1: masters of the universe?

An image which has become a cliché. A foreign exchange dealing room of a major bank in London, New York, or Tokyo. The mainly young men and women who inhabit these rooms for ten or eleven hours at a time are under pressure. They are under pressure to make profits. They are under pressure from their fellow traders – they don't want to lose face by screwing up a deal. They are under the pressure of constant surveillance – from managers, from video cameras, from tape recorders capturing all their calls. Above all, they are under pressure of time. Dealing itself is largely a matter of timing and dealers are expected not only to make profits from their deals but to make them quickly. You are only as good as your last deal.

To cap it all, these dealers are at the sharp end of time–space compression. Their world is a world where telecommunications have become more and more sophisticated and, as a result, space has virtually been annihilated by time. A dealer's world consists of a few immediate colleagues, the electronic screens which are the termini of electronic networks that reach round the world to other colleagues in other cities, and the electronic texts that can be read off the battery of screens (figure 2.1). If there is a space of flows, then this is it.

Certainly, what these dealers can do to the world is, in its own way, quite extraordinary. In any day's work of foreign exchange dealing, currencies are being transmitted backwards and forwards across the world by dealers to the tune of $100 billion in a day. This money is increasingly able to gainsay governments. As one dealer (cited in Kahn and Cooper 1993, p. 10) put it, "its a huge big global casino. If a government steps out of line they get their currency whacked." There are plenty of governments that can attest to this, as the example of the European Exchange Rate System (the ERM) shows only too well.

The ERM, introduced as a means of transforming the European Union into a zone of monetary stability in which the exchange rates of the different European national currencies would vary in an ordered and predictable way, came under an intense series of speculative assaults from foreign exchange traders over the course of 1992 and 1993.

> Highly mobile flows of money began making speculative assaults on currencies believed to be particularly vulnerable to devaluation. Vulnerable currencies were identified as those attached to national economies where the internal costs of retaining existing ERM parities were considered to be unsustainable. The aim was to anticipate a currency devaluation which would generate windfall gains for currency traders. As selling pressure in the foreign exchange markets forced the value of a currency down below its permitted floor value in the system, it was possible to turn a profit by selling

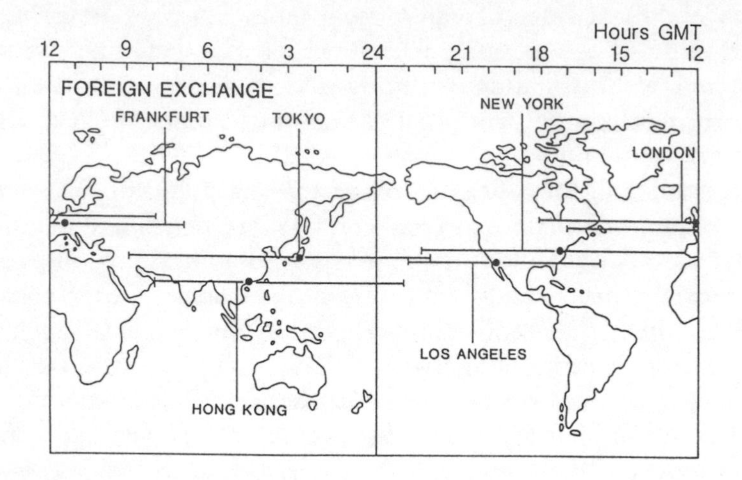

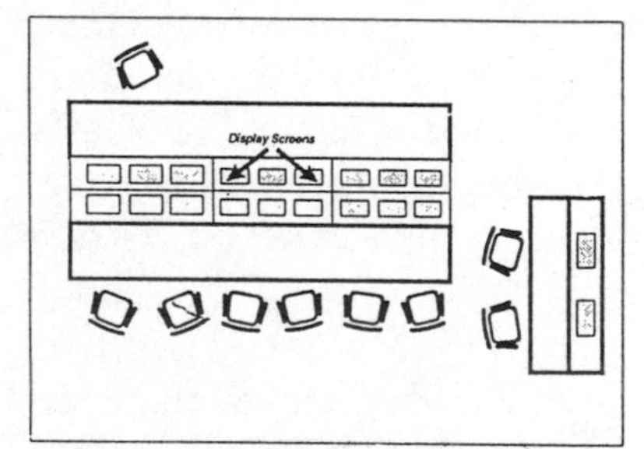

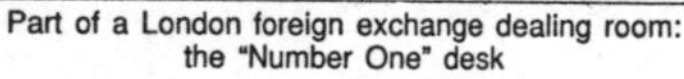

Part of a London foreign exchange dealing room: the "Number One" desk

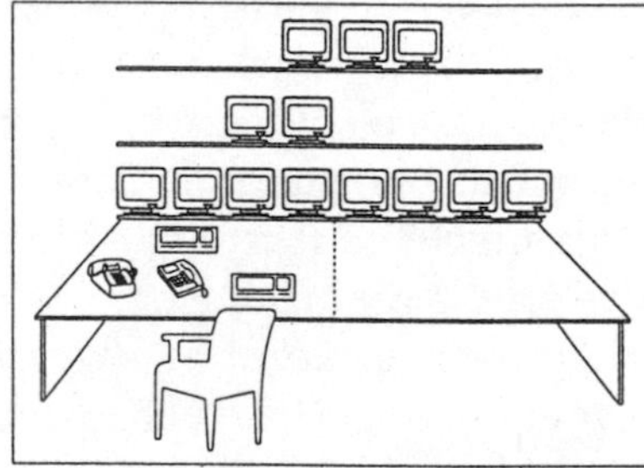

Displays available to the "Number One" desk

Figure 2.1 Foreign exchange trading.

> the currency back at a higher price to the EC central banks intervening in the market to defend the prevailing parities. If governments and central banks could be made to devalue a currency within the system, then sellers of currency would make a profit by selling it back at a far lower price. (Leyshon 1993, p. 1555)

The results of these speculative assaults were devastating. All but one ERM currency was devalued against the deutschmark, the anchor currency of the system; two currencies – sterling and the lira – were forced to leave the system altogether, and the rules of ERM membership were relaxed to such a degree that it became a pale shadow of its former self.

An example like this might be used to suggest that international money can flow where it will without let or hindrance – in other words, we see here a true space of flows. But like those apparently fraught young men and women in the dealing room (according to Kahn and Cooper 1993, they apparently actually suffer less stress than nurses dealing with the mentally handicapped), this is something of an exaggeration. First of all, the "phantom state" of international money is a nomad state (Thrift and Leyshon 1994). That is, it has no permanent spaces to call its own, only a series of transient sites in a few global cities. This constant mobility has its advantages. In particular, the world of international money is difficult to tie down. But it also has its disadvantages. The phantom state is always in danger of being trapped by nation states which control territories, and are able to regulate what goes on within them. Thus this space of flows can be choked off by the rules nation states impose – like capital adequacy ratios, which force banks to set aside a certain portion of their capital, or rules on how or what financial investments to use. Secondly, the phantom state has to be constantly in motion, chasing into all the nooks and crannies of the world economy that might produce a profit. Such a task requires an enormous investment of not only money but also communication. Nowadays money is essentially information and getting that information, interpreting it, and using it at the right time, requires constant human interaction. The result is that the hypermobile world of international money is actually a hypersocial world, a world of constant interchange between people, whether over electronic networks, or in face-to-face meetings, or at the end of often lengthy journeys. In this sense this world of flows is not abstract at all – it is the product of and it is produced by people communicating about what is going on.

So the barometers of illegibility, of a space of flows, and of time–space compression can clearly be seen to be partial, even when an example is chosen which should cast them in a favorable light. But there is one last point that needs to be made that repeats the historical lessons of the first section of this chapter: for the denizens of the world of international money these barometers do not represent some new condition. They are practiced in living with them. Since the international financial system has been in operation, its practitioners have had to live with

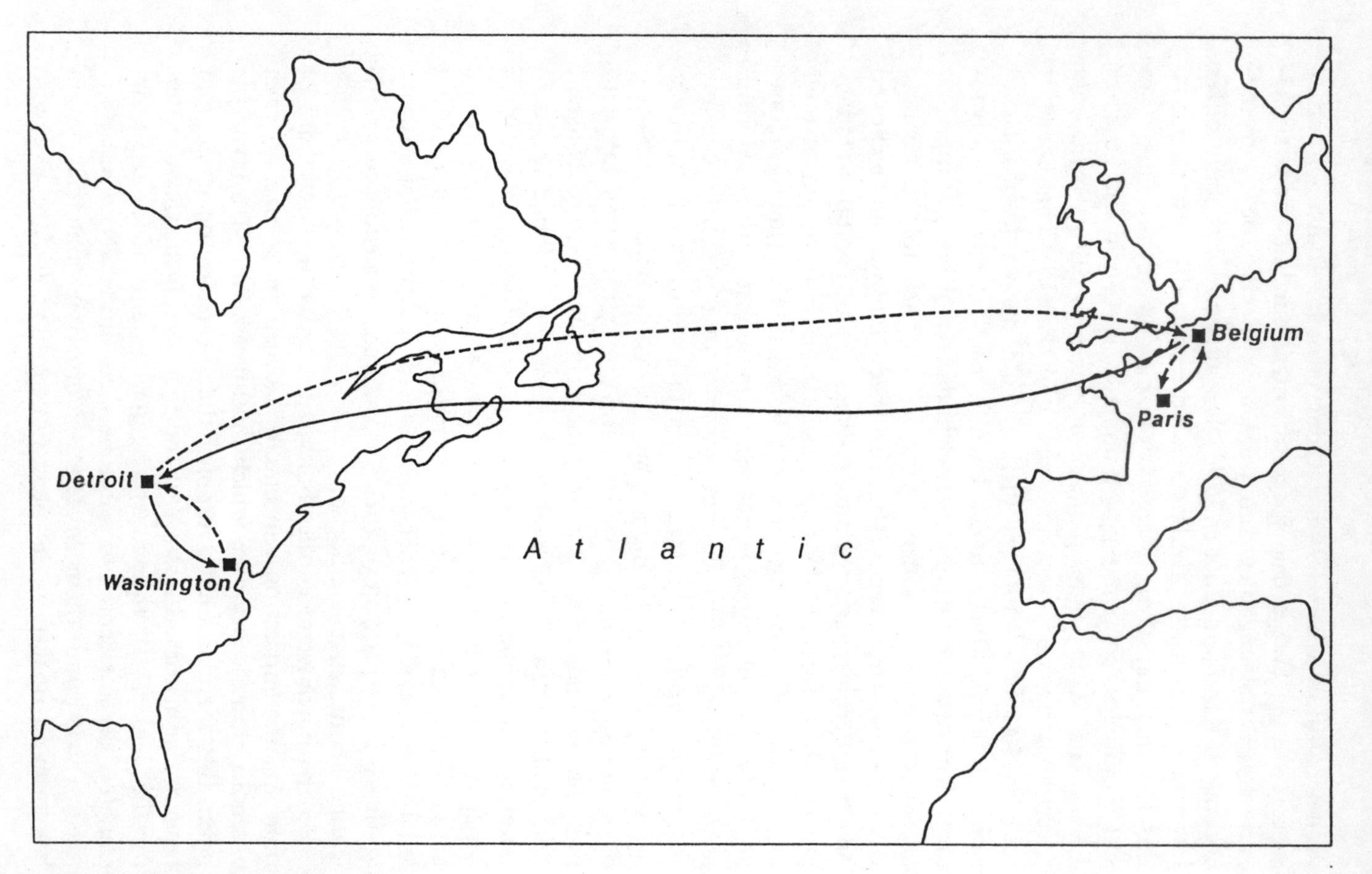

Figure 2.2 Paris to Detroit and back.

uncertainty, using only limited information to assess the risks they run in investing money. Since international financial markets started to coalesce in the late nineteenth century, because of the telegraph and then the telephone, their practitioners have become well versed in living with time–space compression. It is all part of the game they play every day and it is a game they are good at.

Take 2: networks and ghettoes

The dealers in the international financial system live life in the fast lane. But what about those of us waiting for the bus or cursing the late train? For us too, monetary transaction is speeding up. The installation of credit cards, automated teller machines (ATMs), and the like means that life in the slow lane is moving faster (figure 2.2).

> I'm in Paris, it's late evening, and I need money quickly. The bank I go to is closed . . . but outside is an ATM . . . I insert my ATM card from my branch in Washington DC and punch in my identification number and the amount of 500 francs, roughly equivalent to $300. The French bank's computers detect that it is not their card, so my request goes to the Cirrus system's inter-European switching centre in Belgium which detects that it is not a European card. The electronic message is then transmitted to the global switching centre in Detroit, which recognises that there's more than $300 in my account in Washington and deducts $300 plus a fee of $1.50. Then it's back to Detroit, to Belgium and to the Paris bank and its ATM and out comes $300 in French francs. Total elapsed time: 16 seconds.

Increasingly, we are all dependent on the speed and processing power of telecommunications that examples like this illustrate. But the new space of telecommunications is not, in reality, a smooth global space over which messages can flow without friction. It is a skein of *networks* which are "neither local nor global but are more or less long and more or less connected" (Latour 1993, p. 122). To think otherwise is to mistake length or connection for differences in scale level, to believe that some things (like people or ideas or situations) are "local" whilst others (like organizations or laws or rules) are "global."

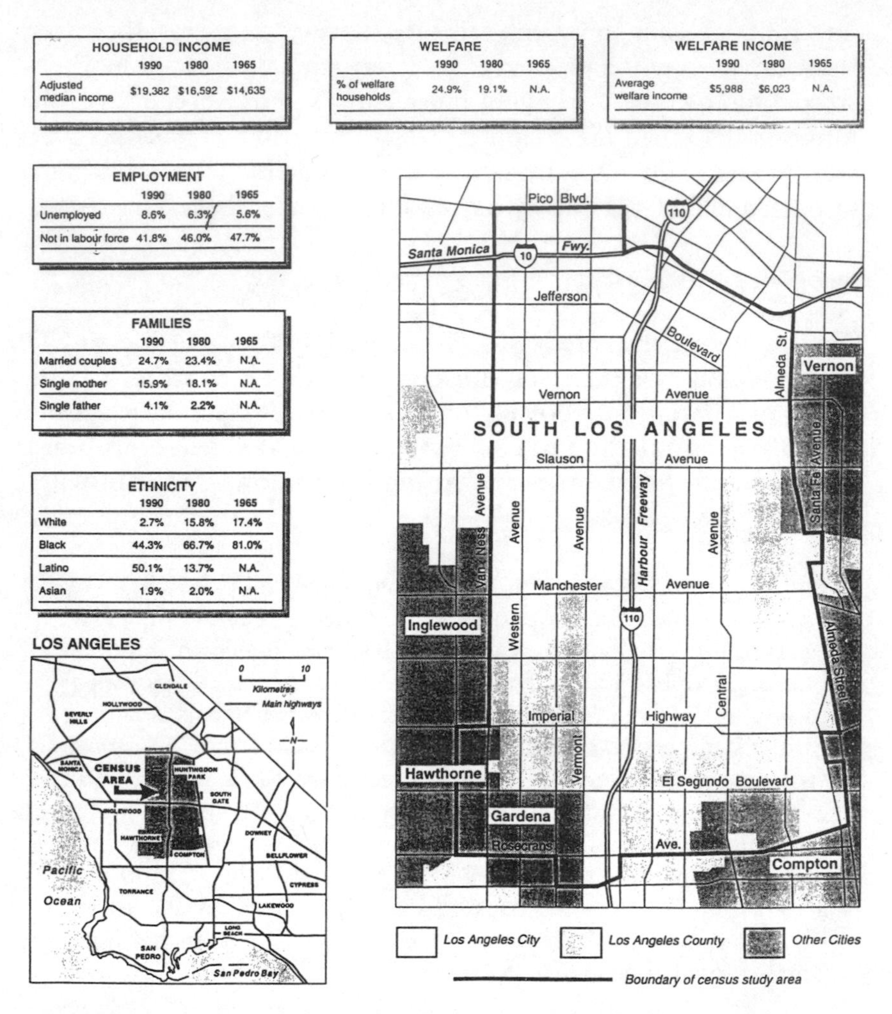

HOUSEHOLD INCOME	1990	1980	1965
Adjusted median income	$19,382	$16,592	$14,635

WELFARE	1990	1980	1965
% of welfare households	24.9%	19.1%	N.A.

WELFARE INCOME	1990	1980	1965
Average welfare income	$5,988	$6,023	N.A.

EMPLOYMENT	1990	1980	1965
Unemployed	8.6%	6.3%	5.6%
Not in labour force	41.8%	46.0%	47.7%

FAMILIES	1990	1980	1965
Married couples	24.7%	23.4%	N.A.
Single mother	15.9%	18.1%	N.A.
Single father	4.1%	2.2%	N.A.

ETHNICITY	1990	1980	1965
White	2.7%	15.8%	17.4%
Black	44.3%	66.7%	81.0%
Latino	50.1%	13.7%	N.A.
Asian	1.9%	2.0%	N.A.

Figure 2.3 South Central Los Angeles.

The late twentieth century is covered by a lattice of networks. Public and private, civil and military, open and closed, the networks carry an unimaginable volume of messages, conversations, images and commands. By the early 1990s, the world's population of 600 million telephones and 600 million television sets will have been joined by over 100 million computer workstations, tens of millions of home computers, fax machines, cellular phones and pagers. (Mulgan 1991, p. 6)

But when we see this world of telecommunications as indeed a world of networks, we can also see other things. First of all, telecommunications networks still rely on many hundreds of thousands of people and machines all around the world who build, monitor, repair, and use them, just like the navvies and the telegraph engineers did of old. Second, modern telecommunications networks may be hybrid systems, in which machines and people are increasingly mixed together in queer combinations, but this does not make them more abstract or more abstracted. They break down. They stutter. They pause. They make errors. Whether it is a case of a vandalized ATM, a line fault or an atmospheric disturbance, these networks are not self-sustaining. Third, these networks can be organized in different ways which are more or less effective (Mansell 1993). Fourth, not everyone is connected to these networks. The new telecommunications networks have produced "electronic ghettoes" (Davis 1992) in which the only signs of globalization that can be found flicker across television screens, endlessly mocking their viewers by producing in front of them the lives and possessions that they will never be able to obtain. This is not even life in the slow lane. It is life on the hard shoulder.

In the electronic ghettoes the space of flows comes to a full stop. Time–space compression means time to spare and the space to go nowhere. It is all horribly legible – as the example of the South Central area of Los Angeles shows only too well. There, access to normal monetary transactions and credit, represented by a network of facilities like ATMs, bank branches and the like is in decline as these facilities have been shut down by banks whose bottom line is under pressure, further fuelling South Central's economic decline (Dymski and Veitch 1992) (figure 2.3). Its inhabitants are forced back on an informal system of cheque-cashing services, mortgage brokers, credit unions, and cash. Yet just a few miles from South Central are the recently constructed corporate towers of Los Angeles's financial district, a place with all the necessary connections with telecommunications networks that are long and interconnected all around the world. It is a district that is nearer in this network time and space to New York, or London, or Tokyo than it is to South Central.

Conclusions

When we actually look at the space of flows, instead of taking it as a given, we find nothing very abstract or abstracted. What we find, as the example of the world of money shows only too well, is a new topology that makes it possible to go almost anywhere that networks reach (but without occupying anything but very narrow lines of force). But what we also find is a system where people are often in interaction with only four or five other people at a time – on a trading floor, or in a bank branch. What we find, in other words, are networks that are always both "global" and "local."

The problem is that writers like Jameson and Castells and Harvey, with their ideas of a global capitalist order typified by barometers such as an illegible globalization, a space of flows and time–space compression, are in constant danger of simply reproducing modernist views of an increasingly frantic commodified world filled with decontextualized rationalities like capitalism, various kinds of organizational bureaucracy and markets – rationalities which are soulless and relentless and shiveringly impersonal. But things, as they say, ain't necessarily like that;

> An organisation, a market, an institution are not supralunar objects made of a different matter from our poor local sublunar relations. The only difference stems from the fact that they are hybrid and have to mobilise a greater number of objects for their descriptions. The capitalism of Karl Marx or Fernand Braudel is not the total capitalism of (some) Marxists. It is a skein of longer networks that rather inadequately embrace a world on the basis of points that become centres of profit and calculation. In following it step by step, one never crosses the mysterious lines that should divide the local from the global. The organisation of American big business described by Alfred Chandler is not the organisation described by Kafka. It is a braid of networks materialised in order slips and flow charts, local procedures and special arrangements which permit it to spread to an entire continent so long as it does not cover that continent. One can follow the growth of an organisation in its entirety without ever changing (scale) levels and without ever discovering "decontextualised" rationality. . . . The markets . . . are indeed regulated and global, even though

> none of the causes of that regulation and that aggregation is itself either global or total. The aggregates are not made from some substance different from what they are aggregating. No visible or invisible hand suddenly descends to bring order to dispersed and chaotic individual atoms. (Latour 1993, pp. 121–2)

In other words, what we need to produce geographies of global change is less exaggeration and more moderation.

But what would the more modest accounts look like? There are three closely related ways in which we need to ring the changes. First of all, we need to change the way that we do theory. In particular, that means recognizing that large-scale changes are always complex and contingent. They must be seen as:

> a multiplicity of often minor processes, of different origin and scattered location, which overlap, repeat, or imitate one another, support one another, distinguish themselves from one another according to their domain of application, converge and gradually produce the blueprint of a general method. (Foucault 1977b, p. 138)

Secondly, we need to recognize that all networks of social relation, whether we are talking about capitalism, or firms, or markets, or any other institutions, are incomplete, tentative, and approximate. They are constantly in the process of ordering a somewhat intractable geography of different and often very diverse geographical contexts. "And ordering extends only so far into that geography. The very powerful learn this quickly" (Law 1994, p. 46). Accounts of these networks therefore need to recognize that it is a struggle to keep them at a particular size. There is no such thing as a scale. Rather, size is an uncertain effect generated by a network and its modes of interaction. A network of social relations has no natural tendency to be a particular size or operate at a particular scale; "some network configurations generate effects which, so long as everything is equal, last longer than others. So the tactics of ordering have to do, in general, with the construction of network arrangements that might last a little longer" (Law 1994, p. 103). Usually, this trick depends upon the invention and use of materials that can be easily carried about and retain their shape, "immutable mobiles" like writing, print, paper, money, a postal system, cartography, navigation, ocean-going vessels, cannons, gunpowder, xerographics, computing, and telephony. Electronic telecommunications can be interpreted as the latest of these mobiles, yet another means of allowing certain networks of social relations to retain their integrity by ordering distant events. Accounts of these networks also need to recognize that new forms of connection produce new forms of disconnection. We will never reach a totally connected world. As the example of the new electronic ghettoes shows, new peripheries are constantly being created. Thirdly, and finally, we need to beware of confusing our theoretical ambitions with the reality that we are located *in* the world. Too often, the works of authors like Jameson, Castells and Harvey seem to assume that there is a place, like a satellite, from which it is possible to get an overview of the whole world. Yet, as numerous feminist commentators, have made clear, this assumption now looks increasingly like a classic masculine fantasy, a dream of being able to find a vantage point from which everything will become clear and a way of refusing to recognize that the world is a mixed, joint, commotion (Morris 1992; Mack 1990). Once we see that this assumption is a fantasy then we are also able to take account of all the subjects which in the past were often considered to be, somehow, "local" – like gender, or sexuality, or ethnicity – which recent work has shown are crucial determinants of global capitalism. In other words, one could understand "global capitalism" the better and, at the same time, realize that not everything can be explained by studying "global capitalism" (Walker 1993).

So now how might we see barometers like illegibility, the space of flows and time–space compression? To begin with, we can recognize that a globalizing world offers new forms of legibility which in turn can produce new forms of illegibility. For example, electronic telecommunications provide more and new kinds of information for firms and markets to work with but the sheer weight of information makes interpretation of this information an even more pressing and difficult task (Thrift 1994). The space of flows is revealed as a partial and contingent affair, just like all other human enterprises, which is not abstract or abstracted but consists of social networks, often of a quite limited size even though they might span the globe. Finally, time–space compression is shown to be something that we have learned to live with, and are constantly finding new ways of living with (for example, through new forms of subjectivity). It might be more accurately thought of as a part of a long history of immutable mobiles that we have learnt to live through and with. After all, each one of us is constructed by these "props," visible and invisible, present and past, as much as we construct them.

For those looking for big answers to the big questions that challenge us now in a globalizing world, this level of provisionality may all seem a bit frustrating. Yet the history of the last one hundred years or so suggests that big answers founder when they come up against the messy, contingent world that we are actually landed with. Worse than this, the big answers sometimes become a part of the problem, as their proponents force order on the world by applying ordered force. In other words, the same history suggests that, if we are going to try to clean up the uglier bits of our messy, contingent world, big answers are not the solution. Of course, this means that our actions are likely to be modest and sometimes mistaken, but the fact of recognizing this state of affairs is in itself empowering; it means that the future is open and we can all do something to shape it. There is positive value to be gained in "striving to incorporate the problems of coping with that openness into the practice of politics" (Gilroy 1993, p. 223). To put it another way, there is nothing definite about the geographies of global change, and we do not have to be definite to change them.

3

From Farming to Agribusiness: the Global Agro-food System

Sarah Whatmore

Introduction

Food is a basic condition of human life, but for most people in the advanced industrial countries of Western Europe, North America and Australasia today, it has become a taken-for-granted facet of daily consumption. Stacking a trolley in the supermarket is an everyday chore; getting a take-away, a commonplace convenience; eating out, an integral part of many business and leisure routines. These consumer experiences of food are quite profoundly distanced from the social and economic organization of agriculture and the contemporary processes of food production. Milk may still come from cows and apples grow on trees (don't they?) but how does farming, the anchor of commonsense understandings of food production, fit into the creation of oven-ready meals; genetically engineered plants and animals; or synthetic foodstuffs? The prevalent representation of such experiences as the mark of "consumer choice" belies a diminished understanding of, and control over, what it is that we are eating and the social conditions under which it is produced. Moreover, the language of "choice" is at odds with the still widespread realities of food scarcity and uncertainty faced by those without the money-income to secure their basic needs through the market, particularly in the so-called "Third World" and parts of Eastern Europe, but also amongst the growing numbers living in poverty in the West itself.[1]

These divergent experiences of the political economy of food are intimately connected; bound together in highly industrialized and increasingly globalized networks of institutions, technologies and products, constituting an *agro-food system*. The OECD (Organization for Economic Cooperation and Development) has defined this system as

> the set of activities and relationships that interact to determine what, how much, by what method and for whom food is produced and distributed. (OECD 1981)

It is most readily symbolized by the worldwide presence and cross cultural potency of such food icons as the McDonald's hamburger or the Pepsi Cola drink. However, it would be misleading to take these high-profile cases as in any sense representative of the complex and highly uneven processes of industrialization and globalization which have been reshaping food production and consumption in the postwar period. This chapter outlines some of the key features of this restructuring process, focusing on the technological and socioeconomic transformation of food production, its impact on farming, and the consequent re-mapping of the geography of agriculture emerging from recent research.

Shifting Research Horizons

Increasingly, over the last 10 or 15 years, researchers have sought to understand and inform these transformative processes in terms of what has been called a "*new* political economy of agriculture"; a term reflecting both the changing organizational structure of capitalist agriculture in the postwar period and a refocusing of research questions in geography and other disciplines. As proponents see it,

> the present situation is one in which the connotations of "farming" – in particular, rurality and community, but also other categories that are limited to national economies, nation-states, and national societies – are giving way to vertically and horizontally integrated production, processing and distribution of generic inputs for mass marketable foodstuffs. (Friedland 1991, pp. 3–4)

The distinguishing feature of this "new" research agenda is that it takes the analysis of agriculture beyond the farm gate in two directions. The first has been to look at the wider organization of *capital accumulation* in the agro-food sector, focusing on the social, economic and technological ties between three sets of industrial activities, those of food raising (i.e. farming as a rural land-use); agricultural science and technology products and services to farming; and food processing and retailing. The second direction has been to look at the role of the agro-food sector in the wider institutional fabric of *social regulation*. The focus here has been the political and policy processes by which national, and supra-national, state agencies underpin agricultural markets by regulating the terms of trade and the food component of wage costs in the wider economy. In simplified terms, these two dimensions of current research stress the dynamics of agro-food sector restructuring generated respectively by the search for profit by private capital, and by state concerns with securing social order.[2]

Efforts to comprehend these expanded parameters of an agro-food system beyond the farm have led researchers to adopt a number of unfamiliar terms which flag a variety of new conceptual approaches. Several such terms are in current circulation in the literature and it seems appropriate to establish some reference points by beginning with a few basic definitions.

One set of approaches focusses on the reorganization of capital accumulation in the agro-food sector, tracing particular agricultural goods through the sequence of processes they undergo before reaching the consumer, and analyzing the social and economic relations within, and between, each stage in this sequence. An early version of this kind of approach adopted the concept of *agribusiness* to mean

> the sum total of all operations involved in the manufacture and distribution of farm supplies; the production operations of the farm; storage, processing and distribution of farm commodities and items made from them. (Davis and Goldberg 1957, p. 3)

In practice, this term has largely been applied to a particular business configuration, in which a single (usually transnational) corporation coordinates industrial activities in each of these spheres through subsidiary companies; a configuration known as vertical integration, and characteristic of the US agro-food sector in the 1960s and 1970s.

Another influential version of this kind of approach – the idea of agricultural *commodity chains* – emerged in the United States or, more precisely, in California in the late 1970s. It too emphasizes the industrial character of agriculture, treating agricultural products and businesses in the same way as those in, say, the steel or automobile industries.[3] A good example is the study of the Californian lettuce industry by Bill Friedland and his collaborators called *Manufacturing green gold* (1981). Despite criticisms about the extent to which such terms can be generalized beyond forms of industrial agriculture particular to certain US contexts, such analytical models have become abstracted and widely used.

A second set of approaches places a quite different emphasis on the regulatory institutions which have evolved to underpin strategies of accumulation in the agro-food sector. There are two related concepts of importance here. The term *agro-food complex* was coined by Harriet Friedmann (1982) to describe the industrial relations of the production and consumption of specific foodstuffs which became dietary standards, such as beef, and canned, frozen, or otherwise "durable foods," in association with distinct *agro-food regimes*. This latter term signifies the regulatory apparatus sustaining world agricultural markets and food prices, as these articulate with the efforts of nation states to regulate the social conditions of capital accumulation within their borders during particular periods in the development of the capitalist world economy (Friedmann and McMichael 1989). In this chapter we shall be most concerned with the consequences of the decline of the second of two "agro-food regimes" identified by Friedmann, which dates to the period 1945–73 and is associated with what has come to be known as a "Fordist" mode of capital accumulation and regulation.[4]

The concepts of agribusiness and, more particularly, agro-food "chains," "complexes" and "regimes" can be seen as complementary in the sense that they throw light on different aspects of the institutional structure of the contemporary agro-food system. Their composite perspectives are represented in a simplified way in figure 3.1.

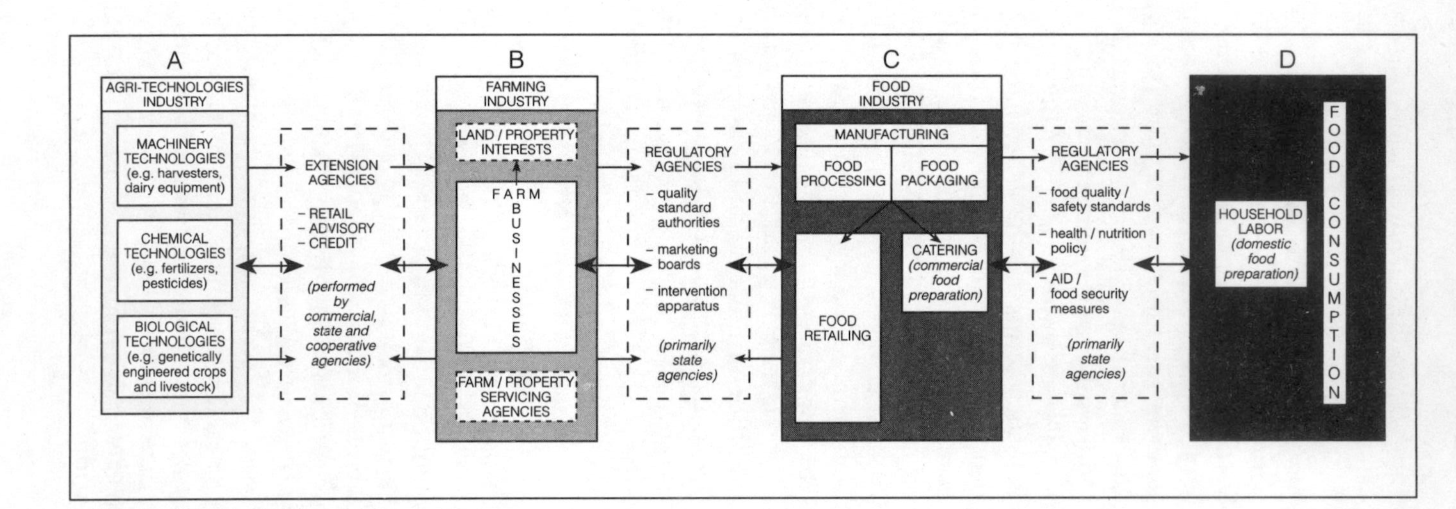

Figure 3.1 An outline of the contemporary agri-food system.

Within geography these developments have proved difficult to locate because of established divisions between agricultural geography, with its traditional focus on agricultural land-use (i.e. farming), and industrial geography, which has a strong political-economy tradition, but has tended to regard agriculture as an anomalous industrial sector outside its remit. As a result, the approaches and terms outlined above entered geographical research relatively late, in the mid-1980s, with a series of agenda-setting papers from North America (Wallace 1985), the UK (Marsden et al. 1986), and New Zealand (Le Heron 1988). While such approaches can be seen as the core of a reinvigorated agricultural geography, their real significance lies in their contribution to eroding the divisions between agricultural and other industrial geographies, and generating a more creative engagement with agro-food system research in other disciplines.

The Making of an Industrial Agro-food System

Goodman and Redclift (1991) argue that:

> to understand accumulation in the agro-food system it is essential to recognise that industrialisation has taken a markedly different course from other production systems. (p. 90)

This distinctiveness is centered on the biological foundations of agricultural production in which the growth and reproduction of plant and animal life that are its products are bound to a series of "natural" processes which have resisted direct and uniform transformation by capitalist relations of production. These biological constraints on accumulation are not fixed but they have made agriculture as a land-based activity unattractive to the direct involvement of industrial capital and its characteristic business forms and divisions of labor. In addition, agriculture has long been seen to face a market handicap in that expenditure on food tends to fall as a proportion of income, as average incomes rise. As a result of these constraints, the capitalist restructuring of agriculture has centered on reducing the dependence of agro-food accumulation on "nature" and increasing the market size and value of agro-food products through a process of industrialization. This strategy has been pursued through the technological modification of biological processes in farming itself, and by valorizing agricultural products "off-farm" in the manufacture of technological farm inputs and the increased processing and packaging of food products after they have left the farm.

An influential study by David Goodman et al. (1987) suggests that the rapid growth of these "off-farm" sectors has taken place through two discontinuous but persistent processes:

- *appropriationism* – in which elements once integral to the agricultural production process are extracted and transformed into industrial activities and then re-incorporated into agriculture as inputs, and
- *substitutionism* – in which agricultural products are first reduced to an industrial input and then replaced by fabricated or synthetic non-agricultural components in food manufacturing (ibid. p. 2).

Current developments in the application of biotechnologies herald a potentially more radical recomposition of the socioeconomic and spatial organization of the agro-food system. Such genetic and biochemical manipulation techniques fracture the integrity of biologically defined agro-food chains centered on particular types of crop or livestock. They permit, for example, the transfer of genetic features across species barriers as "inputs" into agricultural production, and the transformation of agricultural products into generic raw materials, say starch or protein sources, in the manufacture of industrial foodstuffs. To illustrate some of the implications of these technological processes for the recomposition of the agro-food system, figure 3.2 represents four configurations of the potato commodity chain.

The postwar expansion of these "off-farm" sectors of the agro-food system has seen them become much larger, in terms of market value and employment, and more concentrated, in terms of business and market structure, than farming itself. In the UK, for example, just four companies supply 77 percent of the agro-machinery market and the same number accounts for 50 percent of the food retailing market, compared with over 144,000 units which make up 90 percent of farm product markets (Munton

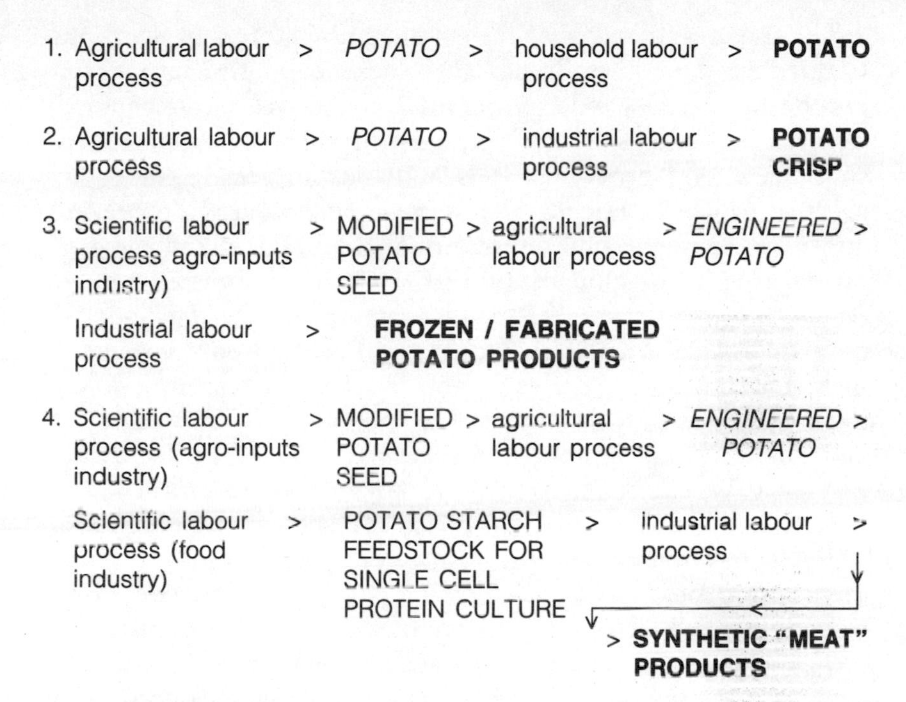

Figure 3.2 Industrialization of the potato commodity chain.

Key:
agricultural product leaving the farm
food product reaching the consumer

Source: S. Whatmore (1994). In A. Amin and N. Thrift (eds), *Holding down the global*: 77. Oxford University Press.

1992). Contrary to the agribusiness model of agricultural industrialization, corporate agro-food capitals have become *directly* involved in the farm production sector to only a limited extent. Instead their influence has primarily been exerted *indirectly*, through networks of marketing contracts, technical services and credit arrangements with independent farm businesses. These relations take a number of forms according to the particularities of different commodities, farm business structures, and local political contexts. In concert, they bring into play an important distinction between the corporate profile of today's agro-technology and food industries and the family-based organization of agriculture which prevails, in different forms, in both advanced and developing economies.

The rise to dominance of these "off-farm" sectors of the industrialized agro-food system has been inextricably tied into a complex of state policies and agencies including: public agro-technology research and extension services, agricultural price supports, and subsidized food programs. It is activities and institutions such as these which, in Friedmann's terms, constituted the regulatory apparatus of the second international food regime and played a pivotal role in the globalization of this industrial agro-food system in the postwar era. This regime centered on the consolidation of technological modernization and political corporatism which placed agro-food industries at the heart of agricultural policy-making institutions, first in the US and then, via US technology and food "aid" programs, in Western Europe in the 1950s and the Third World in the 1960s (Goodman and Redclift 1989).

Perhaps the most dramatic and lasting result of the globalization of this agro-industrial system has been the reorientation of much of the food production capacity of developing countries away from staple food crops supplying local markets. Agriculture in these countries has become increasingly tied into Western markets for unseasonal or luxury primary foods, or bulk feed crops for intensive livestock production. Simultaneously, they have been turned into net importers of staple and processed foodstuffs from the US and other Western states (Susman 1989). However, the competitive development of this agro-industrial model in other advanced industrial countries, notably those within the European Community (EC) and Japan, had begun to undermine the hegemony of the US in the international agro-food system by the mid-1970s.

The institutional foundations of this mode of regulating food production and consumption were further destabilized by the escalating fiscal costs of maintaining it which had come to a head for national governments in all three major trading blocs by the mid-1980s. A final blow came from the growing influence of financial interests in the agro-food sector, through speculative trading in agricultural "futures" and the disciplinary effects of the stockmarket on corporate investment and takeover strategies, which eluded the machinery of regulation (McMichael and Myhre 1991). Cumulatively, these elements have compounded a crisis in the industrial agro-food system which we are still living through today.

The Changing Place of Farming

Making sense of the restructuring of the farming sector of the agro-food system in terms of "commodity chain" and "food regime" approaches is more problematic. Despite their many important insights into the processes of industrialization and globalization, such approaches are inclined to represent them as the logical progression of some inexorable law of motion, popularly translated as "market forces." Their holistic and apparently seamless accounts have limited purchase on the uneven and unstable geography of the integration of farming (and consumers) into this industrialized agro-food system and all but eclipse the significance of the interests and actions of farmers, and others, in shaping the process and pattern of change. While the global strategies of transnationals in the "off-farm" sectors of this industrial system certainly exercise a powerful influence over the terms on which other actors are drawn into the agricultural restructuring process, they do not determine its outcomes in any straightforward or uniform way.

Above all, and despite forecasts of its imminent demise dating back to Marx's time in the mid-nineteenth century, the farming sector retains a highly diverse but distinctive family-based structure. This centers on dynamic, and culturally variable, forms of kinship and household relations in which social divisions by gender and generation are integral to the organization of "family" land, capital and labor (Whatmore 1991). It also sustains very diverse and locally sedimented farming cultures, political identities, economic strategies and land-use practices which are active ingredients in the restructuring process. As a result, the articulation between farming and corporate sectors of the agro-food system is a more complex and contested process than the above accounts allow. A good illustration is the tortuous progress of the recent round of GATT negotiations on agricultural trade, resulting from the political embeddedness of farming interests, as well as those of transnational agro-food corporations, in the machinery of agricultural regulation in most nation states and regional trading blocs.

The restructuring of farm production in the postwar period is marked by one dominant feature, *differentiation* in the extent to which farmers have been integrated into the industrial agro-food system; a feature which also marks trends in food consumption. The leading-edge technological and market relations outlined above, which set the contours for farming in particular commodity sectors of the industrial agro-food system, are realized unevenly. They are concentrated in particular regions and sectors which are characterized by large farm businesses (although many still "family" owned) employing intensive production methods and highly integrated into the global networks of the agro-technology and food industries through the ties of technological dependence, debt and production contracts. For example, 80 percent of the European Community's agricultural production derives from just 20 percent of agricultural businesses, which are concentrated in agro-industrial "hot spots" such as the Paris Basin, East Anglia, Emilia Romagna, and the southern Netherlands (Commission of the European Community 1988).

Conversely, the same example implies that much of rural Europe, many of its agricultural products, and some 80 percent of its farmers (concentrated in the south) are in some sense "marginal" to these global networks of agro-food accumulation and effectively left out of the "food regime" and "commodity chain" accounts. Similarly, it has been estimated that $10,000 represents the minimum GDP (Gross Domestic Product) per person necessary to "trigger" a consumer transition to the basic ingredients of an "industrial" diet (*The Economist* 1993, p. 15). The highly uneven social distribution of national income seriously undermines the significance of such estimates but, even on their own terms, is does not take an economist to calculate that those "marginal" to this mode of consumption, particularly in developing countries, exceed the numbers integrated into it. Making sense of the livelihood strategies of households within these rather substantial "margins" of industrialized farm production and food consumption draws attention to the broader significance of social and political processes beyond the scope of frameworks defined in terms of the dynamics of industrial processes or commodities.

From the perspective of these "margins", the globalization of the industrial agro-food system can be seen as a lattice, or network, of locally embedded nodes in which transnational corporate agro-food capital strategies are variably, rather than uniformly, articulated with farmers (and consumers) through specific regional compromises and institutions, and sectoral regimes (FitzSimmons 1990). Farmers and consumers "marginal" to the industrial agro-food system occupy the interstices of this network. But rather than seeing these interstices as the "black holes" of the global agro-food system, these tangential spaces represent sites of alternative strategies which build on traditional production practices centered on subsistence and "informal" market networks, or are bound up with new social movements associated with non-agricultural or non-food issues.

Two brief examples must suffice to illustrate this latter point. The first centers on consumer resistance to industrialized food products, and the cultural limits placed on the market strategies of the corporate agro-food industries. This takes a number of different forms, but represents an important basis for new political alliances and marketing networks between consumers and farmers in different parts of the world which sustain ecological or traditional agro-food practices. The second centers on the environmental consequences of industrial agriculture, which has become another important focus for the politicization of the agro-food system. This has seen both market and regulatory machinery in the agricultural sector increasingly reorient towards the production of "environmental goods," whether in the form of sustainable farming practices, which add value to their products through quality control schemes which certify them as organic or "conservation grade"; or in the commoditization of particular habitats or landscapes through conservation subsidy schemes and as a feature of the tourist and leisure industries.

The place of farming in the increasingly globalized industrial agro-food system is thus one of growing differentiation. On the one hand we see a rapid concentration of industrial farming methods and their products onto a smaller number of much larger farm businesses which are bound to the corporate sectors of the agro-food sector by a variety of economic and knowledge ties. On the other, we can trace a diversification of farm livelihood and business strategies which link the rest of the farm sector more closely into wider struggles over the cultural and material resources of the rural environment.

Conclusions

The industrial agro-food system, which has emerged since the Second World War as an increasingly globalized network, has dramatically altered the contours of the production and consumption of food. Simultaneously, biological time–space rhythms like crop gestation, seasonality and regional specialization have been disrupted, while the social distance between producers and consumers at the heart of the system and those on its periphery have been magnified. As this account has tried to emphasize, this agro-food system is both uneven and unstable, and is characterized today by a number of overlapping crises. The first is a crisis of *production* which centers on the problems of surplus production and indebtedness amongst agricultural producers faced with escalating input costs and declining product prices. The second is a crisis of *regulation*, resulting from the growing political and institutional tensions in a policy apparatus which regulates global agro-food trade, on the one hand, and national farm incomes, on the other. The third is a crisis of *legitimation*, which centers on the politicization of concerns about the consequences of industrialized agriculture for food security, food safety and the farmed environment at national and local levels.

Extending research beyond the farm gate and into other sectors of the agro-food industry and its regulatory framework represents a necessary and important contribution to a re-mapping of the geography of agriculture. But the accounts which have come to define this new map generate their own silences and blind spots. The challenge now is to make space for a broader canvas of social struggles and political alliances over rural resources and identities within which this industrial story is couched. We could do worse than to begin by giving greater conceptual and empirical prominence to consumption processes and the cultural politics of food in refashioning the social and environmental relations of agriculture.

Notes

1. For example, United Nations data document a growing polarization of income between 1960 and 1990 with a staggering 30 percent of the world's

population experiencing hunger, including a suprising 10 million people in the USA.

2. This simplified account of the relationship between the dynamics of capital accumulation and social regulation must be placed in the context of a much disputed literature on "regulation theory." A geographer's guide to these debates is provided in Tickell and Peck (1992).
3. A parallel body of work was generated independently in continental Europe, particularly France and Italy, at around the same time using the terminology of agro-food (agro-alimentaire) "filières."
4. This Fordist periodization is not unproblematic, but a critique is beyond the scope of this chapter.

4

Multinational Corporations and the New International Division of Labor

Richard Barff

Introduction

In the early 1970s, Bulova Watch movements were made in Switzerland. These semi-finished goods were then transported to Pago Pago in American Samoa, where assembly with cases, bands, dials, and other components took place, before they were shipped to the US for final sale. Then corporate President Harry Henshel commented on this production geography at the time as follows: "We are able to beat the foreign competition because we are the foreign competition" (quoted in Bluestone and Harrison 1982, p. 114). How typical is this story? For how many years has international production of this sort taken place? Is this kind of production network stable over space? How do we explain such a geography of production? What is the future of the geography of international industrial location? By focusing on the strategies of multinational corporations and the concept of a spatial division of labor at the global scale, this chapter seeks answers to these and other derivative questions.

The multinational corporation (MNC) is the most complex of several archetypes of the production unit (Hymer 1979, pp. 146–7). The first and simplest – the workshop – involves a small number of people working together with very little specialization by worker. In the second, the factory, a large concentration of people work together but on separate and specialized tasks: this involves a fine division of labor both within the laboring classes and between those who plan and those who work. Third, the national corporation comprises many occupations, markets, units, and places within a country. In this type of enterprise, forward integration from single-function firms draws in, say, wholesaling activities: i.e. those activities further along in the sequence of operations. Backward integration (upstream in production to supplying activities) incorporates, for example, mining or wellhead operations. Such vertical linkages produce, at least potentially, multilocation, and multimarket firms. It was this type of enterprise that resulted from the US "merger movement" in the early twentieth century.

These corporations evolved, with the development of the market, into a fourth manifestation – the multidivisional corporation, wherein the horizontal integration and subdivision of corporations was based on the introduction of new products or on differentiation within a product line. Accompanied by the creation of a middle level of administration to coordinate activities of a division within a country, management remained to concentrate on strategic planning. Multinational corporations simply entail the international location of a division or part of a division of a "multidivisional corporation." International involvement can be seen as perhaps a natural extension of transferring corporate power from across divisional space and operations to across political space. If the international wings of the corporation are sufficient in size or complexity, these divisions may be organized under separate international divisions.

During the last one hundred years, international investment by corporations has increased steadily to the extent that the multinational corporation at the end of the twentieth century is one of the most important forces creating large-scale shifts in global economic activity. Explanations of the forces driving these shifts in foreign direct investment (FDI) include ownership-specific advantages and location-specific factors. The first is generally linked with oligopoly and market power, and frequently associated with the existence of competitors (indigenous or otherwise) who can enter the industry if the multinational corporation does not. While indigenous firms might possess a better understanding of the local business environment, MNCs entering the market gain advantage through access to finance, credit, and new technology by virtue of their size and oligopoly power. The second explanation is associated with variations in economic and political conditions over space. Market size and competition vary across space, as do national and supranational government policies. Production costs, or factor supply conditions, also provide for "locational advantages" – such as labor or other factors of production (after Hymer 1979 and Dunning 1981). As it traces the evolution in the geography of these investment patterns, this chapter will show that the explanatory power of these two forces varies considerably over time and space.

In the last fifteen years or so, high labor costs plus union militancy and intransigence in technologically advanced nations (i.e. associated with particular "locational advantages") have influenced the geography of intra- and international investment patterns. As a reaction to these location-specific factors, some corporations have moved manufacturing operations offshore in search of lower-cost, non-unionized labor in the developing nations of the global periphery, creating what has become known as a new international division of labor (NIDL). MNCs control economic activities in more than one country, and therefore can take advantage of geographical differences between countries and regions in factor endowments (including government policies) by virtue of their flexibility – such as their ability to shift resources and operations between locations around the globe (Dicken 1992, p. 47). According to one group of commentators, "[t]he development of the world economy has increasingly created conditions (forcing the development of the new international division of labour) in which the survival of more and more companies can only be assured through the relocation of production to new industrial sites, where labour power is cheap to buy, abundant, and well disciplined; in short, through the transnational reorganization of production" (Frobel et al. 1980, p. 15). The overall goal of the chapter is to evaluate this thesis in the context of the evolution of the multinational corproation.

The Evolution of Multinational Corporations

Although the number of transnational networks of production accelerated after the Second World War, the first signs of multinational enterprise originated around the turn of the century, mainly in the United States. Improvements in communications and transportation at this time, coupled with massive increases in scale economies in industrial production, provided the foundation for the initial development of multinational operations (Vernon 1992). By 1914, US firms had invested $2.5 billion in other countries (Hymer 1979, p. 209). The United States, along with France, Germany and Britain, as origin countries, accounted for 87 percent of total global FDI in 1914. As recipients, the total was much smaller – less than 30 percent. Most pre-First World War FDI targeted countries in Asia and Latin America (Dunning 1983).

Two interwoven strategies accounted for this geography of investment. One, associated with resource-based investments, developed backward linkages to supply home producers with primary products in the form of either raw materials or food. For example, the investment decisions by leading European firms to invest in countries with which their home governments had colonial relations explain part of this geography. The other strategy, associated with market orientation, occurred mainly in the form of manufacturing investment in other developed countries. US-based companies made most of the early moves in establishing bridgeheads overseas. Companies like Ford, General Motors, and General Electric, for example, started foreign affiliates to establish footholds in the European market. They were not alone, however, in seeking to develop international operations: Fiat, for example, opened manufacturing branches in Austria in 1907, the United States in 1909, and Russia in 1912 (Fridenson 1986). Merck pharmaceuticals, which grew out of an apothecary shop founded in 1654 near Frankfurt, set up an affiliate in the United States in 1887. Bosch had factories producing automobile parts in France and Britain in the early years of the twentieth century (Hertner 1986).

As late as 1939, the geography of global FDI had changed little from a quarter of a century earlier. Developing countries in Latin America and Asia received almost 67 percent of total global foreign investment. Twenty-one years later, however, technologically advanced nations received two-thirds of global FDI. US companies accounted for almost half of the total international investment (Dunning et al. 1986). Global investment stagnated during the 1930s and most of the 1940s and tremendous and hugely significant changes in the rate and direction of flows took place in the 1950s.

Part of the explanation for this shift is associated with the rapid revival of the European and Japanese economies. A fuller explanation of the shift in the destination of global FDI requires, as in all debates about the structure of international operations, a consideration of "ownership" advantages as well as "locational" advantages. Before the Second World War, state regulatory mechanisms restricted ownership advantages by US firms in Western Europe – especially the UK. The war changed all that. The war devastated the European economy, providing opportunities for investment that simply were not there before 1939. Even before the war, many countries in Europe had economies with significant structural problems and most of the economies of the larger industrialized countries were operating at capacity. The devastation of the war exacerbated these weaknesses. Also, markets loosened after 1945. For example, Roosevelt continually lobbied Western European heads of state throughout the early 1940s for the end of discriminatory tariffs, especially in the UK and its colonial economies. Relatedly, the establishment of the 1944 Bretton Woods Accords contributed to US industries moving into Europe. Under the Bretton Woods agreement, each nation was responsible for keeping its currency within 1 percent of its par value. To keep within this tight range, each country had to buy or sell its own currency in foreign exchange markets. The US consistently ran large deficits in its balance of international payments, forcing foreign central banks to buy excess dollars with their own currencies. US multinational corporations (and other players) were provided with European currencies, which were invested in the UK, France, West Germany, and elsewhere on the continent (Bluestone and Harrison 1982).

Vernon (1992) notes additionally that a "follow-the-leader" strategy drew firms to set up operations offshore as a hedge against threats posed by rivals: once a corporation in an oligopoly began production in Europe, other US-based firms in that oligopoly subsequently tended to establish affiliates in the same country. More generally, as Hymer (1979 p. 210) observed, subsidiaries of

MNCs "once established, tend to grow in step with that industry in that country except where interrupted by extraordinary events like war."

Decolonization, or the threat of independence by colonized countries, also probably discouraged investment by European-based firms in Latin America, Africa, the Middle East, and Asia, and helped reorient the global geography of FDI away from developing countries and toward (Western) Europe, where new fierce competition was erupting with American affiliates. Last, but not least, the formation of the European Economic Community in the late 1950s helped reorient foreign investment as US multinational corporations sought market access through the establishment of affiliates in Western Europe.

Since 1960, the internationalization of US production has continued apace (although the late 1960s is the benchmark of the peak proportion of FDI originating in the United States). The volume of US foreign investment grew from a little over $30 billion in 1960 to about $250 billion in the mid-1980s. This rate of growth is matched by UK and French corporations, but has been outpaced by MNCs in Germany and Japan (UNCTC 1983; 1988). Furthermore, in the 1980s multinational corporations have emerged with bases in so-called newly industrialized countries (NICs), such as Hong Kong, South Korea, and Taiwan. Nevertheless, patterns of MNC transnational investment are still concentrated (and increasingly so) in the global core, and the MNCs based in the United States are as significant as ever. The evolution and growth of the European Community (EC) has been significant in this regard as US-based enterprises continue to seek market access in Europe. The development of the Single European Market and the opening up of Eastern European economies will probably affect the size and geography of US investment in Europe.

One of the most noteworthy shifts in global FDI since the mid-1970s has been the change in the position of the United States in terms of inward and outward direct investment. While the ratio of outward to inward investment grew by about 20 percent in Western Europe and Japan between 1975 and 1983, the ratio for the United States has declined by about two-thirds. As table 4.1 indicates, the USA is becoming a significant host country to global FDI. Almost paradoxically, however, since the 1970s, as new

Table 4.1 The balance of inward to outward direct investment: 1975 and 1983

	FDI Ratios*	
	1975	*1983*
Western Europe	1.20	1.56
United States	4.48	1.66
Japan	10.65	12.32

* FDI Ratio = Outward Stock/Inward Stock
Source: Dicken (1992, p. 55), after Dunning and Cantwell (1987, Table B1).

foreign competition has challenged US economic supremacy, profits of some major corporations have been squeezed, and inward investment to the US by non-US MNCs has expanded rapidly; many producers in the global core have sought lower wage labor sites in developing countries in the global periphery to reduce production costs. The term "new international division of labor" was coined to explain this drift of work from the core to the periphery.

The next part of this chapter expands on the reasons why a new international division of labor came about, and then questions its utility in the light of global patterns of investment by MNCs that still tend to favor core-country locations.

The Old and the New International Division of Labor

The term "new international division of labor," popularized by Hymer (1972; 1976) and Frobel et al. (1980), has become central to the discourse on global industrial restructuring. The NIDL refers to a spatial division of labor at the international scale and can be considered part of a bigger family of labor divisions, which includes social divisions of labor, divisions of labor between production and exchange, and spatial divisions of labor at scales other than the international (see Sayer and Walker 1992). The idea of a new international division of labor opposes that of an old international division of labor (OIDL). Under the OIDL, the global periphery was seen and theorized as the provider of many primary goods and raw materials for processing in the core countries of Western Europe, North America, and Japan. In exchange for these materials, the periphery received finished goods manufactured in the core.

The economic activities in the global core and periphery are changing under the development of the new international division of labor. A process of vertical uncoupling and subdivision of production results in the periphery developing low-skilled, standardized operations such as manufacturing assembly or routine data entry, while the global core retains high-skill knowledge- or technology-intensive industries and occupations. Through deskilling labor, and the functional and physical separation of various tasks in the corporation, this process creates "roles" for places in the world economy (cf. Massey 1984).

According to Frobel et al. (1980), countries in the global periphery have transcended their old role, beyond the search for low-cost, union-free labor environments, for three main reasons.

1. The introduction of capital intensive green-revolution agricultural methods, especially in Southeast Asia, has liberated hundreds of thousands of people from what was largely a subsistence rural life. Much of this labor force is inexpensive and productive by Western standards. Moreover, because of the size of this labor supply, multinational corporations (and other companies) can afford to select their employees according to age, skill, and particular disciplinary factors. Notably, the NIDL is especially augmented by gendered labor markets and the underpayment and devaluation of female wage labor (Sayer and Walker 1992).
2. The increasing subdivision of labor processes lends itself directly to the substitution of minimally-skilled workers for those who have had more training or education. The fragmentation of productive tasks in manufacturing has developed to such an extent that the execution of the simplified tasks of perhaps a very sophisticated overall process requires only a brief training period. Knowledge functions are therefore extracted from the production process, creating a situation in which skilled workers commanding relatively high wages can be replaced by less expensive semi-skilled or unskilled workers.

3. New, permissive technology extends to the realm of transportation innovations like containerization, that now present the possibility to produce finished or semi-finished goods at virtually any site in the world. In other words, while some places might be more practical than others, physical space per se is a far less limiting factor in the industrial location calculus than it once was.

The NIDL's basic geography, of the advanced industrialized countries on the one hand and the rest of the world on the other, can be modified in several ways, as much labor-oriented global industrial production has evolved beyond that simple organization. A regional core and periphery has developed within Southeast Asia, for example, with the territorial differentiation of Japan, Taiwan, South Korea, Hong Kong, and Singapore in a core from other countries in a new periphery (Lipietz 1986; Donaghu and Barff 1990). In an intranational context, producers seek low-wage locations in peripheries of the global core: Converse, the athletic footwear producer, for example, employs mainly female American Indians (from a tribe unrecognized by the US Bureau of Indian Affairs) to assemble shoes in Lumberton, North Carolina, in the poorest county of one of the lowest-wage states in the country.

MNCs in the NIDL

Joining MNCs and the NIDL in the discussion almost automatically centers on an explanatory emphasis of international corporate behavior on the locational advantages of offshore production. More specifically, fusing MNCs and the NIDL centers attention on the influence of labor on international location. In one respect, this is entirely appropriate. In terms of labor costs, hourly earnings in manufacturing varies hugely across space. Earnings in manufacturing industries can be as much as 75 times greater in certain core countries relative to some of the poorest developing nations (Dicken 1992, p. 134). Add the costs of health and social security benefits and the gulf widens even more.

These location-specific advantages associated with labor affect a particular set of economic activities associated with routine, mass-produced, highly differentiated production, such as manufacturing industries like computer electronics assembly, textile and clothing production, and shoe manufacturing, and non-manufacturing activities like routine data entry/processing. For example, Xerox Corporation, based in Rochester NY, manufactures copying machines at various sites around the globe. Large, complex machines, however, are produced exclusively at their New York plant; smaller, simpler machines are assembled in developing nations in Southeast Asia and South America, taking advantage primarily of low-cost labor in the global periphery.

Shifts to offshore production sites in Third World countries are frequently highly publicized. Examples occur regularly, from the auto plant closures of General Motors in Michigan and their new investments in Mexico, to the shifting of production of Levi's Dockers from their South Zarzamora Street factory in San Antonio, Texas, to Costa Rica, and the move of Smith Corona typewriters from rural New York State to northern Mexico.

Expanding the argument to subcontracting relationships, we find similar global patterns of production. Nike, the athletic footwear marketer, used to own manufacturing plants in the United States and the United Kingdom, but presently subcontracts 100 percent of its production capacity to suppliers in South and East Asia. The geography of Nike's production partnerships has evolved over time, a change powered in large part by changing labor costs in Asia (Donaghu and Barff 1990). Initially, production of Nike shoes took place in Japan. Soon, subcontracting arrangements diffused to factories in South Korea and Taiwan. Presently, those partnerships are diminishing in importance as labor costs rise and new networks of subcontractors become established in Indonesia, Malaysia, and China where workers involved in shoe production are paid about one-thirtieth of the wage their counterparts make, working for other companies, in the United States (Barff and Austen 1993).

Commentary and Criticism

While many multinational corporations take advantage of low-wage labor in places around the globe, and MNCs play a significant part in the development of what has become known as a new international division of labor, the NIDL model explains only a portion of MNC operations and their global strategy, and a small portion at best. The geography of core-based MNC investments tends to be in developed countries, because forces driving the international investment decision extend far beyond a singular consideration of labor-cost minimization. Understanding MNC behavior hinges on the issues of the importance of ownership-specific versus location-specific advantages, with labor costs being only a part of the latter. Multinational corporations invest offshore not only to utilize low-wage labor in what Dicken (1992) calls "rationalized specialization," but also for many other reasons such as access to resources and markets, development of import-substituting manufacturing, and so on. At this point, perhaps we should return to Bulova Watch – a small MNC that at the end of the twentieth century no longer produces timepieces in Pago Pago (or the West Indies, or any other developing country) but now has parts production and assembly only in countries in the global core or some of the NICs. Labor cost variation apparently now plays only a small role in Bulova's international locational calculus.

Because of the exclusive focus on labor, other problems occur with the NIDL approach, Schoenberger comments that the NIDL model understates the extent to which unemployment and unstable employment in countries of the global core result from labor-saving technological change: "This omission reinforces the impression that employment gains in one location are made at the expense of jobs in another. . . . It also encourages workers in the industrial countries to view low cost foreign labour as the primary threat to their well-being" (Schoenberger 1988, p. 116).

Note also that not all developing countries in the global periphery offer the same opportunities for labor exploitation. The geography of multinational corporate foreign investment favors, of course, developed countries primarily, and South America and South and East Asia secondarily. Within the global periphery, multinational corporations invest heavily in locations possessing an Export Processing Zone (EPZ), places where the main lures include not only relatively inexpensive wage labor, but also land, electricity, and water at cheap rates, as well as government grants, tax breaks, tariff/duty reductions, lax pollution controls and less rigorous health and safety standards in the workplace. Furthermore, even when all of the factors of location are taken into account, including access to primary resources and agricultural products, the continent of Africa is largely excluded from investment consideration. In terms of the NIDL, this whole continent plays only a very minor role. For example, the small island of Mauritius has a workforce of 62,000 involved in EPZ activity, which represents about 40 percent of the continent's total EPZ employment (Marshall 1991).

One other criticism centers on the idea at the heart of the NIDL thesis of dividing the world into two parts. Lawson and Klak (1993) argue that the social science literature is pregnant with binary categories, such as new international division of labor/old international division of labor, modern/traditional, developed/underdeveloped, core/periphery, First World/Third World, and North/South. They point out that these binary labels have at least two severe shortcomings. Dualistic theorizing can undermine our making sense of reality. Simply put, new and old international divisions of labor homogenize large areas of the world. Moreover, places, in their words, are "othered," which creates an unequal power relationship between the namer and object of the labeling. The dualities are not only opposing, but also unequal. In terms of the new international division of labor, Lipietz suggests, in a related argument, that as geographers we should be far more sensitive to spatial variation than the NIDL model implies: "The Third World today appears as a constellation of particular cases with vague regularities constituted out of fragments of the logic of accumulation (which proceeds badly or well depending on local circumstances) and [production] tendencies that rise and fall over a few years" (Lipietz 1986, p. 34).

Conclusion

A recent lead article in *The Economist* (14 April 1993) entitled "The Fall of Big Business" held that corporate giants "that once bestrode the globe" are in trouble. The spectacular failures of IBM, General Motors, and others signify how vulnerable the large, multinational firm has become. The declining ownership advantages of MNCs through the opening up of international markets has made it easier for smaller firms to sell their products in overseas markets. The growing efficiency and internationalization of capital markets allows relatively small firms to raise capital in much the same way large firms used to do exclusively. The opening up of foreign markets (via the removal or simplification of legal and tax complexities of international business) allows small and medium-sized companies to operate globally. The use of computers, by narrowing economies of scale in manufacturing and distribution, has made possible inexpensive, small-volume production, further diminishing the advantages of corporate size.

Some might hold that a company like CORPORATION X is an exemplar of a portion of future global economic activity. CORPORATION X is a maker of relatively expensive brand name textile and apparel goods (see Clark 1993). The company is not a multinational corporation, because CORPORATION X "has not made significant investments or built significant organizational units outside its home base." It is simultaneously a global and a local company: global because of its international sourcing, marketing, and design networks, and local by virtue of its sensitivity to local variation in tastes, fashions, and revenues. Different products are marketed in different places, and experience shows that utilization of local knowledge about place-specific tastes and styles is key to success.

Mass markets around the globe still exist, however, and will continue to do so. It follows that multinational corporations serving mass markets that need to take advantage of the NIDL might go relatively unaffected by niche market servicing. Remember also that new mass markets are emerging, in Eastern Europe and China for example. There should be a continuing place for the MNC and the NIDL in our contemplation of international investment patterns, so long as that understanding is part of a larger theoretical framework accounting for revenue maximizing and other cost-minimization strategies.

Acknowledgment

Thanks go to two recent graduates of Dartmouth College, Mike Donaghu and Jon Austen, for inviting me to think on my feet.

5

Trajectories of Development Theory: Capitalism, Socialism, and Beyond

David Slater

Introduction

Ours is an era of endings and beginnings. The contemporary literature is pervaded by a sense of farewell or departure from a constructed past; debates are punctuated by the prefix "post" – we can move from poststructuralism to postmodernism, or from post-Marxism to post-development. At the same moment, we are faced with the signposts of a radical closure; hence we read of the posited "end of history," or the "end of the social." Conversely, in a symptomatic reaction to assertions of endings, we are encouraged to believe in new beginnings or new affirmations. Critically juxtaposed to the "end of history," it has been contested that we are living the "beginning of geography," or the emergence of a new globalism in human relations, or a new age for democracy. Equally, in the more specific context of development theories, and analysis of the North–South divide, it is possible to discern a connected juxtaposition, whereby ideas concerning the end of development, or "post-development" (Latouche 1993), exist next to the desire to re-think and re-problematize what can be meant by "development." Furthermore, the contemporary scene is characterized by the continuing imposition of an orthodox Western vision of development for the non-West, most clearly reflected in notions of "structural adjustment," deregulation of the economy, privatization and more recently "good governance."

Such a vision, which is a continuation of an apparently endless Western will to develop the world, has been increasingly questioned; and within a newer current, development, *tout court*, has been rejected outright as a series of unacceptable Western practices for subordinating the Third World (Marglin and Marglin 1990; Sachs 1992).

What is now clear is that together with the post-1989 dissolution of the Second World, the accelerating tendencies of globalization, and the explosive surfacing of a variety of acute social tensions and conflicts, there has also been a resurgence of interest in the state of North–South relations and in the ways in which development is conceived. This resurgence has also been reflected in the domain of geographical scholarship, where contributions by, inter alia, Corbridge (1993), Peet and Watts (1993) and Schuurman (1993) have connected development thinking to debt and environmental issues, as well as to more general questions of social and political theory.

In the present climate of intellectual as well as sociopolitical turmoil, it is even more necessary than before to trace the breaks and connections in the formation of development discourses. And it is not only the history but also the geopolitics of ideas that is so crucial, as I shall seek to demonstrate below. An active politics of memory, carrying on and revivifying what is considered to be significant and relevant, is accompanied by a politics of forgetting that can be used to insinuate the idea of a new truth, the roots of which may actually stretch far back into the past.

In this chapter, I intend to discuss two major perspectives on development, perspectives which are counterposed and essentially incompatible. The first I shall simply refer to as the orthodox Occidental vision, which has been expressed by both modernization theory and neo-liberalism. In opposition to such a vision the postwar period has witnessed the rise and decline of a Marxist-inspired theorization of development, which despite a degree of internal heterogeneity, has possessed a certain conceptual and political regularity. In the third and final section of the chapter I shall refer, if only briefly, to the outlines of today's situation within which a critical re-casting of development theory has acquired an increasingly global meaning.

Develop and Rule: a Western Project

It was Enlightenment discourse which originally gave meaning to concepts of the "modern." The West became the model, the prototype, and the measure of social progress. It was Western civilization, rationality, and progress that were proclaimed and bestowed with universal relevance. At the same time, such an enunciation was intimately connected to the creation of a series of oppositions – for example, "civilized versus barbaric nations," or "peoples *with* history and those *without*" – which were reflections of the need to generate a non-West other so as to ground a positive identity for the West itself.

The nurturing of a positive identity for the West found a key expression in the United States of the nineteenth century. Already by the 1850s it was firmly believed that the American Anglo-Saxons were a separate and innately superior people who were destined to bring good government, commercial prosperity and Christianity to the American continents and to the world. Thomas Jefferson, for example, wrote that England and the United States would be models for "regenerating the condition of man, the sources from which representative government is to flow over the whole earth" (quoted in Horsman 1981, p. 23). The sense of mission, the aura of geopolitical predestination, were captured in the phrase "Manifest Destiny," which first appeared in the 1840s. This belief in an ethic of destiny, anchored to a particular religious conviction, did not remain restricted to the territories of North America, but extended south into Central America, the Caribbean and in some cases to the whole of the Americas. Furthermore, throughout the nineteenth century, Anglo-Saxon assumptions about US civilization being the highest form of civilization in history took firm root, as US attitudes toward other nations and inhabitants came also to be increasingly based on a well-defined racial hierarchy (Berger 1993).

Adherence to a sense of mission, and to Anglo-Saxon supremacy, carried within it a driving desire to bring the posited benefits of a superior way of life to other less fortunate peoples and societies. Frequently the desire to "develop," the will to "modernize" another society went together with a belief in the need for order, as well as with a grand sense of civilizing zeal. At the beginning of the twentieth century, President Theodore Roosevelt, turning his eye south to the Latin American world – this "weak and chaotic people south of us" – wrote that "it is our duty, when it becomes absolutely inevitable, to police these countries in the interest of order and civilization" (Niess 1990, p. 76).

The idea of a modernizing, developmental, civilizing project, that was legitimated as part of a wider mission of imperial destiny, was given a practical political realization in a whole series of occupations. Before 1917, and the birth of what came to be seen as the "Communist threat" to Western freedom and civilization, the United States occupied and administered the governments of the Dominican Republic (1916–24), Haiti (1915–34) and Nicaragua (1912–25 and 1926–33), as well as maintaining a Protectorate role over Cuba from the beginning of the 1900s. In the case of the Dominican Republic, the imposition of development through occupation went together with a five-year guerrilla war against the forces of the US military government, and in other instances too, especially in the Nicaraguan case, there was no absence of resistance. Whilst preserving order and imposing a geopolitical will to power over these other societies, the United States also introduced a series of related social and economic programs that were the precursors of contemporary development projects; there were initiatives to expand education, improve health and sanitation, create constabularies, build public works and communications, establish judicial and penal reforms, take censuses, and improve agriculture. In the case of Cuba, for example, which was occupied and ruled by the United States from 1898 to 1902, public school reformers built a new instructional system with organization and texts imported from Ohio; in 1900 Harvard brought 1,300 Cuban teachers to Cambridge for instruction in US teaching methods, and protestant evangelists established around 90 schools in Catholic Cuba between 1898 and 1901. At the same time, serious efforts were made to "Americanize" the systems of justice, sanitation, transportation, and trade, whilst the institutions of the Cuban independence movement – the Liberation Army, the Provisional Government and the Cuban Revolutionary Party – were disbanded by the US military government. US investments were encouraged and teams of experts from the United

States placed the mineral, agricultural, and human resources of the island under their scientific gaze so as to determine the proper means for harnessing the country's wealth.

These geopolitical interventions entailed projects for the modernization and development of other societies. The interventions and penetrations were portrayed and underwritten in terms of order, civilization, and destiny. To develop another society was also to rule over it, and to restore order was part of a wider project of civilizing the Latin South. It is important to remember these origins because in the post-Second World War period the growth of modernization theory had key roots in these previous histories of North–South relations. There were significant discursive continuities, but equally the postwar period witnessed the emergence of a series of crucial geopolitical changes.

In the years since the late 1940s, two related but far from identical discourses of Western development came to be constructed. First, during the 1950s and 1960s, in a time of the Cold War and the coming into being of a whole series of newly-independent nation states, conceptualizations of the modern became central. Modernization theory, as it came to be called, took root in the academic citadels of the West and found expression across a broad spectrum of disciplinary domains. It was multi-disciplinary and more multi-dimensional than the econocentric Marxist analysis of the 1970s. It encompassed questions of economic growth, social institutions, political change and psychological factors. Its tenets found a home in geography as well as history. Essentially, modernization theory was constructed around three interrelated components: an uncritical vision of the West, largely based on a selective reading of the history of the United States and Britain; a perspective on the non-West or traditional societies that ignored their own histories and measured their innate value in terms of their level of Westernization; and an interpretation of the West–non-West encounter which was based on the governing assumption that the non-West could only progress, become developed, throw off its backwardness and traditions, by embracing relations with the West. The posited dichotomy between the "modern" and the "traditional," ideal types of a Weberian vintage, was also replicated within the so-called traditional societies of Africa, Asia, and Latin America. Here, the researcher was encouraged to observe a duality between modern urban centers of growing Western innovation and traditional rural peripheries; "development" would come about, would be engineered, through the diffusion of innovations (capital, technology, entrepreneurship, democratic institutions, and the values of the West).

Such a brief sketch remains unavoidably incomplete. For example, it needs to be stressed that modernization theory went through two phases, especially visible in the political science and sociological literature. In the initial phase, that lasted until the mid-1960s, there was a sense of optimism, a firm belief in the potential success of modernizing and Westernizing the traditional society. If only the pre-modern society could find ways of accepting and adapting to the spreading tide of modernization, its future long-term development would be assured. But with the rise of radical nationalist movements in the Third World, the onset of the Vietnam War, and the growth of social protest inside the leading modernizer itself, notions of modernization in the United States increasingly came to be associated with discussions of political order, societal breakdown and "diseases of the transition," as ostensibly exemplified by communism. Finally, by the beginning of the 1970s, belief in the universal applicability of the modernizing imperative was clearly on the wane. This can be explained partly as a result of the growing realization in the West that the societies of the developing world were far more heterogeneous and complex than originally depicted by the protagonists of modernization theory, but also because of the upsurge of resistance to the West's and especially the United States' project of developmental rule. Defeat in Vietnam was emblematic; the will to modernize and control the recalcitrant other had been broken, albeit temporarily.

In a second wave of developmental doctrine, frequently considered as neo-liberal, and customarily couched in the terminology of structural adjustment, privatization, deregulation, free trade and market-based development, an apparently new model from the West was prescribed for the Rest, as their model too. Economies had to be opened as never before, state structures, a rigid barrier to successful development, had to be rolled back and streamlined, financial discipline was to be strictly imposed, and the logic of the market was to be given full reign for the benefit of all. Initially and fundamentally rooted in earlier currents of economic liberalism (Slater 1993a), the doctrine of the 1980s has been amplified to incorporate notions of good governance, fiscal decentralization, participation in development and the strengthening of civil society. The terrain of intervention has been extended and the project is thoroughly global in reach. In the words of a recent OECD (1992, p. 49) report, solutions to the domestic problems for which policy makers in the West have responsibility "are increasingly associated with the economic and institutional functioning of other societies . . . and this creates new scope for mutual understanding and synergy among policy makers in donor governments as they tackle development as part of achieving a global agenda."

In comparing these two waves of Western development theory, it is important to understand the convergences as well as the points of difference. One important break characteristic of neo-liberalism has concerned the treatment of relations between the public and private sectors. In the 1980s the private sector was championed, whereas the public sector, and more specifically the state, has been envisaged as a brake on development, a site of inefficiency, and institutional stagnation. A supreme belief in the benign opacity of market forces, the sanctity of private ownership, and the superiority of achievement-oriented individuals has permeated the orthodox texts of neo-liberalist doctrine. In contrast, modernization theory gave greater weight to the nation state in developing countries, and stressed the importance of a greater degree of balance between the public and private sectors. At this time the influence of Keynesian ideas was still an important factor, and overall the state's role in economic development was seen in a more pragmatic light.

In terms of commonality, both theorizations shared a belief in the universal relevance of Western models of development. Their points of departure both drew on an idealized construction of Occidental history and geopolitics, and their recommendations for development and progress all assumed that the North–South encounter was intrinsically beneficial to the South. The ethnocentric universalism of modernization theory provided one of the main targets of the dependency critique of the late 1960s and early 1970s (Slater 1993b), but in contrast, the neo-liberalism of the 1980s and beyond has tended to escape a similar interrogation. To some extent this may be related to a weakening of critical thought in development studies during the 1980s, but also the encompassing political climate has clearly favored the reassertion of Occidental hegemony in matters of development and modernization. And it is precisely in this context that we need to stress the fact that it is the political ascendancy of those who believe in the supposed superiority of the orthodox Western model of development, combined with their institutionalized power and enormous resources, that explains the widespread effectiveness of the "development as Westernization" discourse.

The Waning of the Marxist Challenge

The emergence of a Marxist and neo-Marxist challenge to orthodox, or what were customarily referred to as capitalist, views of modernization and development dates from the 1960s, and, with reference to the imperialist countries, especially from 1968. Dependency perspectives, which have been discussed in great length elsewhere (Kay 1989), prepared the ground for a move on to a more clearly-inscribed Marxist terrain. The rise of Marxist thought was a generalized phenomenon in the 1970s world of the social sciences, and in the domain of development studies, theoretical interpretations of modes of production, unequal exchange, world systems, and class conflicts figured prominently. Whilst primary conceptual importance was given to the relations and forces of production, to capital and wage labor, to the internationalization of capital as a social relation, and to class structures and struggles, in a more directly political language, capitalism was denounced as a system of exploitation that had to be replaced by socialism.

Although it is certainly the case that the Marxist diagnosis of the social and economic issues of development retains a role in the critical geographical and social science literature, I would argue that its influence is in decline. Apart from the impact of the events of 1989 and the disintegration of the erstwhile Second World, the fading intellectual and political influence of Marxist approaches can be related to three interrelated problems.

1. First, there has been a traditional tendency, although now much less marked, to assume that the economy would always be determinant in the last instance. In other words, it was presupposed that the logic of capitalist economic development governed the outcome of social and political processes. In one such reading the nature of the state was read off from the dynamic of this underlying logic. Moreover, it was frequently the case that not only was the economic structure centralized within the overall explanatory frame, but additionally subjects or social actors came to be absorbed within this determining structure. This analytical tendency represents an undiluted example of Marxist economism.
2. Second, when there has been an examination of sociopolitical change, the key subject has always been a class subject, and the class struggle has been interpreted as the defining historical struggle. There have been two interconnected difficulties here. Overall, there has been a failure to analyze the ways in which different forms of social subjectivity come into being. The processes through which individuals in society are constituted as social subjects or agents have been neglected, since the overriding concern has been with class subjects. Instead of viewing the class category as one possible point of arrival in an examination of social subjectivities, it has been taken as a pregiven point of departure. In addition, it has been assumed that in the construction of social consciousness, the point of production is central and determinant. Consequently, the understanding of the heterogeneity and complexity of social consciousness has been severely circumscribed. But also, since the social subject at the point of production has been interpreted as a unified subject, centered around the experiences of the workplace, it has been less possible to begin to comprehend the barriers to mobilization at this site of potential conflict.

The major problem with Marxist class analysis, particularly as it has been deployed in development studies, concerns the failure to theorize subjectivity and identity. This failure is in its turn conditioned by the belief that what classes do is spelled out by their situation in the relations of production, which precedes them causally as well as logically. In Marxism, classes are the agents of the historical process, but its unconscious agents, since it is posited that social being determines consciousness. But it can be more effectively argued that the reproduction of material existence, as well as the constitution of social being, must presuppose thought; they are not prior to it.

3. Lastly, permeating so much of the Marxist and neo-Marxist canon has been the belief in the existence of a pre-given, privileged social subject, cast in the role of the historical bearer of the revolutionary rupture from capitalism. Working-class revolution was for so long seen as a species of cure for the diseases of capitalism. But when actual revolutionary breaks occurred, as in a number of peripheral societies, the agents of those ruptures and splits could not be straitjacketed into any category of class belonging. In the cases of the Cuban and Nicaraguan Revolutions for instance, at specific historical moments, a variety of social subjects, unified around a particular political horizon, which combined a range of attitudes, feelings, sentiments and desires – around questions of the nation, of the fight against dictatorship, of the need for social justice and equality, of the struggle against US imperialism – came together and took a series of actions that culminated in the moment of revolution. These revolutions and others in the societies of the South were not engineered by an insurgent working class, but were brought to fruition by an aggregation of forces sharing a common political imagination. And such a sharing was precarious, temporary and unfixed. What these kinds of revolutionary upheaval demonstrated was the emergence at given moments of a collective will to overturn an existing order. Crucial to these processes of change was the fusion of a will to overthrow an established and unjust order with a desire for national dignity and independence.

Further to these three analytical problems of the Marxist reading of development and political change, we have witnessed the institutionalized embodiment of Marxist, and more precisely Marxist–Leninist, ideas, in highly authoritarian, one-Party states. Cuba is a striking example of such a phenomenon. With its one-Party system there has been a clear lack of any division of powers within the political space of the state. The notion that the Party is a synthesis of society, or the earlier belief dating back to Ché Guevara that in the post-Revolutionary period there ought to be a total identification between society and government, capture the meaning of a desire for total power. Within such a system, Marxist–Leninist thought was converted into a state doctrine, and used to portray the idea of a society without antagonisms, a society in which the class conflicts of capitalism had been transcended, a society in which political struggle had been successfully concluded. The unwillingness to accept difference, and the lack of any institutionalization of the means for the expression of effective difference, or alternative strategies, has sealed into place a rigid structure that carries within it little potential for constructive renewal.

The development achievements of the Cuban Revolution are well known. Transformations in the systems of health and education, impressive improvements in the utilization of agricultural and mineral resources, the installation and extension of public utilities and services, and sharp reductions in the degree of social and economic inequality. The meeting of basic needs was always a key priority of the Cuban Revolution before it became a catch phrase for international organizations. But, unlike the Nicaraguan Revolution, the Cuban process has not been characterized by any attempts to create new ways of combining different democratic practices. The longevity of the US trade embargo and general geopolitical isolation in the Western hemisphere ushered in a protracted reliance on the former Soviet-bloc countries. With the dissolution of the Soviet bloc, and a severe reduction in the amount of financial aid, concomitant with sharp changes in post-Soviet strategy, the Cuban polity looks even more fragile in its rigidity. The lack of political plurality and the absence of institutional channels for the expression of difference throw a long shadow over the basic-needs achievements of the Revolution. For many years the debate on Cuba was polarized between the uncritical supporters of revolutionary change and those who obdurately refused to countenance any positive evaluation of Cuban development. One of the contemporary dilemmas takes shape around the following problem: how will it be possible to preserve those social and economic achievements of the post-revolutionary period, while at the same time opening up the political system to a variety of different currents? In the Cuban case not only is the primacy of the political so apparent but equally the future trajectory of the island's development will be intimately connected to the changing impact of geopolitical circumstances.

Development and the Geopolitics of Knowledge

As Marxist perspectives have begun to fade away, the more recent critical currents of theory have been particularly concerned with questions of identity, difference, subjectivity, knowledge, and power. In development studies, and as exemplified in the Marglins (1990) and Sachs (1992) contributions, attempts have been made to construct a radical critique of the discourse of "development," seen as a hegemonic form of representation of the Third World. This new critical current, which has been characterized as expressing an "anti-development" viewpoint, carries within it certain shared assumptions and concerns. Prominent among these are the following: an interest in local knowledge and cultures as bases for redefining representation and societal values; a critical stance with respect to established scientific knowledge, and the defense and promotion of indigenous grassroots movements.

One of the encouraging and stimulating facets of today's critical research is the opening up of a series of interconnected pathways of analysis. Encouraged and enriched by the influence of feminist theories which have established not only the significance of gender, but also the centrality of questions of identity and difference, new work on the possible meanings of development has increasingly come to include a key gender dimension. In the connected analysis of new social movements and their relevance for rethinking the purposes and ends of development in neo-liberal times, the role of women in these struggles has been paramount, just as the investigations carried out by women social scientists and activists have been so fundamental to the theorization of social movements (Radcliffe and Westwood 1993).

Environmental issues, and the movements that have emerged to counter the damaging effects of conventional development projects, have received increasing attention and have been linked to the relevance of indigenous knowledge for the protection and sustainability of vital resources. When considering environmental politics, the links with the relevance of indigenous movements leads to a broader discussion of the objectives of development within which a challenging of the orthodox Western definition of knowledge for development can take place. At the same time, as environmental issues are also global issues, the framing of the agenda and the selection of priorities directly connects with the geopolitics of North–South relations. Whilst in the North, for example, there has been a tendency to concentrate on the rural aspects of environmental policy, in the South more urgent issues relate to the *urban* nature of environmental deterioration. Also the attempts by the North to define the environmental agenda and monitor the policies of Third World governments evoke crucial problems of ethics and international justice.

The processes of self-reflexivity, of stretching out toward new themes of dialogue, learning and re-thinking, can well be regarded as a positive and enabling element of the postmodern sensibility. In the North, there is the need to reinvent ourselves as other, which can be set in the context of taking historic responsibility for the social locations from which our speech and actions issue. The process of this reinvention requires the will and desire to learn from the South, not in a romantic or uncritical vein, fueled by an unconscious sense of culpability, but as a way of better understanding the North itself and with it the South. The life of the mind does not begin and end inside the Occident; the enclosure of Occidental thought needs to be fissured and broken open to other currents of thinking and reflection. If we are to develop a genuine global expansion of knowledge and understanding, the West's self-enclosure within ethnocentric standards will have to be transcended. Conventional development theory and practice has been one expression of power over another society and economy – a reflection of a belief in the West's manifest destiny. The Marxist challenge sought to bring into being another form of power based on socialist principles, but here too there has been a tendency to underwrite a privileged position for the West's standards and meanings of development. In today's discussion of the significance and dispositions of development, the politics of the production and deployment of knowledge has become an increasingly pivotal question. In an increasingly global world, the geopolitics of knowledge and power is a theme for which geographers can make an important analytical contribution, and, in the field of development studies, a critical geopolitics can begin by considering the power relations involved in the history of the North–South divide.

PART II

Geopolitical Change

Introduction to Part II: After the Cold War

Among the categories of global change covered in this volume, the geopolitical seems, at first, to be quite distinctive in its pattern of global shift. In the 1990s nobody doubts that recent geopolitical changes have fundamentally altered the meaning of world politics, but a decade ago even thinking of such a likelihood would have been deemed fanciful. During the 1980s, while other scholars were reporting critical changes such as the end of Fordism, the Aids pandemic, the postmodern celebration of cultural diversity and the depletion of the ozone layer, international political scholars were confronted with the seemingly unchanging politics of the Cold War to study. The pattern of international relations appeared frozen into a bi-polar world of USA and friends versus USSR and friends where the only changes were ones of degree: "freezes" and "thaws," in the jargon of the time. In the early 1980s President Reagan, viewing the USSR as the "evil empire," led the creation of the so-called "second cold war." This was to be the last "freeze" phase and was a reaction to US suspicions of the previous "thaw" phase, the *détente* of the 1970s. But whatever the political "temperature," the basic pattern of conflict was not changed, although as the conflict intensified the world was periodically made a more frightening place. The "thaw" that followed the coming to power of Gorbachev in the USSR in 1985 was interpreted by contemporaries, quite reasonably, as merely a second *détente*. As late as 1987 Edward Thompson could write about the Cold War seeming to be "an immutable fact of geography," with Europe "divided into two blocs which are stuck into postures of 'deterrence' for evermore." With our immense power of hindsight we know that all was soon to change. The policies developed by Gorbachev, *perestroika* (restructuring) and *glasnost* (openness), did not create a new phase of the Cold War but set in train processes that were to bring about its demise.

The political upheavals between 1989 and 1991 constitute a *geopolitical transition* (P. J. Taylor 1993b) from one world order, the Cold War, to another that, despite early US proclamation of a "new world order," has yet to be clearly defined. These transitions separate one relatively stable pattern of international politics, a geopolitical world order, from another. Despite the fact that the changes are very large, with old friends become new enemies and vice versa, the change-round is typically very rapid. Hence their hallmark is surprise as the political world is "turned upside down" in just one or two years. All this is true of 1989 and its aftermaths. The end of communism as a force in Europe (1989), the reunification of Germany (1990) and the demise of the USSR (1991) are each massive world events which together have completely changed contemporary world politics. Post-Cold War geopolitics are much more fluid and unpredictable and, in the 1990s, have been final confirmation of the general world ambience of great social changes out of control.

The geopolitical-transition pattern of change – stability followed by very rapid alteration – can be viewed as a political time-lag, a delayed response to the reduction in world economic growth rates that was about two decades old by 1989. During such difficult phases of the world-economy, we can expect additional competition between states as they strive to protect their economies from the general malaise. In the rich core countries policies are devised to export economic problems to each other and down the economic hierarchy. In the process, the periphery is economically devastated; but it is in between, in what are termed semi-periphery countries, that economic pressures have their greatest impact on politics (Chase-Dunn 1989). The 1980s witnessed the collapse of semi-peripheral Latin American military dictatorships as the precursor to the collapse of semi-peripheral communist Eastern Europe including the USSR. It is the peculiar role of the latter state, as political superpower but with a semi-peripheral economy, that precipitated a geopolitical transition deriving from the common travails of the semi-periphery.

This interpretation of the 1989 revolutions and their aftermaths shows that the rapid changes of geopolitical transition can be related to more gradual shifts typical in other areas of social change. There is much more to geopolitical change than alternating world orders. There have been important worldwide and gradual political changes throughout recent decades that are at least as important as the 1989–91 geopolitical transition for understanding future geopolitics. For instance, although in Eastern Europe alterations in state functions and the rise of nationalisms had to await 1989, elsewhere key debates about the nature of states and their relationships with national groups had been ongoing for more than two decades. Cutting the welfare state has been a universal response to the slowdown in economic growth, and this has usually been accompanied by a redefinition of the state's role. As markets have been proclaimed to be the arbiters of income and wealth distribution, the old corporatist state of partnership between government, business and labor has given way to a far less interventionist state. There has even been a wide-ranging discussion on whether states as we know them are in terminal decline.

Behind all these changes lies the matter of US relative decline in the world-system (Wallerstein 1984). Often interpreted as the latest decline of a world hegemon (dominant in economics and culture as well as politics), since about 1970 other Western states, notably Japan and Germany, have become economically powerful to rival the US in the world market. From undisputed leadership after World War II, in the 1970s the USA had to settle for a trilateral pattern of power (USA, Japan, Western Europe) in economic matters. The "second cold war" of the 1980s can be interpreted as the USA reasserting its authority over allies by privileging security, where it remained undisputed leader, over economy. However, this just created a huge budget and trade deficit to exacerbate the general relative decline. The result has been the power anomaly of the 1990s, where the USA is the sole remaining superpower but has a very vulnerable economy. Nevertheless we should not overlook the important legacies of US hegemony. In particular, the system of trans-state organization constructed by the US in the 1940s is as salient as ever. If the state is really in terminal decline it is these trans-state institutions centered on the United Nations that are expected to take on the burden of world governance in the future.

6

Democracy and Human Rights after the Cold War

John Agnew

Introduction

With the exception of a few reactionaries such as the Russian politician Zhirinovsky and even fewer unrepentant Stalinists, nearly everyone today claims adherence to "democratic" principles. Even Zhirinovsky calls his political party "Liberal Democratic"! Political regimes of all kinds throughout the world style themselves democracies, irrespective of whether their mode of leadership recruitment or treatment of their citizens conform to the ideals of limited government and popular rule. A claim to democracy bestows a legitimacy on modern political life. Rule by the few is justified when their decisions, rules, and policies can be portrayed as representing the "will of the people." This is a modern phenomenon. It is also controversial intellectually. The majority of political thinkers from the ancient Greeks to today's post-modernists have been critical of both the theory and practice of democracy.

The worldwide commitment to democracy as a justifying ideal, and certainly as a set of established practices, therefore, is both modern and flying in the face of much fashionable Western opinion (as recounted, for example, in Pangle 1992). Yet the "struggle" for democracy is fundamentally related to the historical struggle of subordinated groups for "recognition" and responsibility in governing their lives. The limited geographical spread of democracy and its fragility even where long established point to the tenuousness of its hold. The end of the Cold War does not signify a celebratory moment, even though some commentators have argued precisely this point (in particular, Fukuyama 1992).

After laying out the major terms in which democracy and human rights have been discussed, this chapter addresses how "democracy" was implicated in the Cold War, the dilemma for democracy posed by the increased openness of all territorial states to transnational economic and cultural influences, the challenge to democracy posed by the explosion of "identity politics" based on group affiliation (race, gender, interest group) rather than state citizenship (in particular, regional and ethnic politics), and the specific threats to and major challenges facing democratic achievement and possibility after the Cold War.

Democratic Theory and Practice

Proponents of democracy have differed profoundly over what its practice entails. Dispute has centered on such questions as: What constitutes a "people" capable of self-rule? Is democracy a process for selecting rulers or a set of institutions designed to achieve popular control over rulers? Is there an optimum size of political community for democracy? What is the role of obligation and dissent? Can democracies coerce recalcitrant minorities in the name of a majority? Are there social rights (e.g. full participation in management of the workplace, full employment, welfare rights) as well as political rights (e.g. voting, safeguards against arbitrary arrest and torture, an independent judiciary) that should be fostered in a democracy? Is democracy compatible with systems of hierarchical subordination such as some churches and most large business enterprises?

Modern thinking about democracy developed alongside the expansion of the territorial state in Europe from the sixteenth century on. Democratic theories sought to balance the rights of citizenship against the regulatory and coercive capacity of the state (Held 1987). The relative weight given to the limitation of arbitrary power versus the extension of participation in government differed, respectively, between those with liberal and those with democratic agendas. Eventually a syncretic "liberal democracy" provided the most frequently institutionalized way of resolving the dilemma – representative democracy. This intellectual solution grew out of the experience of, and the challenges to, absolute monarchy emanating from the English Revolution (1640–88), the American Declaration of Independence (1776), and the French Revolution (1789). Perhaps the most important statements of the liberal democratic position can be found in the philosophy of James Madison (1751–1836) and in the collected works of two English liberals, Jeremy Bentham (1748–1832) and James Mill (1773–1836). From their point of view the rulers could be held accountable to the ruled only through electoral choice involving a secret ballot, regular voting, and competition between potential representatives. The question of who would count as a "citizen" was left unresolved. Through often violent struggles, the idea that voting and other citizenship rights should be available to all adults (i.e. including women and ethnic minorities) has been slowly realized as a reliable feature of political life in only a limited number of states. Worldwide, the main achievement of liberal democracy, the periodic selection by the populace of representatives to make political decisions on their behalf, is more often than not either absent or episodic and subject to dramatic reversals.

Another approach to resolving the claims of citizenship against those of the state derives from the Marxist tradition. To Karl Marx (1818–83), and those who followed him, the great universal ideals of "liberty, equality and justice" associated with the French Revolution could not be achieved only at the ballot box and in the market place. Rather, capitalism deepened economic inequalities, and the state only served to protect the collective interests of the capitalist class against the emancipation of the working class. Only in the wake of capitalism's demise would "true" democracy be possible. In this context Marx envisaged a hierarchy of democratic forums extending from local communes to higher-order entities elected directly from the representatives at the lower level. This pyramidal structure was to inspire the Soviet system which emerged in Russia after the Bolshevik Revolution of 1917. A single party, the Communist Party, came to dominate this structure and effectively undermined the "bottom up" vectoring that inspired it. By appealing to the idea of social rights (full employment, etc.), the Communist Party offered compensation for the loss of political rights and an alternative model to that of liberal or bourgeois democracy: "people's democracy."

In Western Europe and North America after the Great Depression of the 1930s a hybrid "social democracy" offered a third approach to the definition of citizenship rights, in which elected governments provided a basic range of social services by means of progressive taxation. Under conditions of economic growth such as prevailed in Western Europe and North America during the 1950s and 1960s social democracy became widely accepted by most major political groupings. But since the 1970s this approach has been questioned, as the "welfare state" has been defined by some influential political leaders as a fiscal burden in an intensely competitive world economy.

The Democracies of the Cold War

The Cold War that emerged between the United States and the Soviet Union in the late 1940s, and which lasted until 1989, was closely tied to competing claims about democracy and human rights. On one side, the United States was represented by its governments as the ideal-typical liberal democracy or "limited" government, whereas the Soviet Union was represented as an expansionist tyranny. On the other side, the Soviet Union was represented by its governments as an embattled people's democracy encircled by the agents of an expansionist world capitalism. The Cold War, therefore, was a conflict of claims about democracy – a conflict between political discourses – as much as it was anything else.

The sudden collapse of Communist rule in the Soviet Union and Eastern Europe, for reasons of economic failure more than lack of living up to Marx's model of political democracy, has stimulated a burst of theorizing about the "victory" of liberal democracy over its counterpart. This implies that both models were somehow real rather than abstracted versions of what actually went on in the United States and the Soviet Union during the Cold War. In this construction, ideological struggle over the "best" form of politics is said to be over and, in one extreme formulation (that of Francis Fukuyama 1992), history itself is declared to be at its end. More temperate commentators note, however, that in most countries with a history of democratic practice, liberal democracy is in a condition of arrested development. Not only has electoral participation shrunk dramatically, most notoriously in the United States itself, but democratic institutions are seen as corrupt and inefficient in ways not seen in Europe since the 1920s and 1930s. Political discourse that was organized largely around the categories and slogans of the Cold War has been unable as yet to find a satisfactory substitute. The collapse of the Christian Democrat–Communist division in Italy as part of the fall-out from the end of the Cold War, for example, has produced an interminable political crisis and increased cynicism about liberal democratic politics. Those issues high on the agenda in the US and Europe as alternative foci for political competition – namely foreign immigration, "civilizational" clashes (particularly Islam versus Western secularism), free trade and regional integration, and geoeconomic competition – do not provide the same capacity for systemic integration as did the Cold War with its manichean contest of "good" versus "evil" and competing visions of the essence of democracy.

The essential distinction of political versus social rights to which the two sides in the Cold War drew attention remains unresolved, irrespective of the outcome of the Cold War. As the welfare state in Western Europe, Australasia, and North America is challenged on the grounds of its undermining economic efficiency and national competitiveness, the defense of social rights is likely to become as important an issue as the protection and enhancement of political rights. The influence of the Marxist tradition's emphasis on social rights will be felt in countries where its manifestation as Soviet Communism was never of much political significance.

Yet, the continuing importance of liberal democracy should not be slighted (Przeworski 1991). During the 1980s the practice of liberal democracy spread widely into world regions where its previous hold had been tenuous at best (see, e.g. Diamond 1992). Almost all the countries of Latin America witnessed some form of transition to a broadly democratic (civilian constitutional) regime; though the hold of liberal democracy is still tenuous in the region, as it is throughout the global "South" of

Figure 6.1 Elections in Latin America.

underdeveloped and formerly colonized countries (figure 6.1). In southern Europe, Africa, and Eastern Europe similar if less noted transitions to liberal democratic rule have been under way since the 1970s. In South Africa and in Israel/Palestine seemingly intractable struggles have produced new openings toward a more democratic future. In many of these settings the transition to democratic rule has had to coexist with economic stagnation or decline and the austerity imposed by the structural adjustment and stabilization policies of the International Monetary Fund and the World Bank. What is most hopeful is that this conjuncture has not invariably produced demands for a reversion to authoritarian solutions. In many countries it was the authoritarian regimes that helped bring about the economic disasters that more democratic governments must now try to manage (for example, in Argentina or Brazil). Unfortunately, in this context social rights tend to be neglected in order to reduce government spending and attract international investment. Of course, previous liberal democratic interludes produced reversions to authoritarian government and this could well happen again.

The Territorial State and Democracy

The debate over democracy's content (the mix of rights and duties) grew alongside the modern territorial state as it evolved in Europe in the nineteenth and early twentieth centuries. The association between liberal democracy and the state has been particularly close given liberal democracy's focus on procedures for ensuring the election of state rulers for limited periods of time. Indeed, the sense of its *necessary* association with democracy gives the state much of its normative appeal. In his classic work *Politics and Vision* (1960), Sheldon Wolin put the dilemma bluntly:

> To reject the state [means] denying the central referent of the political, abandoning a whole range of notions and the practices to which they point – citizenship, obligation, general authority, . . . Moreover, to exchange society or groups for the state might turn out to be a doubtful bargain if society should, like the state, prove unable to resist the tide of bureaucratization.

But the territorial state's relevance to the contemporary world is increasingly in question. Businesses are less and less organized solely to exploit local or regional market opportunities. They are increasingly multinational or global in scope. Unlike in Marx's day, the state appears less the shield of capital and more its latest victim. States must now choose whether they represent the interests of territorially circumscribed populations or the interests of businesses that operate globally but which originated within their confines (see Reich 1991). In fact, in a recent work Wolin himself (1989, pp. 16–17) captures the sense of the moment:

> compelled by the fierce demands of international competition to innovate ceaselessly, capitalism resorts to measures that prove socially unsettling and that hasten the very instability that capitalists fear. Plants are closed or relocated; workers find themselves forced to pull up roots and follow the dictates of the labor market; and social spending for programs to lessen the harm wrought by economic "forces" is reduced so as not to imperil capital accumulation. Thus, the exigencies of competition undercut the settled identities of job, skill, and place and the traditional values of family and neighborhood which normally are the vital elements of the culture that sustains collective identity and, ultimately, *state power itself*. [my emphasis]

The state's failure to protect its citizens from enhanced capital mobility undermines its claim to represent their best interests. This "legitimacy crisis" is compounded by the deadening impact on political life of the disruption of social and political institutions (such as local government and community groups) in which democratic practices have become most strongly established.

In this context, the interest of political theorists is now more, as it were, above and below the level of the state: in transnational formations of various sorts and in the variegated groupings within society that are not orientated to the state per se. Transnationalism is the obvious result of the attempt to match political control with the increasingly globalized workings of the world economy (Held 1991). How can vesting powers in different transnational arrangements (regional groupings of states such as the European Community, and international organizations linking sovereign states, e.g. the United Nations), be organized given traditional reliance on notions of state sovereignty to justify political legitimacy? How democratic can and should such arrangements be? Will a transnational order be composed of states or composed of individual persons liberated from lower jurisdictions? What threat to "traditional" cultures and religions is posed by the defense of individual rights implicit in most transnational organizations?

How can one justify intervention by international organizations such as the UN in the "internal affairs" of sovereign states to protect or enhance individual rights?

The Politics of Difference

There is a growing tension between the aspirations of many globalist programs and the increasing interest in "the politics of difference" involving various local and sectoral (ethnic, gender, behavioral) claims to "group rights." Postcolonial and communitarian theorists defend the mobilization, empowerment and education of people as members of social groups whose identities have been stigmatized, repressed, or ignored or who identify issues (such as the global ecology) that are not readily contained within the territorial confines of state-regulated and state-mandated politics. Feminism and environmentalism are the paradigm examples of the "new social movements." The main contrast is with class-based and nationalist movements which aimed at taking state power by adopting broad political programs. The new movements conversely aim at changing particular state policies and changing the workings of society through "consciousness raising." They are not interested so much in becoming the rulers of a territorial state as in creating spaces for associational life autonomous of state power.

Some groups have proved capable of "crossing the borders" of particular states and have begun the construction of a viable international society beyond the direct control of states. Many of these, such as Greenpeace, Amnesty International, and Oxfam, are concerned with acute environmental, political, and food crises. They have profited from the immediacy of modern televisual representations of such crises. Unlike the Catholic Church or the Communist International, they are not simply out to realize a totalizing vision of the world. They want to save the ozone layer, save women's lives from systematic abuse, or expose the routine use of torture by police and military forces.

The "new" politics is more tumultuous than the old and less contained by state boundaries. One question concerns how it will link with older political movements based upon class and nationality. What also remains to be seen is whether it is more democratic. Certainly, groups that were previously marginal to mainstream politics (such as racial minorities, sexual minorities, disabled people) now appear regularly in the political arena in many countries. But one frequently asserted threat to democracy comes from the elevation of group rights over those of individuals. The worry is that standards of conduct will be imposed upon some individuals because their beliefs, behavior, or language are offensive to a particular group. In the United States this concern with "politically correct" behavior has emerged most strongly in relation to so-called "First Amendment" rights involving the constitutionally guaranteed right to express opinions free of prior restraint. As yet, however, only with respect to community regulations governing pornography and in regard to some universities' codes of speech and sexual conduct have these issues had much actual impact (Hughes 1993).

There is a more general challenge to conventional thinking about and practice of democracy from the "politics of difference." This lies in the valuation of communal or sectoral ties at the expense of participation in larger political communities (such as states). When political identity is totally bound up with a singular social identity then the search for communalities and common understandings among geographical neighbors who are identified as belonging to some other social group is abandoned. Harry Goulborne (1991), for example, worries that in postcolonial Britain ethnic–communal identities are beginning to exclude the possibility of the shared political identity upon which democratic politics is premised. Sikhs and Guyanese, for instance, care more about the politics of their homelands than the United Kingdom's. At the same time, "British" identity is more exclusionary and openly based on an ethnic or "bloodline" conception of nationality.

Regional and Ethnic Politics

The concern is that such long decried practices as "apartheid" and "ethnic cleansing" will escape from the geographical confines in which they have been contained (in, respectively, South Africa and Bosnia) to afflict societies all over the world. The collapse of Yugoslavia into a frenzy of inter-communal violence, and murderous civil wars in such former Soviet republics as Georgia and Azerbaijan, suggest one kind of outcome for an overemphasis on the politics of difference. Movements for ethnic, religious and regional autonomy, however, can argue with some justification to be claiming their democratic rights (Nairn 1993). One feature of democracy and human rights as they developed in the nineteenth century was their connection to the right of national self-determination. People were claiming their rightful place in history when they campaigned for national unification or separation from large and frequently tyrannical empires. The Greek patriots fighting for their freedom from the Turks, and such heroes as Garibaldi fighting to overcome the partition of Italy into disparate kingdoms and principalities, inspired democrats as much as nationalists. Indeed, nationalism appeared as an ideology of freedom in a way that it no longer does. At the end of the twentieth century it appears as parochial and backward-looking as it once appeared universal and progressive. After the bloodletting and concentration camps of two world wars, nationalism has lost much of its earlier sheen.

A case can be made that, contrary to the idea of ethnic nationalism as a worldwide trend, it is in fact a response to regional and local forces. Prime among these is the collapse of centralized Communist (and other) governments which used (and abused) ethnic labels as a tactic of divide-and-rule and to reward certain groups at the expense of others. Other forces would include frustration with corrupt central governments (as with the Lombard League in contemporary Italy), the misalignment of political and cultural boundaries (as in large parts of Africa; see Davidson 1992), and the use of religious and language affiliations as techniques of political mobilization (as in India and Sri Lanka). This is not to minimize the importance of violent ethnic conflicts in threatening democratic politics in many parts of the world. It is simply to suggest that ethnic and religious divisions are not *universal* in the threat they pose to democracy and human rights.

Four Threats to Democracy

At a global scale, four other trends are much more threatening to the past two centuries of fragile democratic accomplishment.

One is the reassertion of centralized authority in such hierarchical structures as transnational corporations and multinational churches (such as the Catholic Church under Pope John Paul II and various Islamic and Christian revivalist movements) which, in their investment and moral decisions, respectively, show little sympathy for the free expression of opinion and popular constitution of authority that lie at the heart of the liberal democratic tradition. Corporate managerialism and theocratic authority fundamentally challenge the open process of negotiation and decision that democracy requires. Efficiency for its own (or profit's) sake and conformity to hierarchically constituted authority have always coexisted uneasily with modern democracy. Hierarchical subordination did not die out with the rise of the modern territorial state. Indeed, a case could be made that in some world regions, such as the Middle East, Africa, and South America, churches, corporations and social networks (such as lineages and chieftainships) are more effective units of rule over people's lives than are states, which frequently fail to deliver on their promises. Yet these other organizations are often anti-democratic in their procedures and goals.

A second trend, exemplified above all in Eastern Europe and the former Soviet Union, but by no means restricted to them, is the increasing confusion between democracy and economic liberalization. This confusion was characteristic of a certain American position during the Cold War in which the terms capitalism and democracy were used synonymously. Although the conduct of open elections was a necessary part of this definitional conflation, open markets were much more central to the calculus. Democracy was essentially redefined as a market-access regime in which the free flow of mobile factors of production was the central attribute. Today such international economic organizations as the International Monetary Fund and the World Bank seem to have adopted this definition of democracy. Little or no attention is paid to how responsive governments are to the demands of their citizens. Rather, the needs of external creditors and potential investors have a higher priority. Sometimes, democracy in its classic sense is seen as a barrier to economic development; something that must wait until economic "take-off" is under way. The empirical research on this connection is equivocal, suggesting that liberal democracy is not the villain it is usually painted as: sustaining the "outrageous" demands of a citizenry without any sense of the economic "realities." In other words, political rights are not a "luxury" conferred by economic development.

A third trend inimical to the prospering of democracy is the growing social–geographical inequality within and between states. At the global scale there is a widening gap in material well-being between residents of some states, particularly those in sub-Saharan Africa, and those in the more industrialized countries. Within states, even relatively prosperous ones, regional and social inequalities have been growing once again, having been much reduced in the period 1945–75 by means of various income and regional policies. These inequalities not only challenge the achievement of the great democratic virtues in the French revolutionary tradition (one is "equality"), they also prevent equal or equivalent participation in political life. Consequently some regions, localities, and social groups are disadvantaged in the political process. Poorer regions (e.g. southern Italy, the former South African homelands) can become prone to the depredations of patronage politics. Richer regions can use their wealth to buy political advantage. Only a reversion to the "discredited" regional investment and income redistribution policies of the 1960s and 1970s could reverse this trend (Denitch 1992).

Finally, a fourth trend threatening democracy around the world is the posing of human rights as particular to cultures rather than actually or potentially universal in nature. At the 1993 UN Conference on Human Rights a major contention of the representatives of several Asian states, including China and Indonesia, was that the understandings of human rights enshrined in the charter of the UN, far from being universal, were the global projection of European understandings of human freedom, the rights of individuals in relation to governments, the duties of states in relation to citizens, etc. From this point of view there can be no such thing as universal human rights. There are only rights *particular* to certain cultural traditions. The ideal of democracy as a universal form of rule, therefore, is rendered moot.

There is certainly some truth in this position. The modern discourse of democracy and human rights is largely European in inspiration. However, most of the world's cultures have democratic "currents" of one sort or another and most "cultures" are now so syncretic in origin that claims for national cultural particularity do not bear close examination. The idea of "oriental despotism" is European; Asia, Europe's classic Other, was always more politically complex than portrayed by Europeans (Springborg 1992). However, European imperialism also worked its effects in contradictory as well as devastating ways, none so ironic as the spread of democratic ideals which eventually helped undermine European claims to rule their colonies without popular consent. It is difficult to take seriously the charges of governments against a universal model of democracy whose own legitimacy rests at least in part on popular consent. Most importantly, the consequences of the "cultural critique" of democracy appear slighter than at first glance. No government, not even that of China, was appealing for the right to torture political dissidents or the right to lock political opponents in prison without trial. The past two hundred years have at least had that effect, minor as it often appears to be.

In aspiration, democracy is an inventive set of arrangements for the arousal, expression, and mediation of popular political sentiments and interests. It is a reflection of the staying power and spreading influence of the idea of democracy that the battle over its essential elements is likely to continue indefinitely. Democratic politics does not admit of authoritarian definitions. This indeterminacy is precisely what disturbs authoritarians.

Challenges Facing Democracy after the Cold War

One key area of dispute connects democracy to the increasing geographical scope and pace of contemporary life. Conventional thinking sees democracy as flourishing in "small places" where active participation can be nurtured. Without extension into the realms of international relations and global flows, however, the possibilities for real or strong democracy in the twenty-first century seem limited at best. Keeping democracy tied to the territorial state seems, like state socialism, to be something of the past in a world of increasing international migration and vastly expanded global economic transactions.

A second key area connects the deterritorialization of democracy to an increased concern for economic democracy. The one (large) area of life where both liberal democracy, people's democracy and social democracy have failed to make much headway is in bringing a degree of democracy into the management of the economy in general and workplaces in particular. The difficulty has lain in the restriction of the terms of democratic debate to the "political sphere." Yet, the arguments that vital decisions should not be made without popular consent, used by classic democratic theorists to justify political democracy, also apply to the economy. The point here is not to advocate a single form of ownership; that road is well traveled and beckons beyond democracy. But only through building a vigorous popular involvement in economic life can democratic politics itself be realized (Putnam 1993). Enhancing the possibility of democratic regulation of economic activities without massive bureaucratization increases the likelihood that democracy can achieve the goal of popular involvement in key areas of life that its advocates have always aspired to.

But the greatest challenge facing advocates of democracy after the Cold War is to combine its historic commitments to universal human rights and increasing the equality of the material conditions of life with a relativistic mission which allows for the different meanings of needs, justice, and ownership characteristic of different societies. A place for "fraternity" and "sorority" needs to be found in a democratic discourse dominated hitherto by liberty and equality (Taylor 1992c). Unfortunately, this is not a task for which we have been well prepared, either by political theorists or by recent political practice.

Acknowledgment

I would like to thank Amy Becker, Peter Taylor and Michael Watts for their help with this chapter.

The Renaissance of Nationalism

Nuala C. Johnson

Introduction

December 1989 was a month of monumental change for the world political map in general and for Eastern Europe in particular. It precipitated the most significant changes in the map of Europe since the Second World War. The Brandenburg Gate, dividing East and West Berlin, was opened on December 22 after 28 years of closure and this symbolic act prefaced the unification of East and West Germany in October 1990. The former dissident Czech writer Vaclav Havel became the non-Communist leader of Czechoslovakia on December 29, 1989 (later to be divided into the Czech Republic and Slovakia), and Romania's dictatorial leader Ceausescu and his wife Elena were executed on Christmas Day of that year. The initiation of *glasnost* and *perestroika* under Gorbachev's leadership of the Soviet Union in the mid-1980s eventually resulted in the break-up of that union, with the establishment of independent states in the Baltic in 1991 and the evolution of the loose confederation of states comprising the Commonwealth of Independent States (CIS).

In long-established states, the 1980s also witnessed a rejuvenated sense of national identity and patriotism, culminating in Britain with the Falklands War and in the USA with the Gulf crisis. As Anderson (1983, p. 12) comments, "the 'end of the era of nationalism', so long prophesied, is not remotely in sight." As the geopolitical blocs that characterized the Cold War have evaporated, nationalism has emerged as one of the dominant discourses of recent times. Although global processes appear to be eliding the role of the national, and postmodernism is emphasizing the fractured basis of political and cultural identities, the national state continues to exercise power as a mediating link between the local and the global. The changing political geography of the "new world order" adds yet more complexity in the cultural–political make-up of the planet.

This chapter will treat nationalism not just as an ideology and practice that has recently experienced a revival (Smith 1986), but as one of the most enduring ideologies that has structured political life over the last two hundred years. As such, nationalism – the desire to bring cultural and territorial imperatives together – will be analyzed by examining key concepts in the lexicon of nationalist discourse, both the symbolic and the literal, to elucidate some of the ways in which we can forge a better understanding of the changes being 'mapped out' in the global political landscape.

Imagined Communities

Although there is a huge literature on nationalism (see Hobsbawm 1990), proffering definitions, typologies, explanations and case-studies of the phenomenon, from a theoretical viewpoint Anderson's claim that the nation is an "imagined community," in which there is an assumed cultural communality in spite of class, or geographical and social distance, continues to be a persuasive framework from which to analyze nationalism. This imagining emerged in the context of the rise of capitalism in the sixteenth century, the replacement of religious and dynastic orthodoxies with new political and social formations associated with the Reformation and the Enlightenment (Anderson 1983). These changes were located in an era when there was an extension of the world capitalist economy which centered on national states (Wallerstein 1974; Agnew and Knox 1989). The production of cheap books and newspapers, written in the vernacular, enabled geographically dispersed peoples to recognize the existence of others with a same language and to unite culturally as a result. While by the sixteenth century the printing press and its products were experiencing a boom, books were still only available to a small, literate minority. It was increasingly with the introduction of mass education in the nineteenth century that widespread literacy was attained, thus making possible the "mass" imaginings underlying nationalist sentiments (Fishman 1972), and allowing for the break-up of poly-vernacular empires and the unification of localized territorial kingdoms.

Nation-Building

The "imaginative discourse" surrounding the nation can assume many forms (Bhabha 1990), variously centered on conceptions of history, collective memory, habitat and folk culture, traditions, poetic spaces and symbolic landscapes (Smith 1986; Williams and Smith 1983; Hobsbawm 1990; Hobsbawm and Ranger 1983; Cosgrove and Daniels 1988). In the drive to achieve territorial and cultural consolidation, exercises in nation-building are necessary and are constantly subject to renewal and rejuvenation. These can be overt exercises (even coercive) such as a state's education policy, or they can be much more subtle processes such as the gradual promotion of particular landscapes as representations of national identity. In either case, cultural legitimacy must be achieved by the national state which, as Gramsci noted more than half a century ago, seeks to "raise the great mass of the population to a particular cultural and moral level" (1971, p. 258). In Eastern Europe the evidence suggests that "national" cultural legitimacy was not achieved under communism and that these states today are struggling to grapple with diffuse political and cultural allegiances (e.g. Bosnia). Similarly in postcolonial Africa the construction of unified nation states has proved difficult; linguistic, religious and ethnic tensions have persisted. The remainder of this chapter will examine, with examples, the process of nation-building and some of the challenges that have been raised against the nation state.

Linguistic Imaginings

> Has nationality anything dearer than the speech of its fathers? In its speech resides its whole thought domain, its tradition, history, religion,

> and basis of life, all its heart and soul. . . . With language is created the heart of a people. (J. G. Herder 1968 [1783])

The cultural definition of identity frequently rests, as Herder asserts, on linguistic differentiation. Fishman (1972) contends that there are four main reasons why language is useful and often intrinsic to the nationalist cause. First, functionally it can arouse ideas of a common identity. Second, it forms a link with the past. It can safeguard the "sentiment and behavioral links between the speech community of today and its (real or imaginary) counterparts yesterday and in antiquity" (Fishman 1972, p. 44). Third, language becomes a link with authenticity. It provides a secular source of mass symbolism in modern society and yet can lay claim to uniqueness. Finally, a vernacular literature can allow elites to become central to a nationalist movement. The politicization of language requires planning. The standardization of spelling, grammar and so forth, and a mass education system achieves a degree of uniformity, at least as far as the written word is concerned. Language planning is crucial for the breaking down of old and the construction of new spatial barriers at the scale of the state. Where language planning is unsuccessful tensions between linguistic communities forming a state can lead to a separatist politics.

A classic example of linguistic tensions is to be found in Belgium. Created as a state in 1830, with a merging of Flemish areas in the north and French-speaking lands in the south, Belgium has experienced a series of constitutional and linguistic crises centered around divisions between the Flemish and Francophone communities (Senelle 1989; Frognier et al. 1982). Although the state has endeavored to overcome some of these divisions by moving toward a federal system of government and by the allocation of limited autonomous powers to the four regions making up the state, a crisis continues to haunt Belgium, despite Brussels's status as the administrative capital of the European Community. The cultural conflict has a micro-geography as well as a regional geography. This micro-geography is expressed in the suburbs of Brussels, where Flemings fear for their linguistic future as Francophone Bruxellois move to the suburb of Overijse. This trend has stimulated the local Flemish MP to try to legally quell

Francophone migration to protect the linguistic and economic well-being of the Flemish population living there: "it is regional not national politics which attracts the Belgian political talent, the language question dominates and permeates all political questions" (*Independent on Sunday*, 10 October 1993, p. 15). The future of the state is in a precarious position given the salience of the language question. Similarly the constitutional status of Quebec, in the Canadian context, has dominated the politics of that region since the 1980s (Laponce 1984). The results of the 1993 general election confirm the salience of linguistic issues in the Canadian political system. According to Lucien Bouchard, leader of the main opposition party, the Bloc Québecois, "No longer, will English-speaking Canada be able to pretend the constitutional quagmire is a thing of the past" (quoted in *The Independent*, 27 October 1993, p. 14). Although historically language served as one of the defining characteristics in nation-building projects, today language is an important cause of cultural conflict within many states where communities speaking minority languages are challenging the hegemony of majority language cultures (e.g. Spain, France).

In Eastern Europe, since the liberalization of the political regimes, the rights of linguistic minorities have firmly re-entered the political agenda. In Slovakia, for instance, the Hungarian minority have claimed that the Slovaks have systematically pursued an assimilation policy of "educating Hungarians into state-patriotism towards the Czechoslovak nation state, and a campaign for Hungarians to learn Slovak" (Carter 1993, p. 247).

Postcolonial states have also had to make the choice of "whose" language to use in the context of independence. The revival of a local language can be an important legitimation for independence. In the case of Ireland's independence from Britain, the state declared Irish the national language and instituted a series of policies to maintain and extend its usage as a central cultural marker (Johnson 1992). In the case of a multilingual state such as India, the adoption of an official language (Hindi) has presented great problems, allowing English to remain the official language of administration and government (Seton-Watson, 1977). Not only may the language of the colonizer be entrenched in the social relations of independent states, but the landscape can be named in the fashion of the colonizer. Duncan (1989), in the context of Sri Lanka, has highlighted how after independence the Sri Lankan nationalist party renamed the streets of Kandy in the vernacular, an exercise in "symbolic decolonization." Decolonization, however, is rarely a simple process, and just as the naming of streets, towns, and villages in the language of the colonizer is part of the colonial endeavor, the response of the colonized to negotiate their own cultural space after independence is fraught with difficulties. The place of Russian in the context of the Commonwealth of Independent States is a case in point.

Heroic Pasts

While nationalism has a territorial imperative in the acquisition of space or territory for a "nation" to inhabit, the imagined community also has a temporal dimension, in which previous periods of "past glory" are rejuvenated for present self-aggrandizement (Smith 1986). The process of the "invention of tradition" eschews the chronology of the historical imagination in favor of a focus on particular historical or quasi-historical epochs, when the cultural effervescence of the "nation" was particularly heroic. The geographical and the historical merge in the context of these imaginings as particular places and landscapes become centers of the collective cultural memory (Hobsbawm and Ranger 1983). McCrone (1992), in the context of Scottish nationalism, argues that the tartanization of Scottish culture involved a selective reading of the "national" past which ignored the real political forces that produced such imaginings. The popularization of the tartan kilt and of tartan patterns associated with different clans corresponds with a period of changing economic fortunes in Scotland in relation to the British and the world economy. Not only is the constitution of Scottish identity place-based, but the articulation of separatist politics is embedded in the histories and geographies of particular regions in Scotland (Agnew 1987).

The compelling appeal of "invented traditions" is not confined to minority groups within larger states, but is also found in well-established majority cultures. The contemporary nostalgia for the past has been tentatively linked to public dismay in the purported anonymity of high modernity, where identities are increasingly being fragmented in favor of more situated and flexible ones (Urry 1990). The abandonment of some older myths and traditions associated with national cultures has seen their replacement with new or reconstructed ones. Tourism and the heritage industry frequently reinforce images of a heroic national past, packaged for public, popular consumption (Hewison 1987; Wright 1985). New technologies, new forms of museum display, and the "theme-parking" of historical narratives may have replaced the older traditions of popular ritual (e.g. parades), but the history purported to be represented through these installations often merely anchors conceptions of national identity on new terrains (Lowenthal 1991). While, in the nineteenth century, national languages, school history textbooks, and religion formed the nexus of national imaginings, in the late twentieth century the heritage industry and the associated historicizing of interior design and architecture plays a similar role.

In Britain the resurgence of national feeling in the 1980s is attributed to a number of factors: to the decline of Britain in the world economy as the stockmarkets of New York and Tokyo overtake that of London; to challenges from Scotland, Wales and Northern Ireland to English cultural hegemony; and to immigration from non-European states (Samuel 1989). One reaction to these processes is a reassertion of "national" cultural values. The Prince of Wales, in *A Vision of Britain*, published in 1989, articulated this loss of national prowess in postwar Britain and especially in London. The image envisioned by the prince "is a London restored, re-visioned as landscape, framed by lavish reproductions of eighteenth- and nineteenth-century oil paintings" (Daniels 1993, p. 11). Using Canaletto's painting of London as a blueprint and looking today at St Paul's cathedral and environs, where towers disrupt "the symbolic exchange between St. Paul's and the City, the solid image of a community of interest which framed London's supremacy as the centre of world trade" (Daniels 1993, p. 13), the prince is imagining the city being recast in the image of London as glorious, and by inference Britain as supreme. Daniels illustrates that the cathedral had been subject to various interpretations over the centuries and that it was not always eulogized on architectural or civic grounds, yet the prince's view eschews the historical record for a more heroic view of St Paul's in the history of the nation. The emergence of royal interest in the architecture of the city corresponds with the revival of royal pageantry in the 1970s and early 1980s. The furore generated in the USA in 1991 by the exhibition "The West as America: Reinterpreting Images of the Frontier 1820–1920" at the National Museum of American Art, underlines the unwillingness of many, and of state representatives, to have their nation-building mythologies challenged. As Watts (1992b, p. 116) claims, the exhibition's "unflinching account of the brutality of the frontier – a space that, . . . has been ideologically formative in the construction of a particular national identity – did not lie well with the jingoistic and nationalist sentiments rampant on the Hill [Capitol Hill]."

The potency of a heroic history rejuvenated for current political reasons can be particularly skewed in the context of civil strife. The historical memory is continually jogged to assert cultural legitimacy and it is incorporated into the iconography of everyday life. In Northern Ireland, where there has been a failure to establish an agreed history and an agreed interpretation of the past, history is a source of constant dispute. Northern Ireland is not so much an exceptional case but an example of more general processes writ large. Varying interpretations of history are popularly consumed through graffiti, street parades, carpet painting of pavements, and wall murals adorning the housing estates of Northern Ireland. While street festivals such as London's Notting Hill Carnival may be celebrations of local cultural resistance to hegemonic groups (Jackson 1988), in Northern Ireland they are heavily politicized dramas of binary opposition carrying important symbolic and material consequences. For the Unionist population of Northern Ireland, the historic moment that links present-day political strife with past antecedents is the accession of William of Orange to the crown. Wall murals of "King Billy" and his victories assert for Unionists a legitimation of their current political existence (Rolston 1991), and the painting of such murals has a long history in Northern Ireland (Loftus 1990). The murals are principally "concerned with the entrenchment of existing structures and beliefs" (Jarman 1992, p. 161); and these beliefs rely upon a particular view of the past.

Mythical figures as well as historical ones can also articulate conceptions of national identity (Smith 1986). In Belfast, members of the Ulster Defence Association (UDA) have appropriated the figure of Cuchulain in their wall murals. Traditionally Cuchulain represented "early valour, miraculous feats, generosity, self-sacrifice, beauty and loyalty – evoked an archaic epoch of nobility and liberty, in which the full potential of Irishmen was realized" (Smith 1986, p. 195). The use of Cuchulain in Loyalist wall murals is an attempt, according to Rolston, "to retrieve a history beyond 1690, and the Battle of the Boyne" (quoted in *The Irish Times*, 19 January 1993, p. 6). Rather than viewing this mythical figure as a Gael, Loyalists have redefined him as one defending Ulster from the Gaelic queen Maeve. Ironically the interpretations awarded to "golden ages" vary across political communities and create tensions in the discursive practice of mythmaking.

While heroic histories form an important part in the establishment of an "imagined community," these pasts are also intimately linked with heroic people in nationalist discourse, and these people are embodied in monuments and statuary that adorn the towns and cities of national states.

Monuments and Nationalism

If nationalism appropriates periods of the past to represent its continuity, it also appropriates historical persons, events, and allegorical figures to reinforce its cultural existence. In the nineteenth century public statuary had firmly entered the public domain (Johnson 1994). Monuments dedicated to important figures in the nation's history began to emerge on a large scale, and such a process was fraught with dissension and disturbance. In Eastern Europe historic figures associated with the evolution of these states are being gradually replaced. Statues to Lenin and Stalin have been systematically removed from the public sphere in towns and cities of the former Eastern bloc – "Moscow has set up a Commission on Cultural Heritage to deal with the statues. . . . Few [of the 123 Lenins] will be kept; most will go to the Museum of Totalitarian Art" (quoted in Bonifice and Fowler 1993, p. 126).

The erection of public monuments often is an intrinsic part of the nation-building process. As Mosse (1975, p. 8) effectively posits, "The national monument as a means of self-expression served to anchor national myths and symbols in the consciousness of the people, and some have retained their effectiveness to the present day." They commemorate real historical figures such as political leaders, writers, adventurers and military leaders. Statues also commemorate war or are used to personify for the national community abstract concepts such as justice and liberty (Warner 1985). Statues articulate in a material and ideological fashion the collective memories of a nation's past, but the choice of "whose" heroes and "which" events to commemorate reveals the process by which groups achieve hegemonic positions within a state. The geography of monuments articulates a hierarchy of the sites of memory within a nation and the relative power of different groups.

While the nation state remembers its founding heroes, it also commemorates its wars – they can be wars of independence, international conflict or even civil strife. The war memorial varies in iconography depending on whether it is commemorating defeat or victory. Smith (1986, p. 206) notes that "Creating nations is a recurrent activity, which has to be renewed periodically," and the construction of public monuments is one way in which this renewal and re-interpretation of the national past is articulated.

War memorials that personify particular military leaders tend to be heroic in proportion and may use iconography from previous eras to convey the strength and national importance attached to an individual. The Nelson Column, in London's Trafalgar Square, adopts a design drawn from the iconography of ancient Roman imperial victories (Mace 1976). While it took a long period finally to execute the design of the column and the ancillary figures surrounding the monument, the square itself has frequently been the focus of political protest in London. As a site for lobbying collective memory Trafalgar Square has simultaneously played an ambiguous role as a centre of public protest where the national state has been challenged (Mace 1976).

Memorials to the First and Second World Wars vary considerably in scale and iconography. They can be simple commemorative plaques placed in towns and villages listing local people who were killed, or they can be colossal national monuments located in prestige positions within capital cities. Annual pilgrimages to these memorials and the laying of wreaths at their foot reinforces the states' recognition of the importance of war and the losses it incurs, but these occasions also enable the mass of the population to participate openly in a national event (Mosse 1975). The tone of the inscriptions on war memorials replicates those found on headstones in cemeteries, but unlike the latter they tend to conceal the class and gender divisions of the people they represent. In this sense they function as "nationalized" monuments.

Not all war memorials are heroic in their symbolism, nor is popular support for their construction universal. Where the support for war is contested and ambiguous, the building of a memorial can highlight some of the underlying fissures within the national community. The case of the Vietnam Memorial in Washington is a notable example. As a result of the public's antipathy to America's involvement in Vietnam the conception of a monument proved difficult and raised several issues for the collective memory (Wagner-Pacifini and Schwartz 1991). Rather than being a unifying process, the debates surrounding the construction of a memorial reflected in the popular consciousness the dissension from and ambiguity toward the state's policy in Vietnam. The eventual design of the monument conspicuously deviated from other memorials. First the architect, chosen through public tender, was a female student of Chinese–American descent. She was not an architect with an established record, she was female and the choice of someone with her ancestry was questioned given that the memorial was commemorating a war which took place in Southeast Asia (Sturken 1991). The politics underlying the commission reflected differing attitudes toward the war. Although the positioning of the monument on the Mall in Washington reinforced it as a national icon and part of the nation's history, the iconography of the design deviated radically from that of other memorials in the capital. The listing of each individual soldier killed in the conflict, and represented in chronological order rather than in order of rank, made it non-hierarchical in conception. Both these features offered an interpretation of the war which treated each casualty as equal in significance and which suggested that the war was not an heroic event in the nation's history (Sturken 1991; Wagner-Pacifini et al. 1991). The monument functioned at both an allegorical and a literal level, and in nationalist discourse it served to heal the wounds of a nation in mourning, a nation that did not offer unequivocal support for the state's actions. In contrast to more heroic depictions of war the iconography and ritual associated with visiting the monument centered the story of Vietnam on individual suffering within the broader framework of a national army. It differs therefore from memorials to the "unknown soldier" where the collective loss is embedded in the anonymity of a dead soldier. The design of the Vietnam Memorial did not satisfy all interests in the US and a more conventional war memorial was erected beside the "wailing" wall (Sturken 1991). Three bronze statues of soldiers, two white and one black, framed the wall in a more orthodox form, masculine, heroic, and anonymous. The recent addition of a new monument dedicated to the women/nurses in Vietnam underlines the gendered division of labor in a war context, and reclaims a more active role for women in the war effort.

The nationalist discourse in which war memorials are conceived is confirmed by the fact that they rarely acknowledge the loss experienced by the "enemy." They are not memorials to all those lost in war, but are interpreted in terms of "national" losses and "national" geopolitical considerations. War memorials commemorate "our" dead not just "the" victims of conflict.

For separatist groups within larger states statuary also articulates the divisions within the national polity, and competing public monuments can be erected to reflect this division. Pierre Nora (1989) argues that there are dominant and dominated sites of memory. The dominant "spectacular and triumphant, imposing, and generally imposed – either by a national authority or by an established interest. . . . The second are places of refuge, sanctuaries of spontaneous devotion and silent pilgrimage" (quoted in Hung 1991, p. 107). In China the symbolic meaning of Tiananmen Square and its monuments encompasses the ongoing history of China itself and the competing political ideologies that the state has experienced. The inscription on the monument to the People's Heroes links "separate historical phases into a continuum" (Hung 1991, p. 99) from the older revolutions of the 1840s to the new revolution of the 1940s. The recent protest by students in

Tiananmen Square and their erection of a statue of the Goddess of Democracy "signified consecutive stages in a pursuit for a visual symbolic of the new public" (Hung 1991, p. 109). The suppression of that protest and the removal of the monument confirms the hegemony of the state over sections of the population within its territory (Hershkovitz 1993).

Gender and Nationalism

To a large degree analyses of nationalism have been presented as gender-neutral discourses. The imperatives of creating a national "imagined community" have excluded a discussion of gender cleavages. Yet, as Warner (1985) points out, pictorial representations of nations have generally been expressed through female allegorical figures; Britannia, Marianne, and Lady Liberty are all examples. Female iconography frequently features in public commissions "because the language of female allegory suits the voices of those in command" (Warner 1985, p. 37). That the female body has been used to personify concepts such as justice, equality, and liberty is ironic given womens' lack of access to such freedoms, especially as nations emerge. Today "feminine" characteristics continue to be ascribed to powerful female political figures. In the case of Margaret Thatcher, where the label "Iron Lady" was used to highlight her tough, resolute and determinedly "masculine" approach to politics, she was simultaneously depicted as – mother, housewife, and her *father's* daughter (Warner 1985).

In nationalist discourse men are active agents and women are typically passive onlookers. Although feminist critiques and histories of nationalist movements have emphasized the role of women revolutionaries as active, determined participants in particular contexts (Ward 1983), dominant theories of nationalism continue to ignore the ways in which gender relations inform conceptions of national identity. The ways in which womens' voices are frequently silenced in nation-building projects, despite their role in the achievement of independence, underlines the conventional hegemony of the male voice. That women's role as homemakers was enshrined in the 1937 Irish constitution aptly demonstrates this silencing process. Although, in Eugène Delacroix's painting

Liberty Leading the People, a female allegory is pointing the path to freedom in revolutionary France (Agulhon 1981), in analyses of nationalism women are seldom leading but loyally following their male protagonists. While the achievement of political independence in no way necessarily leads to the emancipation of women (Kandiyoti 1991), the dearth of studies that deal with gender relations and nationalism makes generalizations difficult.

Conclusion

As we move toward the twenty-first century, the decline of the nation state as the basic structure of global political organization would appear to be nowhere in sight. While local and global processes are increasingly challenging the national state, and the notion of "multi-culturalism" is gaining some ground, the evidence suggests that political and cultural identities are still articulated broadly within a national "imagined community."

This chapter has emphasized the ways in which the "imaginings" that consolidate the nation state occur and can change through time. Whether the basis of nationalist imaginings be linguistic, historical or symbolic (or combined), the global restructuring that has taken place since the end of the Cold War appears to have raised nationalist discourse more profoundly than ever on the global political stage.

Acknowledgment

Thanks to Michael Watts and Peter Taylor for their useful comments on an earlier draft of this chapter.

8

Global Regulation and Trans-state Organization

Susan M. Roberts

Introduction

In most atlases one of the first maps included is a "World Political Map." What is on such a map? Usually the patchwork of countries. The borders and capitals are carefully drawn and the whole of the earth's land surface appears covered by named territorial units – countries. Thus the World Political Map is a map of the world of states. Some are sovereign, and others (with another country's name in brackets after their own) are not. Cartographers have been busy redrawing bits of the World Political Map as countries unite (Germany); split (Slovakia and the Czech Republic); shatter (the former USSR); or are torn apart (Yugoslavia). The map, however continuously it is redrawn, remains one of states.

In contrast to the strengthening global views of environment, economy, society, and culture, the power of states has precluded much serious discussion of a global polity or a world government. The world political map seems to deny the very possibility. However, there are organizations that are trans-state in nature and such organizations are the focus of this chapter. In particular, we will consider trans-state organizations that act as global regulators. This entails surveying the institutional framework of post-World War Two global mode of regulation and then examining a recent example of a trans-state organization's regulation of one aspect of the international economy. The case study is an attempt to consider how we might grasp the workings and implications of the relationship between an economy that has seemingly burst the boundaries of the state that held it in and a polity that remains coalesced around the boundaries of states (old and new) as the containers of political life. Economic globalization "from above" is related in various ways to political globalization (such as formal attempts at global regulation) "from above." However, there are also identifiable movements aimed at trans-state organization through international alliances of various grassroots and regional social movements: globalization "from below."

Global Regulation and Trans-state Organization "From Above"

This part of the chapter will focus on trans-state organizations that seek to regulate some aspect or aspects of the contemporary international system and the states therein "from above." The phrase "from above" refers to an arena of regulation at a scale higher than that of the state. Regulation is a term that is used in a number of different ways (see Jessop 1990). For those in the so-called "Regulation School" capitalism is understood not only in economic terms as a "regime of accumulation" but also in social and political terms as a corresponding "mode of regulation" which comprises the "institutional forms, procedures and habits which either coerce or persuade private agents to conform to its [the regime of accumulation's] schemas [of reproduction]" (Lipietz 1987, p. 33). Here the concept of the "mode of regulation" has a meaning that encompasses formal legal systems of rules but extends to include elements of the regime of accumulation such as social processes and cultural norms which also order and "make regular" capitalism's inherently unstable course. In this chapter the case study is focused on a type of formal regulation, but within the context of the wider mode of regulation. Modes of regulation are never completed wholes, but are "contingent" and "provisional" – themselves contested and implicated in social and political struggles (see Jessop 1990, p. 177). Modes of regulation are therefore dynamic: seen to be changeable over time and space – usually configured as a series of national regulatory spaces

(Clark 1992). In this chapter we will see that the mode of regulation at the global scale is constituted in large part by the actions of states. Contemporary trans-state organizations arose out of attempts to establish an institutional regulatory framework as World War Two (1939–45) came to a close, and so we will concentrate on developments of the last 50 years. Given this concentration, three reminders are in order.

First, it is important to remember that trans-state organization did not begin in the mid 1940s. Indeed, those who were involved in setting up the post-World War Two system of global regulation were well-aware of the failings of the institutions set up after World War One (1914–18), notably the League of Nations.

Thus, secondly, it should be remembered that even though there was undoubtedly the beginning of a new era in 1944–5, the slate was not wiped clean. Just as the New World Order being contested today is being built upon parts of the preceding Cold War World Order, so too, many elements of the institutions set up as World War Two came to a close grew out of previously existing arrangements. Certain organs of the United Nations (UN) have their origins prior to the founding of the UN itself in 1945 (see table 8.1).

Thirdly, it would be a mistake to think that once institutions of trans-state organization have been set up they remain unchanged. Rather, the four major postwar institutions (the UN, the World Bank, the International Monetary Fund, and the General Agreement on Tariffs and Trade) have not existed unproblematically. They have changed their character and operations in several significant ways. The two major developments in the postwar world that affected the trans-state regulatory institutions were first, the internationalization of capital, and secondly, the independence of many former colonies. Both of these factors have been significant in rendering problematic the positions and roles of post-World War Two trans-state organizations.

Having noted these three reminders, we can now summarize the formal institutional framework of the international mode of regulation set up as World War Two ended. The regulatory framework resulted from several meetings held by representatives of the countries that would emerge as victors in the war. From 1 July to 22 July 1944, representatives of 45 countries met at Bretton

Table 8.1 Trans-state organizations.

NAME	*Date Estab.*	*Purpose*	*HQ Location*	*Members*
IAEA International Atomic Energy Agency	1957		Vienna, Austria	130
ILO International Labor Organization	1919	Formation and promotion of international standards through International Labor Conventions	Geneva, Switzerland	160
FAO Food and Agricultural Organization	1943	To increase food production, raise nutrition ad eliminate hunger	Rome, Italy	160
UNESCO United Nations Educational, Scientific and Cultural Organization	1946	Promoting collaboration between countries through education, culture and science	Paris, France	164
WHO World Health Organization	1946	"the attainment of all peoples of the highest possible level of health"	Geneva, Switzerland	
IMF International Monetary Fund	1945	To promote international monetary cooperation; exchange rate stability and expansion of international trade	Washington, DC, USA	167
IBRD (WORLD BANK) International Bank for Reconstruction and Development	1946	To provide funds and technical assistance to facilitate economic development	Washington, DC, USA	
IDA International Development Association	1960	Lending agency for projects in poorest countries	Administered by World Bank	
IFC International Finance Corporation	1956	Financial and other assistance to private enterprises	Administered by World Bank	143
ICAO	1947	Coordinates safety	Montreal, Canada	173

International Civil Aviation Organization		standards, navigation, and promotes international cooperation		
ITU International Telecommunications Union	1932 (merger of earlier orgs)	Promotes international cooperation in telecommunications	Geneva, Switzerland	166
UPU Universal Postal Union	1875 (as General Postal Union)	International cooperation between postal services for reciprocal exchange of mail	Berne, Switzerland	168
WMO World Meteorological Organization	1951	Encourages coordination of collection and transmission of meteorological information	Geneva, Switzerland	168
IMO International Maritime Organization	1959	Through Conventions, to encourage cooperation regarding merchant shipping (especially safety, pollution and traffic)	London, UK	137
GATT General Agreement on Tariffs and Trade	1948	Multilateral treaty negotiated in "rounds" to guard against protectionism	Geneva, Switzerland	103 contracting parties; 29 de facto
WIPO World Intellectual Property Organization	1974 (took over BIRPI estab. 1893	Through Conventions, to protect intellectual property	Geneva, Switzerland	131
IFAD International Fund for Agricultural Development	1977	Agricultural and rural development projects in poor countries	Rome, Italy	147
UNIDO United Nations Industrial Development Organization	1986	Promotes industrial development in poor countries	Vienna, Austria	

Woods, New Hampshire, USA, to discuss the regulation of the international economy after the war. The debates revolved around competing proposals put forward by John Maynard Keynes (UK) and Harry Dexter White (USA). At Bretton Woods it was the White proposals that triumphed – a point which is looked upon as signaling the end of British hegemony and the confirmation of US hegemony in the world of 1944–5. Two significant trans-state organizations were established – formally coming into existence in 1945. These were the International Monetary Fund (IMF) and the International Bank for Reconstruction and Development (IBRD) – together with the International Finance Corporation (IFC) and the International Development Agency (IDA) – known as the World Bank.

From August 21 to October 7, 1944, another series of meetings was held at Dumbarton Oaks, Washington DC, USA, with delegates from China, the UK, the US and the USSR. The conference built upon the Atlantic Charter – a declaration signed by UK Prime Minister Churchill and US President Roosevelt in 1941 – which established political and economic priorities for the postwar world. The Charter called for an international system based on security to guarantee free trade and economic growth. The meetings resulted in a draft of the United Nations Charter. This document then formed the basis for a larger conference (with delegates from 51 countries) held from April 25 to June 26, 1945, at San Francisco, USA. By October 24, 1945, enough states had ratified the Charter and the United Nations officially came into existence.

In 1944 and 1945 then, the Allied powers had set up the key institutions for regulating the postwar world. The commitment to international organization has to be seen in the light of events after World War One. The memories of terrible economic depression, a chaotic international financial system, territorial aggression on the part of certain states, and other factors precipitating a second brutal "world" war, were uppermost. It was envisaged that international organizations could and should be created to prevent such events ocurring again.

The postwar framework was refined. A key development was the General Agreement on Tariffs and Trade (GATT), set up in Geneva, Switzerland, in 1947. The GATT was intended to be a stop-gap arrangement to ensure open trade through a series of bilateral tariff concessions written into a final agreement while negotiations for the International Trade Organization (ITO) were underway. The ITO's charter had been agreed upon at a UN conference in 1948. The ITO was intended to be the third pillar (together with the IMF and the World Bank) of the international institutional regulatory structure. However, the ITO was never set up because debates in the US over protectionism versus free trade were won by the protectionists and the US Senate declined to ratify the ITO Charter. This was not to be isolated demonstration of American ambivalence about its economic role as hegemon. However, American participation in the GATT does not require Senate ratification and the GATT has persisted as the organizing frame for world trade. It has proved a cumbersome and complicated institution involving long rounds of negotiations between member states. The latest round, called the Uruguay Round because it commenced in Uruguay, was concluded in 1993 only after long years of dispute and disagreement.

The most significant institution of trans-state organization remains the United Nations. The UN currently has about 180 member countries and is organized (as stated in its Charter) around six major "principal organs." The two major ones are the General Assembly and the Security Council. All UN countries belong to the General Assembly and each gets one vote therein. The General Assembly discusses a range of matters – and often holds special sessions to consider particular issues (for example, on Palestine in 1947 and 1948; on the economic situation in Africa 1986). The Security Council consists of 5 permanent and 10 non-permanent members, each with one vote. The non-permanent members are elected by a two-thirds majority of the General Assembly and serve a two-year term. The permanent members are the victors of the Second World War: France, Russia, the UK, the US, and China. The Security Council of the UN is the body that concerns itself most directly with maintaining peace and security. It is the Security Council, for example, that calls on UN members' militaries for involvement in "peacekeeping operations." There is disagreement over the relative powers and different structure of the General Assembly and the Security Council. The selected membership of the Security Council, arguably the more powerful of the two organs, is contested by many Third World leaders and by those who argue that it does not reflect the present world. Why, for example, should France and the UK each be permanent members? Why not one membership for the European Community as a whole instead?

The Economic and Social Council reports to the General Assembly and is concerned with issues of human well-being – notably development and human rights. However, the role of the Economic and Social Council is notorious for being vaguely (un)defined in the UN Charter. The Economic and Social Council's members are elected by the General Assembly. The regional economic commissions (such as ECLAC – the Economic Commission for Latin America and the Caribbean) as well as other specialized commissions (such as those on human rights and the status of women) operate under the auspices of this Council.

The third UN organ to be noted is the Trusteeship Council, and this also operates under the General Assembly. It was set up to supervise "Trust Territories." There have been 11 Trust Territories since 1946, but only the US-administered Republic of Belau [Palau] in the Pacific remains.

The International Court of Justice (or World Court) is one of the "principal organs" of the UN according to the Charter, although it is somewhat independent. This feature is evident in the location of the Court in The Hague, Netherlands, although this location is also a legacy of the 1899 and 1907 Hague Conventions establishing the Permanent Court of Arbitration – a precursor of the present Court. All members of the UN are "parties to the Statute of the Court" (Statesman's 6). The judges (15) are chosen through elections in the General Assembly and the Security Council. The Court has not escaped controversy: it has been criticized for having a very light caseload, for being slow and expensive, and for being hampered by having to secure consent of all parties (states) involved. States are, in general, very reluctant to give up their jurisdictional integrity. After all, this is a key characteristic of the modern state.

The fifth organ of the UN Charter is the Secretariat, or office of the Secretary-General. The Secretary-General is the chief administrative office of the UN and is appointed for a five-year term. The Secretariat staff of over 25,000 persons performs all the day-to-day administrative operations of the UN.

In addition to the "principal organs" several of the UN's programs are important actors on the global scene. These include the UNDP – the United Nations Development Program; UNICEF – the United Nations Children's Fund; UNFPA – the UN Population Fund; and the UN World Food Program. A significant UN body is the UN High Commission for Refugees. This organization was established in 1951 by the General Assembly to provide protection, to seek asylum, and to find lasting solutions for refugee populations. Agencies, such as the Food and Agriculture Organization (FAO) and the UN Educational, Scientific and Cultural Organization (UNESCO) are technically autonomous but work closely with the UN. The full list is to be found in table 8.1.

The UN and related bodies, including the World Bank, the IMF, and the GATT, are contested institutions. That is, there is considerable and varied criticism of their roles, workings, and objectives. Many of the criticisms derive from experiences of those living in the so-called Third World – in countries such as those that became independent in the 1960s and 1970s and were incorporated into the already-formed structures of trans-state organization. The most pervasive criticism faced by each of these institutions is that they reflect and reinforce the highly uneven global distribution of political and economic might. Indeed, the locations of the headquarters of each of the bodies listed in table 8.1 reflect the political world in 1944–5 and definitely not the contemporary distributions of member states or total population. However, more is at stake here than the location of headquarters offices. For many observers argue that institutions such as the IMF are actively used by rich countries against poor countries. The IMF's response to problems of Third World indebtedness has been to advocate and enforce "structural adjustment" policies in many developing countries – thereby subjecting large numbers of people to the rigors of "economic reform" which often include worsening standards of living for the majority (see Popke 1994). Powerful countries are never subject to the same "discipline" by the IMF. The UN itself is also under fire – both literal and metaphoric – as it faces a new world, less defined by the simple oppositions of the Cold War. The "success" of UN operations in Somalia and the former Yugoslavia is contested from many standpoints.

The UN is an example of global-scale trans-state organization attempting to adjust and shape the so-called New World Order. There are many other trans-state organizations that act to regulate in some way but that are less inclusive than the UN. Regional economic alliances such as the Association of South East Asian Nations (ASEAN) or the European Community (EC) are becoming increasingly important elements in the global political-economy. Military strategic alliances such as the North Atlantic Treaty Organization (NATO) remain significant despite challenges arising from post-Cold War political realignments. Such organizations may not be global regulators in the formal sense but are significant trans-state organizations which regulate the world system in the sense of making its operations smoother and more orderly.

Case Study of Global Regulation From Above

Finance would appear to be a much more footloose economic activity than production, with its necessary fixed capital investments in factories and infrastructure. Indeed the international financial system, prompted by deregulation and technological innovations (telecommunications especially), has undergone tremendous changes during the late 1970s and 1980s. Money appears to fly around the globe as markets are linked by round-the-clock, round-the-globe trading. Several commentators have claimed that the international financial system has truly "globalized" to an extent no other economic sector has. Therefore, if we are interested in the ways in which formal global regulation might arise, we should look to this most globalized sector. The case-study presented here focuses on international banking and its regulation by a trans-state organization – the Bank for International Settlements (BIS).

The BIS is based in Basle, Switzerland, and was originally set up in 1930 to supervise Germany's reparations payments. However, the BIS statutes give it a broader mandate: in addition to supervising international financial settlements (such as reparations), the BIS is charged with promoting and facilitating the cooperation of countries' central banks. The BIS is often referred to as the central banks' central bank. The BIS has recently assumed a more visible role in regulating the world's financial markets, as it appears to be the only existing institution able to play the role of international banking supervisor and regulator. It is to this role and to the only piece of banking regulation that is truly international that we now turn.

Regulators are concerned not only with banks that are fraudulent (e.g., the Bank of Credit and Commerce International or BCCI) but also with international banks that are increasingly subject to, and generative of, various financial risks. In fact, the event prompting the initial discussions about international banking regulation was the so-called Third World Debt Crisis as it appeared to threaten the solvency of major banks. The Debt Crisis revealed the fragility of international financial capital – specifically, the multinational banks. For many analysts this fragility is seen as rooted in the peculiar and complex relations between the state, financial capital, money and credit in capitalism. Aglietta (1987, p. 337) has noted that the "solvency of the banking system becomes the knot where all the contradictions of capitalist accumulation come together" and Harvey (1989) has argued that this "knot" has taken on new and distinct forms and has become a defining feature of contemporary capitalism. From a policy imperative, and with a realization of the systemic risks in the international banking system, the BIS "Committee on Banking Regulations and Supervisory Practices" (called the Basle Committee) was charged in the mid-1980s with examining banks' balance sheets with an eye to assessing capital adequacy. Capital adequacy is a way of checking the ratio of capital (classified as shareholder equity) to assets (in a bank's case these are loans and now many other financial products). The Basle Committee was to: "report back with recommendations for assessing the comparability of different measures of capital adequacy employed by member countries and with proposals for the attainment over time of comparable and adequate minimum international capital standards" (in Kapstein 1992, p. 275).

A sensible question might be: why would the BIS focus so exclusively on capital adequacy? It is a valid question considering that there is no consensus among analysts that this is the best way, or even a very effective way, to ensure bank soundness. Even among those who agree that increased capitalization is a good idea there is no agreement as to what the appropriate minimum level (or ratio) is.

In economic theory there are two reasons given for favoring capital adequacy regulation. These are, first, that such regulation gives banks a buffer against losses. Less likelihood or actuality of bank failure indirectly affects public perceptions of the safety and soundness of the banking system as a whole, so that capital adequacy standards are seen as essential for maintaining public faith and thus contributing to the wider mode of regulation. The second rationale is that if there are no mandated standards, weakly capitalized banks will be able to undercut stronger banks and force them, through competition, to reduce their strong capitalization, and thus the operation of competitive forces will result in a tendency for banks to be capitalized at less than socially optimal levels (see Kapstein 1992 for further discussion). These two points are, of course, related. However, to answer the question "why the focus on capital adequacy?" we have to step out of the realm of economic discourse and into the world of the workings of geopolitical economy, and take this second rationale to the international arena. In fact, in the world of international banking and national banking regulation of the 1980s, banking regulators worried that they were acting in competition with each other – with certain countries' regulators suspicious that they and their countries' banks were over-capitalized in comparison with other countries' banks. The more weakly capitalized banks were able to out-compete the more strongly capitalized banks.

This is indeed what was going on. Essentially, the BIS discussions of international standards of capital adequacy came about because regulators from the US and Britain were looking for ways to halt the tremendously rapid growth in the international presence of Japanese banks in the mid to late 1980s. Indeed, Gerald Corrigan, then President of the Federal Reserve Bank of New York, said in testimony before a committee of the US Congress in 1987: "the single item on which I place greatest emphasis relates to bank capital adequacy standards and specifically the goal of moving Japanese bank capital standards into closer alignment with international standards" (in Kane 1990, p. 34). It seems clear that the BIS capital adequacy regulations were in large part a reaction to the sustained redistribution of financial market shares toward

Japanese financial institutions, which now dominate any list of the world's largest banks and securities houses (for examples, see *The Banker*, July 1993).

In addition to US and British wishes to curb Japanese banks' expansion, there is some evidence that Britain also was concerned to set up rules prior to any European Community attempts to do the same – attempts that might be more guided by the wishes of German and French banks and their regulators than by those of Britain. Indeed, that has happened, and although the EC (for 1992) does now have its own "solvency ratio directive" it has tended to fall in line with the BIS Accord. The BIS Accord was stalled for a while in 1986 and it was really the threat of a bilateral agreement (on capital adequacy standards) between the US and Britain that jump-started the multilateral negotiations leading to the eventual 1987 Accord. So the convergence of interests between two powerful states was crucial to formulating this international regulation. Like the UN then, the BIS is a trans-state organization embedded in assymetrical geographies of power and in a world where states are still the primary unit of political and hence formal regulatory organization.

Although the BIS capital standards are international in scope, in implementation the power of particular states, and of the state in general vis-à-vis a trans-state organization, show clearly. This is because the banking regulators of each country are in charge of setting guidelines for compliance (crucially in deciding the degree of risk of various specialized financial instruments and activities) and in checking banks for adherence to the standards. Because accounting practices, bank structures, and even bank cultures vary considerably from country to country, the ways in which the BIS standards have been implemented also has varied. Further, in each country the regulators have had pressure from the banks to be lenient and not implement the standards in such a way as to "handicap" them in their competition with banks from elsewhere. This just reminds us of the competitive aspect of country-based regulation.

In some ways the capital adequacy standards may be seen as affirmation of the potential for global regulation; on the other hand the difficulties in implementing them in a straightforward and harmonious manner might well be seen as evidence that truly international banking regulation is impossible. The BIS Accord is a less well known case of trans-state organization and global regulation but is significant in terms of its attempts to manage risk in the international banking system and for the light it sheds on the tension between an economy that has gone global and a polity (which shapes the mode of regulation) that revolves around states.

Trans-state Organization "From Below"

Another way of conceiving of trans-state regulation is to focus on organizations that are international but not supra-national. That is, there are attempts at trans-state organization which are not aimed at formally regulating states from above, but rather aimed at uniting groups in various different countries around common causes. Such groups are sub-state (as opposed to supra-state) and thus represent trans-state organization of a quite different nature from that of groups such as the UN or the BIS. Examples of trans-state organizations of this type might include Greenpeace – the international environmental activist group – and Amnesty International – an organization aimed at monitoring human rights and fighting human rights abuses across the globe.

What is the relationship between trans-state organization of the sub-state type and the trans-state organizations operating "from above"? Some attempts have been made to connect the two types of trans-state organization. One example is the "International Dialogue on the Transition to a Global Society" begun in September 1990. This annual dialogue is aimed at discussing "themes relevant to the advance of a global society" and is concerned to work within existing institutions – particularly the UN (see Bushrui et al. 1993). Other trans-state organizations work outside the existing arrangements – in some cases directly against institutions such as the World Bank or the UN. For example, Elkins (1992) criticizes the recent influential UN Reports by the three independent commissions. The Brandt Commission (established in 1977) on "International Development Issues"; the Palme Commission (1980) on "Disarmament and Security Issues," and the Bruntland Commission (1984) on "Environment and Development." Elkins argues that such top-down policy approaches misdefine, and are wholly inadequate in the face of what he terms the "global problematique": war and militarization; persistent poverty; the denial of human rights; and environmental destruction. Elkins sees democratic popular mobilization as the only real prospect for tackling the global problematique (see Elkins 1992 and Woodhouse 1990 for examples).

Other approaches may not be directly organized to counter trans-state organizations but seek consolidation, in the form of a trans-state organization. In addition to international confederations of trade unions and labor groups, new networks of alliances of labor, environmental organizations, and a wide variety of other grassroots social movements are being constructed. Indeed, one group of efforts proclaims a manifesto for "globalization from below." They state:

> Globalization from below, in contrast to globalization from above, aims to restore to communities the power to nurture their environments; to enhance the access of ordinary people to the resources they need; to democratise local, national, and transnational political institutions; and to impose pacification on conflicting power centers. (Brecher et al. 1993, p. xv)

Such efforts are examples of a type of trans-state organization not captured in the formal structures of postwar modes of regulation, but that nonetheless seek to regulate the activities of states, capital and trans-state organizations themselves.

Conclusions

The contemporary world is characterized by a series of spatial disjunctures. The formal regulatory agencies of the post-1945 world – such as the UN – have been challenged by recent political shifts and ongoing processes of economic globalization. All these changes have been uneven and have occurred within the context of a deeply unequal world of power relations between populations and between the states in which they live and around which the trans-state organizations are built. The organizations set up as

World War Two came to an end are under pressure and strain – in particular the UN in its role as peacekeeper.

Trans-state organizations are not extra-state organizations. They do not mark the eclipse of the state. The formal structures of the post-1945 world have been creations of states and reflect the unequal power relations between states. The challenges of economic globalization have been (in part) met by states through trans-state organization. Although at first glance this might appear as a transfer to the trans-state arena of some powers that were previously vested in states, the BIS case indicates that global regulation is built out of the strategies and needs of particular states. The world political map of states does not tell the whole story of political organization, but it depicts a most significant element.

9

The Regulatory State: the Corporate Welfare State and Beyond

Joe Painter

Introduction

This chapter focuses on one particular type of state – the corporate welfare state – and considers the factors that led to its rise in a number of prosperous countries in the twentieth century and the problems which it subsequently faced. After some contextual remarks about processes of state formation, it outlines the character and distribution of the corporate welfare state and its limitations. It then introduces a conceptual framework – regulation theory – which provides a helpful interpretation of changes in the welfare state. Finally, it discusses the possible future directions for the state in those countries where the welfare state has been dominant.

State Formation in the Twentieth Century

In the process of the development of human societies, states are comparatively recent phenomena. The *modern* state, organized on clearly bounded territorial lines and with a claim to authority distinct from the person of a monarch, is more recent still, dating from the seventeenth century and confined initially to part of Europe. The growth in the number of such states to the point at which they account for the entire land surface of the globe (with the partial exception of Antarctica) is one of the most significant features of modernity.

This historical perspective provides important clues about how we should understand the process of state development. First, states are not inevitable features of human existence. Organized societies existed for many thousands of years without them. Today's patchwork of states is thus not a natural and inevitable order, but the product of social processes. Second, the forms and functions of states vary through time and across space. Third, explanations of state restructuring and change must be historically and geographically sensitive.

From its emergence the state has been the focus of both hopes and fears. Throughout the complex processes of the formation of states around the globe, a wide variety of social groups has turned to state institutions as potential sources of progress. Of course, what counts as progress has varied immensely across time and space and depends on the interests of the group concerned. Whatever the "progressive" aim, however, the state has most often been the means by which the appropriate strategy has been prosecuted. Throughout the modern age, until the recent rise of the transnational corporation, states have been seen as the only institutions with the resources and organization capable of producing widespread, deliberate socioeconomic changes.

Yet it is precisely the state's control over resources and organization which simultaneously generates its reactionary side. The turmoil bequeathed to Europe by the First World War saw the rise of fascist states in Germany, Spain, Italy and Portugal, and of authoritarian state socialism in Russia. It has often been assumed by liberals that authoritarian or totalitarian states are exceptional: an essentially distinct and aberrant form of the state which has no implications for the development of "normal" liberal democratic states. The sociologist Anthony Giddens has suggested that, on the contrary, a tendency to totalitarianism is inherent in the modern state, constituting one of its defining features. This derives, according to Giddens, from the concentration of "administrative power" (1985, pp. 172–81) in the state apparatus generated by its control over resources and organization.

"Administrative power" refers to the ability of states to monitor their territories and populations. The development of technologies of surveillance and information storage, from the invention of writing to today's sophisticated microelectronics, has seen states insert themselves ever more thoroughly into the activities of their inhabitants. The state is an increasing presence: regulating and sanctioning; forbidding and allowing; monitoring and recording. Birth, marriage and death; employment and unemployment; collective assembly, politics and expression; business and commerce; sex and relationships; religion and culture: fewer and fewer social activities remain wholly private. The administrative gaze of the modern state is increasingly "panoptic," and, Giddens insists, thus tends inherently toward totalitarianism.

Stuart Hall (1984) suggests that the emergence of unambiguously authoritarian states of left and right in the 1930s marked the beginning of a more critical attitude to the state. The power of the state was not always seen as malign in itself, but the potential for abuse had been demonstrated. Yet, following the Second World War the state was again the focus of hope. In Latin America and in the new countries of the now decolonizing continents of Asia and Africa it was to be the mechanism for economic "development" (Corbridge 1993). In China, home to one-quarter of the world's people, the new communist state promised to sweep away injustice and establish a peasant-oriented socialist society. And in Western Europe new "welfare states" were to be the generators of reconstruction and modernization and the providers of a guaranteed minimum quality of life.

The Corporate Welfare State

The welfare state is popularly thought of as a *part* of the state: those state institutions which provide "welfare" services. Social scientists often prefer to think of the welfare state as a particular *type* of state, in which the state guarantees (in theory) a minimum standard of welfare "from cradle to grave" for its citizens. These guarantees form one defining feature of a particular type of state and an important source of its claim to legitimacy. Seeing the welfare state as a specific type of state is helpful, because it shifts the focus away from the changing *quantity* of welfare services provided and onto the changing *role* of welfare provision.

Table 9.1 Central government expenditure per capita on health and education, ca. 1990.

Country	*US$*
Australia	837
Austria	1642
Belgium	no data
Canada	401
Denmark	962
Finland	1923
France	1853
Germany (old Fed. Rep.)	1308
Ireland	1322
Italy	1604
Japan	no data
Netherlands	2087
New Zealand	1492
Norway	2093
Spain	679
Sweden	966
Switzerland	no data
United Kingdom	996
USA	796
OECD unweighted mean	1310
All developing countries	58

Note: Expenditure for developing countries is general (rather than central) government expenditure.
Sources: derived from World Bank (1992, 1993); UNDP (1993).

If the welfare state is in crisis today, this is not because state welfare provision has declined in quantitative terms (in fact it has increased), but it may be because welfare provision is no longer so central to state strategies and state survival.

Nonetheless, quantitative measures are important. They show just how limited and unusual welfare states are. The Organization for Economic Cooperation and Development (OECD) is made up of all the major industrialized capitalist countries. Table 9.1 shows the expenditure by central governments on health and education, two major components of welfare states. The data are not directly comparable between countries, since in some cases health, or education, or both are the responsibilities of sub-central government. However, taken together they provide a useful comparison with the figure for *total* public expenditure per capita on health and education in "developing" countries. There is clearly a massive disparity in government welfare provision between the OECD, which accounts for just 15 percent of the world's population, and the "developing" countries, which account for 77 percent.

Welfare states are thus rather unusual, limited phenomena, in terms both of time (the twentieth century) and space (Western Europe, the United States, the "white" British Commonwealth and, rather differently, Japan). Nonetheless, the inclusion of a chapter on them in this book on *global* change is justified as they have global significance. First, their development and subsequent problems have had a number of knock-on effects on the rest of the world, and have also partly depended on global economic and political relations. Secondly, the idea and ideal of the welfare state has met with approval around the world.

Its advent seemed to herald the end of a socioeconomic system based on the impoverishment of the majority for the benefit of a few. Where individuals and their families were unable to meet the expense of fundamental goods and services, the state would step in to make up the difference. This would either be through direct provision (of schools, hospitals, and so on), usually free or heavily subsidized, or through cash benefits to enable goods like food and clothes to be paid for.

Not only did the welfare state mitigate many of the social problems previously experienced, but it also had important economic and political effects. Economically, it was seen as partly an investment, by providing a better educated and healthier work-force. It also helped to iron out the economic instability which had dogged many countries, because it guaranteed a minimum level of consumption regardless of the level of economic activity. Previously, the absence of a "social wage" had left economies prone to crises of underconsumption, in which an economic downturn could turn into a slump as increases in unemployment reduced the demand for goods and services, which generated further unemployment and so on. By placing a "floor" under the level of popular consumption, the downward cycle could be broken.

Politically, the welfare state was seen as a third way between the "extremes" of unfettered capitalism and Soviet-style state socialism. It acted as a kind of "truce" between the forces of capital and labor, allowing the continuation of private investment and profit-making, while ameliorating some of the social consequences thereof. In many cases the institutional representatives of capital and labor were given a formal role in the formulation of policy, a strategy known as corporatism (hence corporate welfare state). In Britain, for example, the Trades Union Congress (TUC) and the Confederation of British Industry (CBI) were both involved, with the government, in the National Economic Development Council. This fitted well with the economic doctrines of John Maynard Keynes which were applied at the same time, and this form of the state is often known as the Keynesian welfare state. In many industrialized countries a consensus was established, at least between political parties and elites, that the state should play a significant social welfare role, and the 30 years from 1945 to 1975 are often referred to as the postwar "consensus" or "settlement."

The form of this "settlement" varied significantly from state to state, as did the relationship between welfare provision and the wider social system. The nature and causes of the differences are detailed by Esping-Andersen (1990) and summarized by Johnston (1993). Despite these variations, welfare states were highly successful. This success can be measured in a number of ways. The "welfare output" of states is not easy to assess in quantitative terms. This is partly because welfare is to some extent subjective and partly because it is difficult to determine how much the observable changes are due to state activity, and how much they are a product of a general increase in economic prosperity. Tables 9.2 and 9.3 show the coverage of the West European population by state insurance schemes for health care and unemployment respectively. Figures are given for 1925, some fifteen years before the major growth of European welfare states, and for 1975, just before the emergence of criticisms of that growth. They show that there were dramatic increases in state support for the sick and jobless. Table 9.4 gives data over a

Table 9.2 Active members of medical benefits insurance schemes as a percentage of the labor force, selected European countries.

Country	*1925*	*1975*
Austria	47	88
Belgium	28	96
Denmark	99	100
France	21	94
Germany	57	72
Ireland	37	71
Italy	6	91
Norway	54	100
Sweden	29	100
Switzerland	50	100
United Kingdom	79	100

Source: Flora et al. (1983) p. 460.

Table 9.3 Unemployment insurance: members as a percentage of the lobor force, selected European countries.

Country	*1925*	*1975*
Austria	34	65
Belgium	18	67
Denmark	18	41
France	0	65
Ireland	20	71
Italy	19	52
Norway	4	82
Switzerland	8	29
United Kingdom	57	73

Source: Flora et al. (1983) p. 460.

similar period for one key indicator of well-being, infant mortality. The rapid decrease in infant mortality certainly had a lot to do with improved nutrition and hygiene as well as state health care, but nutrition and hygiene were themselves improved by state policies.

A final, albeit ambiguous, measure of success is the growth of inputs to the welfare state in terms of the proportion of national

Table 9.4 Infant mortality (deaths under one year of age per 1,000 live births), selected countries.

Country	*1925*	*1969*
Austria	119	25
Belgium	100	22
Denmark	80	15
England and Wales	75	18
France	95	20
Germany	105	23
Ireland	68	21
Italy	119	30
Norway	50	14
Switzerland	58	15
United States	68	18

Note: The figures for the United States are for white babies. Mortality among non-white babies is about 60 percent higher in each case.
Source: Mitchell (1975) pp. 130–2; Mitchell (1983) pp. 130–2.

Table 9.5 General government expenditure as a percentage of GDP, selected European countries.

Country	*1930*	*1975*
Denmark	13.5	52.7
France	22.1	38.7
Germany	29.4	47.9
Ireland	20.8	50.2
Norway	17.4	49.3
Sweden	14.0	51.4
Switzerland	17.4	27.2
United Kingdom	24.7	49.9

Note: Data for France relate to 1929 and 1971 respectively. Data for Sweden relate to 1930 and 1974 respectively.
Source: Flora et al. (1983) various pages.

resources controlled by governments. Table 9.5 shows the expansion in public expenditure in a range of countries between 1930 and 1975. This measure is ambiguous since it is not self-evident that increasing resources lead to increased welfare outcomes. However, these figures do measure the *political* success of welfare states.

Table 9.6 The social distribution of expenditure on public services in the UK.

Ratio of per capita expenditure on richest 20 percent of population to per capita expenditure on poorest 20 percent

"Favoring the poor"	
Council housing	0.3
Equal	
Primary education	0.9
Compulsory secondary education	0.9
"Favoring the rich"	
National Health Service	1.4
Further education (16+)	1.8
Non-university higher education	3.5
Bus subsidies	3.7
Universities	5.4
Tax subsidies to owner-occupiers	6.8
Rail subsidies	9.8

Source: after Goodin and Le Grand (1987) p. 92.

Problems for the Corporate Welfare State

Notwithstanding these successes and the enormous benefits that the development of welfare states brought for large sections of the population in industrialized capitalist countries, they also had problems and limitations.

For a start, welfare benefits were distributed in highly socially unequal ways. As noted above, many countries were excluded from these developments. However, even within materially rich countries welfare provision was often universal only in theory. "Advanced" societies are marked by social divisions among which three of the most important are divisions of gender, ethnicity, and wealth or class. These divisions are reflected in the distribution of the benefits produced by welfare states.

Work by Goodin and Le Grand (1987) on the UK welfare state shows that most of its benefits actually go to the middle class, rather than the working class (table 9.6). This raises the question of why the welfare state finds such strong support among the working class. However, the evidence from Goodin and Le Grand does not suggest that poorer sections of society have made no gains from the welfare state, only that they have not gained as much as the rich. Furthermore, popular support is discriminating. In the UK, changes to the National Health Service are much more politically sensitive than changes to universities.

Welfare benefits are also differentially distributed according to gender and ethnicity (Pierson 1991; Williams 1989). Most welfare states depend on a "subsidy" of their *public* welfare provision in the form of unpaid *domestic* welfare provision, performed mostly by women. For example, there is considerable variety in the extent to which welfare states provide public child-care for those under school age. However, comprehensive provision is very rare indeed and in some cases there is virtually no provision at all. Similarly, caring for the sick is often done unpaid by female family members as a complement to, or in place of, public health-care. It is also usually women who undertake the paid caring work *within* the welfare state as well, often for wages lower than those typically paid to men. Members of minority ethnic groups also often suffer from discrimination in welfare provision. Immigrants are frequently permitted to enter their host country only on condition that they will not make claims on the social security system. In some countries minority ethnic groups are denied full citizenship rights and required to become "guest workers." This makes it much easier for the state to exclude them from social benefits, particularly on retirement, when they may be repatriated. Even where there is no overt discrimination of these kinds, welfare provision may still be racist. Formally equal rights may produce highly unequal outcomes in practice, where differences of language, culture and custom lead to unintentional exclusion from welfare provision. In the UK for example, documents and claim forms written in English have in the past led to those who speak other languages failing to receive all their entitlements. Similarly, the failure of Western health services to take account of different attitudes to health-care among patients from non-Western cultures has led to members of some minority ethnic groups losing out.

The corporate welfare state thus brought substantial benefits to the middle classes and to sections of the working class, but it did

Table 9.7 Government consumption and total expenditure, selected industrialized capitalist countries.

Country	*GNP per capita (US$)*	*General Government Consumption (%GNP)*		*Government Expenditure (%GNP)*	
	1991	*1965*	*1991*	*1972*	*ca. 1991*
Spain	12,450	7	16	19.8	34.0
United Kingdom	16,550	17	21	32.7	38.2
France	20,380	13	18	32.5	43.7
United States	22,240	17	18	19.4	25.3
Germany (old Federal Rep.)	23,650	15	18	24.2	32.5
Sweden	25,110	18	27	28.0	44.2
Japan	26,930	8	9	12.7	15.8

Source: World Bank (1986, 1993).

so principally in the interests of men in general and white men in particular. In its operations in practice, the postwar settlement seems to have had substantial social limitations. However, a number of writers have begun to suggest that there are also political and economic limits to the operation of the welfare state even in its present, socially discriminatory form.

It is over fifty years since the publication of the Beveridge Report in Britain, the document which outlined the basis of the British welfare state, and the political and economic situation now looks very different. It has become clear that Keynesian economics and the welfare state were not permanent solutions to the difficulties and contradictions of capitalist development. During the 1980s and 1990s, many governments have attempted to reduce (or to limit the growth of) public services. Since the establishment of welfare states their costs have increased, both in real terms and relative to countries' overall production (see table 9.7).

In addition, Keynesian policies have proved incapable in the long run of maintaining full employment, which has led to the re-emergence of the social problems that welfare states were set up to solve. State resources have been squeezed, as cash benefit payments have increased and tax receipts have been undermined.

Writers on both the left and the right of the political spectrum have argued that the welfare state is in crisis. Increasingly governments blame forces beyond their control for economic problems, particularly economic processes operating at the global scale. In a world of dramatically increased economic, political and environmental interdependence, some writers have even argued that the state itself is a spent force and is in terminal decline. However, evidence about changes in the welfare state point to a restructuring and reorganization of the state and its roles, rather than to its demise. The interpretation of these changes is one of the major tasks facing social scientists at the end of the twentieth century.

Regulation Theory, the Corporate Welfare State and the Crisis of Fordism

"Regulation theory" is a conceptual framework which stresses the interdependence of political and economic processes (Boyer 1990). The regulation approach, developed by French economists in the 1970s and 1980s, has attracted considerable attention from geographers. It does not have a fully-fledged "theory of the state," but its central argument is that capitalist economic growth and stability depend on wider social, cultural, and political practices and processes, including, although not limited to, the state. Regulation theory stresses that it is not inevitable that the "appropriate" regulatory mechanisms will emerge. It thus avoids the problems of functionalism in saying that the character of the state must be understood (at least in part) in terms of its *own* history, rather than simply in terms of its consequences for capitalism.

In that history, the form of the state is both an object of, and an arena for, social struggle and conflict. According to regulation theory those conflicts, which are endemic in any complex society, may from time to time end in a kind of truce or "grand compromise" between at least some of the parties (Lipietz 1992b). It is possible, but not inevitable, that the institutions and mechanisms of that compromise may operate in such a way as to promote stable and reasonably long-lasting economic prosperity, at least for the parties in the compromise. When this happens, and it is likely that it is quite rare, regulation theorists refer to the compromise as a "mode of regulation."

According to regulation theory, the period of dominance of the corporate welfare state was a mode of regulation, or rather a series of modes of regulation, one for each state. A mode of regulation is not *necessarily* a coherence secured at the level of the nation state: there could conceivably be regional or international modes of regulation. However, Fordism, the mode of regulation associated with the Keynesian welfare state (Jessop 1992), was to a large degree national in character. Fordism, and the welfare state which was a part of it, thus took different forms in different countries (Jessop 1989).

For regulation theorists, the fate of the welfare state is linked to the failure of the Fordist mode of regulation. The Fordist mode of regulation was complex and contradictory. Thus while it mitigated many of the contradictions of capitalism in the short term, it was *itself* crisis-prone in the long term (Lipietz 1992a, pp. 313–15). The Fordist mode of regulation has been described by Lipietz as a kind of grand historical compromise between capital and labor in the advanced capitalist countries. According to this view, the working class was guaranteed a minimum wage level and regular real wage increases in exchange for workers' acceptance of regular productivity increases. This arrangement was backed up by the corporate welfare state outlined above. This ensured that "wage-earners (indeed the whole population) remained consumers even when they were prevented from 'earning their living' through illness, retirement, unemployment or the like" (Lipietz 1992b, p. 7). The components of the compromise were not present equally in all advanced capitalist countries – different modes of regulation emerged historically in each. But in general there was a virtuous circle which linked productivity increases with a social wage.

Japan, and to a lesser extent the USA, were partial exceptions to this pattern. Japanese Fordism was grounded on domestic and overseas demand for consumer goods and the famous, if sometimes overstated, "lifetime employment model." With a much lower proportion of GNP (Gross National Product) going to public expenditure, the social wage in Japan was, paradoxically, privatized. In essence the Japanese economic growth pattern was so productive and secure that it could function without being underwritten by the welfare state to the same extent as in Europe. However, the state, and particularly the Japanese Ministry of International Trade and Industry (MITI), was crucial in securing that growth trajectory in other ways, particularly through the strategic planning of industrial development (Itoh 1992). The United States' social security systems were never as comprehensive or generous as those in Western Europe, but according to Aglietta (1979) they were still sufficient to secure a virtuous circle of economic growth, when combined with mass consumption ("the American dream").

As I have shown in respect of welfare provision, the fruits of Fordism were distributed unevenly in both social and spatial terms. In addition, there were also temporal limits to Fordism. Because a mode of regulation is a dynamic set of relationships, rather than a static structure, its evolution through time can lead eventually to its basic principles being undermined. Under Fordism, the welfare state depended upon finance, which derived ultimately from the profits of private industry. It was also based on the assumption that economic distress because of unemployment would be short-lived during any given economic downturn. It was therefore rational for governments to borrow money to sustain demand during temporary economic difficulties. In due course, however, it became apparent that the productivity increases on which Fordism relied could not be perpetuated for ever with the existing technological and organizational approaches. In the UK, for example, there have been substantial productivity gains since the mid-1980s which depended precisely upon at least a partial break with the previous ways of doing things. With the emergence of "structural" unemployment in many advanced capitalist countries, the Fordist state faced a double squeeze: falling revenues on the one hand and increased (and increasingly long-lasting) demand for welfare provision on the other. The virtuous circle of Fordism had gone into reverse.

Further problems arose in social terms as well. Many of the social groups which had been marginalized by the Fordist compromise between capital and certain sections of organized labor began to assert demands. The women's movement across the

Western world became increasingly unhappy with the assumption that women should provide a subsidy to the welfare state through their domestic labor. Women were also challenging the assumption that their wages could be lower than the "family" wages paid to their male colleagues doing the same or similar work. Black people, whose acceptance as legitimate participants in Fordism had been limited at the best of times, found themselves on the wrong end of the long-term unemployment generated by the failure of Fordism. This, together with the poor provision accorded to black people in terms of welfare services, led to very high levels of social deprivation in particular residential areas. Violent reactions to this form of racism (especially by young men) has been increasingly evident in many Western countries – it was seen most recently in the Los Angeles riots of 1992.

None of these tendencies *guaranteed* the collapse of welfare states, and indeed, as table 9.7 suggests, they continued to expand even after the Fordist economic expansion had begun to falter. Rather, as Pierson (1991, p. 177) suggests, the crisis of the welfare state is an intellectual crisis as much as a material one. Just as the success of the "Fordist compromise" had to be secured politically, so its problems are at least in part the result of *political* attacks. Alber (1988, p. 194) documents these attacks and the responses to them in a variety of European welfare states. The picture is one of considerable unevenness, and in many cases government attempts at curtailment have increased political support for the welfare state. Serious doubts about the welfare state may not be universal, but their existence at all among governments and political parties is in marked contrast to the heyday of the Fordist mode of regulation. What this suggests is that the welfare state is now vulnerable in a way which it was not before; a vulnerability which stems from the erosion of its economically beneficial functions and an awareness of the partial nature of its social role.

The Future

The perspective provided by regulation theory allows us to see that, while the Fordist state was to some extent functional for capitalism, it was generated out of social struggles (notably those of the labor movement) rather than the logic of the capitalist system. Similarly, while there were logical limits to the mode of regulation that included the Keynesian welfare state, there are also specific political conflicts which are making the outcome of the collapse of Fordism different in different countries. In addition, regulation theory insists that the future form and function of the state, and of other components of the mode of regulation, are not pre-ordained by the necessary operation of the system, but are to some extent up for grabs.

It would therefore be foolish to make grand predictions about what states will look like in the future. However, some tendencies are becoming clear. States have been affected fundamentally by contemporary processes of internationalization. The internationalization of capital, including the location of production facilities, has stripped states of many of their former powers of economic planning and control. The inability of states to deliver on their economic promises has been labeled by the German political theorist Jurgen Habermas a "crisis of state rationality." This produces in turn a "crisis of legitimation" (1976). Some commentators have interpreted this as leading to the decline of the state as an institution (perhaps to be replaced by wider groupings such as the European Union).

However, this argument seems at odds with experience. For many individuals and social groups, state power appears to be increasing, rather than declining. Because states in advanced capitalist countries are losing their capacity to influence economic events and the overall productiveness of their national economies, they are policing the distribution of that which is produced ever more closely. Since the 1970s there has been a continual tightening up on immigration and citizenship policy (often along racist lines), and a shift to the means testing and "targeting" of social benefits. These trends fit well with the tendencies to totalitarianism noted at the beginning of the chapter. Even in Sweden, the most generous (and expensive) of welfare states, recent political changes have seen attempts to curtail welfare state spending. Economic policy has tended to focus around moves to increase the "international competitiveness" of national economies. In Britain, particularly, but also in the EU (European Union) in general, this has often meant trying to increase their attractiveness to foreign investors, perhaps through lower wages, or decreased government regulation, or more rarely through raising the skill level of the workforce or through investment in research and development. However, it remains unclear whether such measures can be effective without attention to the demand side of the equation – without an increase in aggregate global economic demand, increasing supply-side competitiveness is likely to lead to a downward spiral of wage-cutting and deregulation in increasingly desperate attempts to woo a share of a finite pool of investment. (For an alternative, less apocalyptic vision of a possible future, see Lipietz, 1992b.)

In sum, states are trying to deal with the failure of earlier regulatory mechanisms and their consequent legitimation crises by attempting to secure a new focus for legitimation. In the future, the aim might be to secure legitimation not around the state's ability to plan and manage the economy but around its control over who has access to a smaller economic cake. In Britain, recent months have seen the government implying a new separation between the "deserving" and the "undeserving poor," with ministerial comments about the supposed problem of single mothers and about the need to move to a "welfare society" in which welfare would be provided for the most part by individuals, rather than by the state. In Germany the rise of the right and the costs of reunification have led the government to introduce new restrictions on the liberal German immigration policy. (Though neo-fascism is not a problem just in Germany.) A vicious nationalist war is affecting the former Yugoslavia, and many other former Eastern bloc countries are flirting with nationalism. In China neo-liberal economic policies are going hand-in-hand with political oppression. It seems unlikely that these moves toward authoritarian government across the globe can really provide a secure and prosperous future for any part of it. The collapse of any mode of regulation is followed inevitably by a period of intense social and political conflict both within and between states. The hope must be that alternative paths to today's rising social authoritarianism can still be found.

PART III

Geosocial Change

Introduction to Part III: People in Turmoil

Economic, political and cultural changes interact, with themselves and with other aspects of social organization, such as the structuring of social life. As the world's population grows, for example, so do the demands for food, with consequences for environmental use: as realization of environmental constraints develops, so there may be calls for restraints on population growth, with clear consequences for the restructuring of social relations. Within societies, too, individuals and groups challenge the positions to which they are ascribed. The result of all these interactions is geosocial change, a people in turmoil, whose main features include population growth and mobility.

Population Growth

The world's population has grown very rapidly throughout the twentieth century. Many commentators, especially those in the "developed world" where growth has been relatively slow in recent decades, have argued that unless the increase is very substantially slowed the earth's natural support systems will collapse. Extensive birth-control programs have been promoted in many countries; in China, such a program was accompanied by legislation which penalized married couples having more than one child.

Many birth control programs have been at least partially successful, and the number of children being born to fertile mothers is declining. Some argue that this is a consequence not of the programs and propaganda per se, but rather of the perceived material benefits that flow from smaller families in many societies, plus the alternative life-styles offered by education, especially to women (Todd 1987): these are thought to have stimulated declining birth rates in the "developed world" during the twentieth century, and sustain arguments for promoting economic development globally. Nevertheless, so large is the current female population in the child-bearing years that growth will almost certainly continue for several decades yet, albeit at reducing rates.

The earth's growing population will place increased pressures on the environment during the coming decades, therefore. But growth, with its consequences for food production and distribution, is not the only population characteristic which is generating turmoil at the present time.

Hunger, Disease, and Structural Violence

In a world characterized by much political and military strife, many deaths resulting from behavioral (i.e. intentional) violence are recorded each year. Such premature deaths are small in number, however, compared with those which result from what is known as structural violence.

A capitalist economy is strongly characterized by its class structure, across which the benefits of wealth-creation are very unevenly distributed. Those who receive most can live in better housing conditions, consume many more than the minimum number of calories, vitamins and other substances needed for daily sustenance, and obtain access to better systems of health-care. As a consequence, they tend to live longer.

Variations in life expectancy related to class position occur in all countries, and also between countries. The average life expectancy at birth is much greater in Japan (78.6 years in 1990) than in Sierra Leone (42.0 years), for example, because many more Japanese than Sierra Leoneans are in the higher socioeconomic classes. These differences can be expressed in the concept of "lost (or stolen) years." The difference between the highest and lowest life-expectancy figures in those countries is 36.6 years, which will be lost by (stolen from) the average Sierra Leonean born in 1990. If the Japanese can live that long then their West African counterparts should be able to as well. Unequal life expectancies involve inflicting structural violence on the latter, through their position in the power structure that accompanies the map of uneven development.

The concept of lost/stolen years allows measurement of the amount of annual structural violence. If Japan, with the highest life expectancy at birth of any country, represents what is possible, then every child born in Sierra Leone in 1990 is going to live 36.6 years less than possible according to current societal organization: about 200,000 were born there then, producing a total of 7.3 million lost (stolen) years in that small country alone, for just one year's birth cohort. The extent of structural (perhaps better described as silent) violence is thus rapidly indicated: it is many times greater than the extent of behavioral violence (Johnston, Taylor, and O'Loughlin 1987).

Much structural violence occurs in the early months and years of life: infants are the most likely to die because of malnutrition, unsanitary housing conditions, and poor health-care (and many of their mothers die soon after childbirth for the same reasons). Thus variations in infant mortality rates, within and between countries, provide excellent indicators of the extent of structural violence. Much has been done in recent decades to reduce these variations, through programs of health-care, housing investment, food provision, and education designed to bring rates down to the "developed world" levels. But they remain wide (from 7 per 1,000 live births in Hong Kong in 1989, for example, to 173 in Angola and Mozambique), and the gap between some countries is narrowing very slowly, if at all. The same is the case within countries, even "developed" countries such as the UK and the USA, where there are stark differences between the life chances of those born into the "underclass" and of those born into prosperity (Mingione 1993); those differences are being aggravated by the reduction of welfare state provision and the increased reliance on market mechanisms, from which the poor are largely excluded.

The problems of hunger and survival will continue to present daily concerns (if not crises) to many, perhaps a majority, of the earth's population, therefore. Substantial achievements have been recorded in the control of killer diseases, but others – including new ones such as AIDS – remain virulent. The goal of a high life expectancy for all, wherever they are born and whatever their parents' backgrounds, remains a very distant prospect.

Mobility and Conflict

As demonstrated in earlier chapters, we live in an increasingly mobile world, with goods and information being rapidly – even instantaneously in the latter case – shifted around. People are more mobile, too, responding to the push factors impelling them away from some areas and to the pulls of more attractive places elsewhere.

With more people able to move, so new problems are created for the places that attract them. Immigrants can be very desirable for a country, especially one experiencing labor shortages; there are many examples of workers being imported – whether to work on the land (as in the sugarcane plantations of Queensland, Fiji, and Natal), in factories (as in many Western European countries), or in service occupations (as with the UK's National Health Service in the 1950s and 1960s and Filipino maids in much of Western Europe in the 1980s). Some are allowed to remain and can be joined by dependants, but others are sent home once they are jobless – which was the basis of South Africa's Bantustan policy under apartheid.

Those immigrants allowed to remain as permanent residents may be the source of tensions, however: they are readily targetable because they are identifiably culturally different. Tension often increases during times of economic difficulty, when immigrants are perceived as threats to their hosts' jobs and social positions. There have been several national referenda in Switzerland on limiting the number of foreigners allowed to remain there, for example, and many countries have experienced inter-ethnic strife – often fanned by xenophobic political movements. Refugees, of whom there are increasing numbers because of both famine and nationalist fervor, are particular targets for such movements.

Population mobility assists the rapid dissemination of contagious diseases. Migrations and tourism both increase contacts globally, and enhance the probabilities of epidemics. They stimulate calls for careful regulation of such movements in order to protect the health status of some, usually relatively privileged, populations. Freedom of movement, increasingly promoted as part of a "new world economic order," is thus threatened by calls to defend "national interests" against the possible depredations of too many newcomers.

Difference

Every society has norms which underpin its social relations; these include (often implicit but some explicit) definitions of acceptable roles and behavior for individuals and of micro-social organizations, such as household structures. Those norms invariably sanction unequal power relations between groups within society. Such inequalities may be challenged and altered, though usually only after substantial struggle. Often, however, the inequality survives long after it has been explicitly (i.e. legally) removed, as illustrated by the extremely slow, and continuing, implementation of racial equality after the United States' Civil War more than 130 years ago.

Social structures in most of the "developed" world incorporate unequal power relations between the genders which permeate all aspects of economic, social and political life. There have been many organized challenges to patriarchy, but "equal opportunities" have only recently been built into legal norms: even so, feminist movements have yet to achieve anything like full equality – a condition which applies to minority racial groups in most societies also.

Gender is only one of the characteristics used to define an individual's position in the structure of social relations: race and sexual orientation are others. Individuals and groups in these various positions are developing their own social movements with which to challenge their unequal status. Their goal is to build new, emancipated societies which have broken free from those dominated by white males in nuclear families comprising a wife and two children. Many alternative family and household structures are being created, as society becomes more flexible and individual relationships more fluid.

These social movements are but one set of examples of a society in turmoil. Economic globalization and rapid political change, interacting with continued population growth and striving for material well-being, are contributing to a global society whose various components are in considerable flux.

10

Population Crises: the Malthusian Specter?

Allan Findlay

Introduction

Human population growth has been occurring at an unprecedented rate during the latter half of the twentieth century. It took from 1800 to 1930 to add a billion people to the world's population. At the current rate of natural increase (1.6 percent per annum), and given the age structure of the world's present population, it will take only 11 years to add another billion. By 1998 it is estimated that the global population will have reached 6 billion (that is, 6,000 million or 6×10^9) compared with only 3 billion in 1960. Statistics such as these may cause panic, but they do not in themselves imply a population "crisis". They represent only the global averages produced from the aggregate effects on population change of many different demographic regimes. They indicate that, at a world scale, population growth is very rapid, but they say nothing of the changing quality of life of the world's population, nor of whether population growth should be viewed on the one hand as a cause for concern or on the other hand as evidence of human achievement in technological, economic, and social terms.

Despite the caveats listed above, it would be fair to say that the majority of recent commentators have interpreted the current rapid rate of global population growth in a pessimistic fashion (Wilmoth and Ball 1992). Consider for example the distinctly Malthusian tone of Senator Al Gore: "the social and political tensions associated with [population] growth rates like these threaten to cause the breakdown of social order in many of the fastest growing countries, which in turn raises the prospects of wars being fought over scarce natural resources where expanding populations must share the same supplies" (Gore 1992a, p. 308). Similar Malthusian statements have recently been made by many other scientists and public figures with specific reference to the effect of population growth on food security (Ehrlich, Ehrlich and Daily 1993), global warming (Bongaarts 1992), and the environment (Royal Society of London and American National Academy of Sciences 1992). The 1990s wave of Malthusian concern about population growth perhaps reached its peak of scientific credibility in 1993 with a joint statement by 58 national academies:

> The academies believe that ultimate success in dealing with global social, economic and environmental problems cannot be achieved without a stable world population. The goal should be zero population growth within the lifetime of our children. (Academy of Sciences 1993, p. 1)

This chapter starts by examining the patterns of global population growth which have given rise to such concern and considers the role of demographic processes in remapping the world in the twenty-first century. It then proceeds to document the long and recurrent history of human concern about population growth to show that the Malthusian specter is a phenomenon which has been studied and dismissed many times in the past. This gives rise to the key question of whether similar arguments can be used to counter Malthusian concern at the beginning of the twenty-first century, or whether the current "population crisis" contains new elements which justify not only academic debate but renewed calls for policies leading to global population stability.

The Geography of Global Population Trends

For most of the earth's history, the planet has been home to a very small number of human beings. Hunting and gathering economies required vast areas, but as more sophisticated means developed of meeting human needs for food, shelter and clothing, there was also a gradual increase in population numbers. Up until three centuries ago any sustained growth in population was, however, a very uncertain affair. Some writers believe population growth was regularly halted by the Malthusian checks of war, plagues and famines, while others, such as Woods (1989), argue that poor living standards meant there was a high overall mortality regime, and in particular that infant mortality rates were very high. Woods's analysis, based on the English case, also suggests that preventative checks on fertility were important and operated through the effect of social norms on age of marriage.

In the late eighteenth century a small number of areas of the globe began to experiment with new technologies in agriculture, medicine, and industry. There soon emerged in these areas new economic systems associated with what has come to be known as the Industrial Revolution, and with this also appeared a radically new social order. This is not the place to rehearse once again the arguments about the causes of the timing of these critical changes in human history, but what is salient to the theme of the chapter is that these changes in society and economy occurred in parallel with demographic changes. The demographic transition which took place may be described as "revolutionary" since it involved a fundamental change in demographic regimes which over time also proved to be irreversible.

The case of Sweden illustrates this transition. The crude birth rate (number of births in a year per thousand population at mid-year) and crude death rate oscillated around 30 per thousand at the end of the eighteenth century. This high stationary demographic regime provided little population growth. From around 1810 the death rate began to fall steadily and continued to do so throughout the rest of the nineteenth century and the first part of the twentieth century. Crude birth rates remained high until the 1880s, producing a phase of accelerating population expansion in Sweden between 1810 and 1880. This was only partially offset by waves of emigration. Crude birth rates fell from about 1880, only catching up with low death rates by the 1930s, when Sweden can be said to have entered a fourth demographic phase. This demographic regime can be represented as a low stationary one. By the 1930s crude birth rates and death rates were both between 10 and 15 per thousand.

Demographic transitions, similar to that in Sweden, have been reported for many North and West European countries, although with each case varying in the timing and duration of the transition. The description of the transition has been repeated so often that it is scarcely surprising to find that it has become generalized into a schema which many have imbued with explanatory powers (Notestein 1945; Davis 1963; Woods 1982). For global population the result was the creation of a sustained upsurge in world population numbers in the nineteenth century. By the time that countries such as France and Britain were beginning to experience a slackening in their demographic growth rates, other countries had begun their transition, thus eclipsing in absolute global-population terms the effect of lower growth rates in Northwest Europe. At a global level, rates expanded from about 0.5 percent in 1750 to 0.8 percent in the early twentieth century. The subsequent explosion of medical knowledge combined with the increasing integration of the world economy has seen the benefits of medical knowledge spread around the globe, bringing down mortality rates in virtually all countries and seeing population growth expand to close to 2.0 percent per annum by the middle of the twentieth century. Growth at this level, if sustained, results in a doubling of population every 35 years.

Population growth rates vary very substantially in both space and time. Thus by the late twentieth century, even though global population growth was running at very high levels, very low and in some cases zero growth was being recorded in the more developed countries (table 10.1). The most striking feature of table 10.1 is the stark contrast between the so-called "more developed" and "less developed" countries.

By 1993 the more developed world had 32 countries which had demographic regimes approximating to conditions of zero population growth (annual rates of natural increase between −0.4 and +0.4), and some 21 of these countries had rates of natural increase of less than +0.2. By contrast, in the less developed world the average rate of natural increase was 2.1 percent per annum and there were 39 states with growth rates of 3.0 percent or over. A growth rate of 3 percent per annum will double a population in 24 years. The peak rate of population growth was reached in the less developed countries in the mid-1960s, and since then the

Table 10.1 Global demographic indicators, 1994.

Region	*Population mid-1994 (millions)*	*Natural Increase (Annual %)*	*Birth Rate (per 1,000 pop.)*	*Death Rate (per 1,000 pop.)*	*Life Expectancy (years at birth)*	*GNP per capita ($US) 1992*
World	5,607	1.6	25	9	65	4,340
More Developed	1,164	0.3	12	10	75	16,610
Less Developed	4,443	1.9	28	9	63	950
Africa	700	2.9	42	13	55	650
Asia	3,392	1.7	25	8	64	1,820
Latin America and the Caribbean	470	2.0	27	7	68	2,710
Europe	728	0.1	12	11	73	11,990
North America	290	0.7	16	9	76	22,840
Oceania	28	1.2	20	8	73	13,040

Note: Following the United Nations current classification, the developed countries comprise Europe including Russia, North America, Australia, Japan, and New Zealand. All other regions are classified as less developed.
Source: Selected statistics from the Population Reference Bureau (1994) *1994 World Population Data Sheet.*

rate has fallen slightly, but this has been due largely to the effect of enforced family planning policies in China (Jowett 1989).

An important feature of global population growth is the momentum for future growth which exists even in circumstances in which fertility levels are dropping rapidly. Rapid population growth can appear self-perpetuating when absolute population figures are used, for the obvious reason that the number of births in a population is in part a function of the number of women in the fertile age cohorts. Thus high birth rates from one generation produce increased numbers of potential mothers in the next. Even if each woman chooses to have fewer children than her mother, absolute population numbers will continue to rise as long as the number of births per woman exceed the replacement figure of 2.06. Since the populations of the less developed countries in the 1990s have a high proportion of women in the fertile cohorts this means that the populations of these countries have a high potential for further population growth, even in those cases where total fertility rates (the sum of all age-specific fertility rates) are declining.

The momentum of population growth is well illustrated by the case of Brazil. Between 1965 and 1990 Brazil's total fertility rate (TFR) fell from 5.8 to 3.3. The TFR can be thought of as representing the number of children born on average to women across the fertile cohorts at any point in time. The substantial decline in Brazil's TFR during these 25 years sounds impressive, but during the same period the absolute number of births continued to rise because of the youthful population structure. For example, the number of new babies born in the late 1950s was only 2.9 million, while by 1990 it was over 4.1 million. Even by the year 2025, when it is estimated that Brazil's fertility rate will have fallen to replacement level, there will still be close to 4 million births per annum.

Taking on board the importance of population structure in influencing future number of births, as well as considering the factors most likely to influence fertility rates, both the United Nations and the World Bank produce separate short and (less frequently) long-range population projections. The most recent long-range projections based on 1990 data project regional and global populations to the year 2150. The medium variant projections of both organizations produce very similar expectations of the world's population by the middle of the twenty-first century (10 billion), with another 1.5 billion being added in the following 100 years. Clearly such projections make many heroic assumptions, such as the continued decline of mortality in all countries toward a life expectancy of 85, and fertility everywhere heading toward replacement level (for details of the assumptions see McNicoll 1992). The purpose here is not to challenge these or any other assumptions, although detailed evaluation is warranted (Bulatao, Bos, Stephens, and Vu 1990), but simply to make two points: first that high rates of population growth are not expected to continue in less developed countries throughout the twenty-first century, and secondly that existing demographic momentum will substantially change the relative distribution of the world's population.

Turning to the first point, conventional demographic wisdom does not lead to the expectation of an infinite and continuous increase in world population, but assumes instead that the demographic transition, already experienced by more developed countries, will spread to the less developed countries, and that population growth in these countries is therefore a temporary effect just as it was in the nineteenth century in Northwest Europe, arising from the lag between mortality decline and fertility decline. From this perspective the high population growth rates of the less developed countries will by the twenty-first century give way to lower growth and eventually to relatively stable demographic regimes. The implications for global population growth rates are, therefore, that the twenty-first century will see an overall slackening of growth. To quote the famous demographer Ainsley Coale (1982, p. 15) "if 2000 years from now a graph were made of the time sequence of rate of increase of world population, the era of rapid population growth that began a century ago would look like a unique and narrow spike." If these demographic projections are correct, then current concern from ecologists and some other quarters about population numbers are misdirected. Attention should be focused, not on whether a finite planet can support a population with potential for infinite population growth, but on the social, economic and political consequences of a limited phase of rapid population growth.

The second point arising out of current United Nations and World Bank projections of world population is that they imply differential rates of fertility decline, thus producing variable rates of population growth during the twenty-first century and an overall redistribution of the demographic weight of different regions of the world (assuming zero net inter-regional migration). Table 10.2 shows that by the year 2100 the African continent could have become the single largest demographic unit, with more than a quarter of the world's population (nearly five times its present total) and more than twice the size of China. The political and economic implications of these and other aspects of the projections are clearly massive, given that it is the countries which currently have the lowest incomes per capita which seem set to expand in demographic terms most rapidly. There can be little doubt that demographic processes are and will be responsible for a significant remapping of the world during the next hundred years.

So, do the statistics presented above represent a crisis of numbers, or should the real concern raised by table 10.2 be about

Table 10.2 Regional shares in world population, 1990 and 2100 (United Nations estimates and medium variant projections).

Region	*Share (percentage)*	
	1990	*2100*
Asia	58.8	53.5
China	21.5	12.6
India	16.1	16.7
Other Asia	21.2	24.2
Latin America	8.5	9.6
Africa	12.1	26.2
Sub-total (Asia, L. Am, Africa)	79.4	89.3
Europe	9.4	3.9
Former USSR	5.4	3.6
North America	5.2	2.8
Oceania	0.5	0.6
Sub-total (Europe, FSU, N. Am, Oceania)	20.5	10.9

Note: Percentage figures do not sum to 100 due to rounding errors.
Source: United Nations (1992) and McNicoll (1992, p. 339).

how global economic and political adjustments can be achieved to accommodate the global shift in the demographic center of gravity toward the less developed countries?

Population Crisis: Déjà Vu or Something New?

If one accepts the demographic projections of the United Nations and hence takes the view that rapid demographic growth will last little more than another century, then it becomes pertinent to ask whether a crisis of population numbers really exists, and whether other motives and circumstances might underpin waves of population concern. The current wave of concern about the implications of population growth is only the last in a long series. At least five earlier surges of concern can be identified in the English-speaking world, each introducing different nuances to the debate, but each following the same fundamental theme: that population growth rates are too high given the world's finite resources.

The first wave was initiated by Robert Malthus's famous *Essay on the Principle of Population* of 1798. The timing of the essay is far from surprising, coming as it did just following the sustained and unprecedented upsurge in population numbers which took place in England before the wider implications of the Industrial Revolution could fully be comprehended. The timing of subsequent waves of concern is harder to explain in terms of demographic events. Since the time of Malthus, as reported above, population growth has been continuous and rapid. Concern about population growth has, however, been intermittent and seems to have occurred at times of major economic and political restructuring. Often particular demographic or ecological events have been drawn into the debate, but discussion has been fundamentally influenced by the geopolitical context.

For example, the second wave of concern about population growth occurred in the 1890s. The context on this occasion was a political debate between Germany and Britain over the relative merits of agrarian and industrial economies and a simultaneous short-lived scarcity of wheat. The third and fourth waves of concern followed the First and Second World Wars as the victors sought to establish new international realms of economic and political influence. The fifth wave of concern, in the 1960s, coincided with the timing of the peak rate of population growth in the less developed countries. But as Jones (1990) points out, it also coincided with growing political concern in the West over the spread of communism in less developed countries. The communist takeover in China, and US failure in Vietnam, strengthened the view that population growth would join with communist territorial expansion to increase socialist influence (Wilmoth and Ball 1992). There was a growing feeling expressed by leading politicians such as Lyndon Johnston that Western financial aid for economic development could not keep pace with rates of population growth and that money spent on family planning programs was the most effective way to change the population–resources ratio in less developed countries. Population growth in the less developed world was therefore presented as a major problem to be tackled in order to avert political as well as ecological disaster.

The 1990s have brought the most recent wave of concern over

rates of global population growth. For some, such as Ehrlich and Ehrlich (1990), this has been an opportunity to restate earlier arguments (Ehrlich 1968) about nature being smothered by humanity and about the inevitability of population numbers leading to mass famines. But, for most commentators, discussion of population issues in the 1990s was different from that of earlier years. The changing world order of the 1990s, following the collapse of the Soviet Union, provided a new political context for the population debate. Discussion in the 1990s has also focused increasingly on the global environmental dimensions of population growth. And the debate about how to limit population numbers in developing countries was not an unhappy one for Western political leaders located in an increasingly right-wing world, eager to avoid more costly solutions to problems of pollution and environmental degradation.

Still Waiting for Malthus?

Before turning to consider in more detail the nature of the most recent wave of population concern, it is salient to ask whether previous pessimistic views about population growth have had any basis, or whether they were merely debates engineered and sustained for political ends. This requires, first, an examination of the demographic basis of these debates, and secondly, consideration of the logic underpinning the Malthusian and neo-Malthusian positions.

The most sophisticated analysis of population pressure arising out of the 1960s movement was that proposed by the Club of Rome. This group of 30 scientists headed by Dennis and Donella Meadows produced a series of computer-based simulations of the likely interactions between population growth, resource depletion, food supply, capital investment, and pollution. There is no scope to examine here each of the models, but only room to note first, that the conclusion of the authors was that: "the critical point in population growth is approaching if it has not already been reached" (Meadows et al. 1972, p. 191); secondly, that the population projections of their "World Model Standard Run Model 3" come remarkably close to the demographic trends which have subsequently been recorded; and thirdly that this model anticipates population growth finally being halted in the second half of the twenty-first century "by a rise in the death rate due to decreased food and medical services" (Meadows et al. 1972, p. 124).

Although the fit of their model to present population trends is not perfect, it is more than satisfactory and even over the longer run does not diverge vastly from UN projections. For example, the 1972 World Standard Run Model 3 projected a 1990 population of just under 5.5 billion compared, with the reality of 5.3 billion. The model proposed a peak population of around 12 billion by the year 2070, while the United Nations predict a global population of just under 11 billion by this date with growth still continuing throughout the rest of the century. Although the results of Meadows et al.'s models have been disparagingly described as "Malthus with a computer" (Freeman 1973), it remains the case that their population projections were not untenable, and cannot be rejected as a poor fit to reality. Those supportive of the *Limits to Growth* arguments of Meadows et al. would also point to the prediction in the model that food per capita would peak in the 1970s before falling consistently thereafter. The statistics of the UN Food and Agricultural Organization show that there was a 6 percent reduction in grain output per capita between 1984 and 1992.

Failure of the neo-Malthusian views of scientists from the 1960s to come true, in terms of global ecological disasters following from over-rapid population growth, cannot therefore be explained away by inaccuracies in their population projections. This leads to the need to consider the logic of the arguments surrounding concern about "over-population" in order to discover whether there is reason still to fear the Malthusian specter. The fundamental question which must be answered is "are the principles of neo-Malthusian thinking incorrect, or is it just the timing of a global population crisis which has been repeatedly mis-specified?"

The internal logic of Malthus's case is strong and helps to explain the enduring influence of his thinking and writing: "That population cannot increase without the means of subsistence is a proposition so evident that it needs no illustration" (Malthus 1976, p. 79). More controversial is his claim that:

> The power of population is indefinitely greater than the power in the earth to produce subsistence for man. . . . By that law of our nature which makes food necessary to the life of man, the effects of these two unequal powers must be kept equal. This implies a strong and constantly operating check on population from the difficulty of subsistence. (Malthus 1976, p. 71)

In his first essay, from which these quotes are taken, he gave particular emphasis to positive mortality checks on population growth such as "misery and vice." His later, and much less cited, work of 1803 gave a greater role to preventative checks affecting population numbers through reductions in fertility. Malthus envisaged the positive checks operating through a range of mechanisms. For example, in a wage economy, population growth would produce a rising supply of labor, and given no increased demand for labor this would push down wages. At the same time the limited supply of cultivable land would constrain food production in a situation of rising demand, thus raising food prices and causing a downward spiral in the standard of living. The resulting conditions of poverty and malnutrition would result in population growth being checked by rising mortality.

In England the onset of the Industrial Revolution was associated with a change in the relationships between productivity, wages, and population growth. Economic development associated with the emergence of industrial capitalism made rapid population growth possible at the same time as living standards were rising. Colonial trade made it possible to draw on new sources of food production from around the globe as well as to draw on other physical resources from the less developed countries. By the latter part of the nineteenth century fertility rates were also falling despite sustained increases in the standard of living, thus decoupling the capacity for population growth from trends in the means of subsistence.

The effect of technological advance in increasing the means of subsistence on the one hand, and the geographical shift in the sourcing of both food and non-food resources from the local to the global scale on the other hand, may be seen to have averted the timing of the population catastrophes predicted by Malthus. But the logic of his central principle, that finite resources cannot support a population with the potential for infinite growth, has remained attractive to certain groups of scientists. By the 1960s the Malthusian case was being argued once more in terms of the finiteness of global food and mineral resources. The populations of the less developed countries, it was argued, could not turn to other continents to provide their food needs in the way that Europe had done in the nineteenth century. In addition, water, fuelwood and stocks of mineral resources were shown to be being rapidly depleted.

Skeptics of the ecological Malthusian case turned once more to technology as the answer. It was argued not only that the human species could not be compared with other species in the global ecosystem, but that growing population numbers might actually be a spur to innovation (Simon 1986). Several detailed geographical studies have shown, for example, that increased population densities in the less developed world have not necessarily led to population catastrophes, but rather have led to improvements in land-use practice and to the adoption of more intensive cultivation methods. Other research by population geographers has shown that societal adaptation to situations of pressure on resources has in some cases been the deliberate and effective reduction of fertility levels. Some examples are given below to illustrate these possible responses.

Turner, Hanham and Portararo (1977), in their classic study of 29 tropical communities, show that a positive correlation exists between population density and intensity of land use. Population densities of less than 4 persons per square kilometer were associated with forest fallow production, whilst inversely, at densities of over 64 persons per square kilometer, short fallow or annual cropping was normal. Evidence of this kind has been used by Boserup (1980) to support the case that rising population density is a catalyst to innovation and that it is the stimulus of population pressure on scarce resources such as land which encourages rural societies in certain circumstances to shift to more productive agricultural practices.

This argument applies not only in an historical context, but has been shown by Tiffen, Mortimore and Gichuki (1994) to be one which has relevance to rural populations of Africa in the mid-twentieth century. Their study, based in Kenya with its high population growth rate of 3.3 percent per annum, shows how the population of the semi-arid Machakos region grew fivefold between 1930 and 1990, yet the environment of the region was improved over this time period through the introduction of terraces and the protection of tree cover. Over the 60-year period, agricultural production per person rose dramatically and new technologies were introduced in response to the increased demand for food both in rural and in urban areas. They concluded that "technological change is both impelled by population growth and facilitated by the increased human interaction to which it gives rise" (Tiffen, Mortimore, and Gichuki 1994, p. 264).

There is a number of ways other than technological advancement in which societies experiencing rapid population growth, but with limited local resources, have avoided the positive Malthusian checks. The most obvious of these is the self-regulation of population growth. Jones (1993), following a series of studies of island populations such as those of Bali, Barbados and Mauritius, has provided evidence of significant fertility decline in these high-density locations. This apparently stems from a widespread acceptance amongst these island populations of the desirability of introducing voluntary constraints on family size. In Mauritius the total fertility rate fell from 5.86 in 1962 to 1.94 in 1986.

In the context of small islands such as Mauritius it would appear that rather than the poor having no interest in family limitation, they have had a heightened awareness of the limitation of resources coupled with aspirations for a better lifestyle. This has stimulated increased contraceptive use. In these circumstances it would appear that a degree of concordance has been achieved between the perceived private and the wider social costs of high fertility. This outcome may be taken to suggest that small islands with certain types of shared historical experience such as plantation agriculture form a distinctive type of demographic regime. This particular type of regime differs from the experience of most less developed countries, where poverty may produce circumstances in which couples perceive there to be net benefits from having a large family, while at a societal level such an outcome has high net costs. The purpose of introducing the example of small-island demography is simply to demonstrate that human populations in situations of perceived scarce resources may opt voluntarily for fertility reduction rather than either following a Malthusian trajectory or depending on technological solutions to resolve the potential resource crisis.

The case of small-island demography highlights the fact that many different pathways exist to demographic transition. While many parts of the developing world still have extremely high levels of fertility, and as a consequence continue to experience rapid rates of population growth, it is important to note the trends toward transition which have emerged in the 1990s even in these states. Africa, often seen as the demographic hothouse of the globe, has experienced some remarkable changes in demographic behavior within the 1980s. For example, as recently as 1989 the total fertility rates (TFRs) in Kenya and Zambia were estimated to be 8.1 and 7.2 respectively. By 1984 the TFR estimates for the same countries were 6.3 and 6.5 respectively, and these figures were symptomatic of much more widespread reductions in fertility levels across the continent. At fertility levels such as these, rapid population growth will continue to occur, but at a pace which is less than in the past and which permits a degree of optimism that these countries too have begun on their own very distinctive pathways toward demographic transition. The timing and pace of any such transition depends, however, on a wide range of social and cultural factors which cannot be reviewed in detail within the scope of this chapter. They include the forces which influence the proximate determinants of fertility such as age at marriage and prevalence of contraceptive use (Bongaarts 1985) as well as the culturally modifiable effects of underlying causal factors such as the effect of female education on attitudes to women's roles in society (Cleland and Hobcraft 1985; Royal Society 1994).

Feed the Rich: Starve the Poor

Returning to the global scale of analysis, it is important to note that despite the recent decline in grain output per person which has been recorded in the late 1980s and early 1990s, total world food production rose in absolute terms by 23.7 percent between 1980 and 1991 and by 2.1 percent per head of population. Net per capita gains of 12 percent were achieved in the less developed countries, although in Africa population increase outstripped food production by 4.7 percent. Of even greater concern has been the change in the type of food produced. The rich countries of the world have, over recent decades, consumed increasing quantities of meat. Global meat production has increased fourfold since 1950. This has implied a relative switch in world agricultural production, with increasing quantities of grain being fed to livestock. For example, in Mexico even though 22 percent of the county's population suffer from malnutrition and many more cannot afford to buy meat of any kind, 30 percent of grain is fed to livestock. Elsewhere the geographical spread of livestock production to meet the demands of consumers in developed countries has led to the destruction of rainforest land in countries such as Costa Rica and Brazil. Market forces have therefore helped to produce an efficient system of food production to meet the needs of the populations of the richer countries, but this has often been achieved at the expense of ignoring the low-profit sector for food in the less developed countries.

In terms of the evidence which has been presented above, the growth of population is not the primary force responsible for so-called population crises. Instead it can be argued that the organization and control of resources, the way in which economic systems are structured and, in particular, the positioning of population groups relative to modes of production hold the key to understanding why malnourishment continues to blight the existence of over 500 million people. In a recent evaluation of the causal structure of hunger and famine, Watts and Bohle (1993) make no mention of population growth, except as a mechanism through which the poverty of those facing a food-consumption crisis is expressed. This is not to suggest that population growth is an unimportant dimension of economic development, but it is an affirmation of the belief that hunger and poverty stem from the nature of the economic system rather than from any given level of population. This conclusion is not at odds with the view of population taken by Marx.

Marx was well acquainted with Malthus's view on population growth, but unlike Malthus he interpreted demographic increase as a key mechanism in creating surplus population and through this the expansion of the labor force to levels exceeding the demand for labor. "Surplus population . . . becomes a condition for the existence of the capitalist mode of production. It forms a disposable industrial reserve army which belongs to capital just as absolutely as if the latter had bred it at its own cost" (Marx 1976, p. 784). Thus population increase serves to undermine the position of labor by keeping wages low within the capitalist mode of production. Where levels of capital accumulation are threatened through rising wage levels, owners of capital seek new reserve armies of labor, either within a country or through international migration from other parts of the globe. Unlike Malthus's position, rising population and falling incomes per head are associated with one another because of the existence of economic and social structures which seek to exploit the weakness of those selling labor relative to those who own capital. This exploitative relationship and its effects on demographic indicators have been explored by a number of authors. For example, Johnston has argued that international inequalities in life expectancies reflect the "core–periphery structure of the world economy, whereby the life chances of the population of the periphery are subordinated to those of the population of the core" (Johnston 1989a, p. 222). Spatial variations in life expectancies are not of course the result of deliberate malevolent actions by people living in the developed world, but arise from the cumulative effects of lowering living standards in the less developed world of economic mechanisms such as the unequal terms of world trade and the impact of severe debt-servicing ratios.

From a Marxist perspective the main challenge of population growth is not how to increase global food production by the 36 percent required over the next two decades to match population growth, but how to tackle the structural forces which are producing uneven life opportunities for the world's population. From this perspective, boundaries on "entitlement" rather than global technical limits will continue to define which groups, both within Western society and in less developed countries, are at greatest risk in encountering population crises. In this respect the writings of Sen (1981, 1990b) are particularly helpful. Sen notes that it is not the existence or absence of food which determines whether starvation of human populations takes place, but rather whether an individual or a group commands the economic power to exchange their labor for food. In positions where a population experiences a reduction in its "exchange entitlement" this places it at risk of famine even in circumstances where food may be available in the market place. Some would argue that the starvation experienced by the populations of Tigre and Eritrea in the mid-1980s and by the populations of Somalia in the early 1990s reflected exactly this position. Sen (1981) shows that many other famines, such as that in Bangladesh in 1974, can also be traced to a decline in the exchange entitlements of the population rather than, in Malthusian terms, to a per capita reduction in food supply.

Watts and Bohle (1993) have sought to extend the analysis of the causes of hunger and famine. By recognizing that malnutrition is due not only to poverty, but more particularly to powerlessness, Watts and Bohle show that the position of those affected by malnutrition cannot fully be understood in terms of economic relationships. It must also be located within societies and regions in terms, on the one hand, of power and institutional relations, and on the other hand of their social relations to production and class. Thus, deprivation and starvation may not only arise because of an individual or group's disadvantageous location in economic space, putting them at risk as a result of entitlement decline, but may equally be because of positioning in political and social space. From this perspective, so-called "population crises" may arise because of the powerlessness of a group in political space, as may be the case with deprivation related to gender in a patriarchal society, or related to class in a situation where the bounding of food resources results from specific exploitative relations enforced relative to particular production relations. Watts (1991b) has mapped the way in which these different forms of vulnerability to famine have emerged across Africa in the 1970s and 1980s.

The Malthusian Specter of the 1990s

The line of argument of this chapter so far has been that population growth around the globe is a real and significant influence in remapping the world for the twenty-first century. The current pace of demographic expansion has rightly attracted the attention of academics and politicians but, as has been pointed out, many previous waves of concern have led to unwarranted pessimistic forecasts. In general, technological advances have increased the resource base faster than population growth, while social and cultural factors have produced a marked drop in population growth rates in the consumer societies of the developed world and a limited demographic transition has been experienced in some less developed countries. Technological advances such as improved medical knowledge have produced tangible improvements in living conditions in the less developed countries. For example, the average life expectancy at birth in the less developed countries has risen from 46 years in 1960 to 62 years in 1990 (Population Reference Bureau 1991). Adult literacy has improved from 43 percent of the population to 60 percent over the same period, and primary health-care has been extended to 61 percent of the population despite rapidly rising numbers. Ironically it is technology and high personal consumption levels which may be as much the cause as the cure of the current population crisis. This is the case not only because medical technology has helped to lower crude death rates while fertility levels have remained high, but also because technological advances have made possible high living standards involving the profligate use of resources and diverse negative impacts on the physical environment.

At a general level one could represent the potentially harmful impact of population increase on the environment as follows:

$$E = P \times I$$

where E is the ecological impact, P is population size and I is the income per capita of the population. Many would argue that it is rising per capita income which is the most harmful influence, since it is the wasteful processes of mass consumption in the wealthier countries of the world which result in the greatest environmental damage, but clearly the number of people sharing in wasteful consumption behavior also has an influence on the overall scale of environmental change. Consider for example the case of global warming and carbon dioxide emissions. The relationship between the total annual emission rate, T, and its five main determining factors can be expressed as follows:

$$T = P \times I \times E \times C + D$$

where P, I, E, and C represent, respectively, population size, income per capita, energy, and carbon intensity (Bongaarts 1992). D measures the amount of carbon emitted as a result of deforestation, but is not related directly to population growth according to Bongaarts's (1992) formula. On this basis the developed world, in 1985, with its high income levels and extravagant levels of energy consumption, accounted for 64.2 percent of global CO^2 emissions, while accounting for only 25.3 percent of the world's population. Population growth and rising income levels in the developing countries will, however, dramatically alter this pattern over the next three decades. If Bongaarts's (1992) projections are accepted, then by 2025 global CO^2 emissions will have more than doubled, with the less developed world accounting for 55.2 percent of the CO^2 total and 82.5 percent of the world's population. Even with the increased emission levels per caput in developing countries which will accompany modest improvements in their standard of living, per capita levels will still lie far below those of the developed world. Despite this, and given the uneven pattern of power relations around the globe, it is not surprising to find Western leaders demanding preventative checks on population growth in less developed countries, rather than accepting the financial and political costs of implementing preventative checks on pollution levels in the developed world.

The position described above has led to a new Malthusianism, not only in terms of calling for preventative fertility checks in less developed countries. It has also underpinned a new and more sinister Malthusianism in terms of identifying ways in which mortality patterns should be redistributed around the globe. For example, some have noted that in relative terms the less developed countries are currently "underpolluted" for their population sizes, and recommend an export of pollution to the less developed countries. This is little more than a twentieth-century version of some of Malthus's statements concerning the logical human actions to be taken when faced with an impending mortality crisis. Contrast

Malthus's eighteenth-century view with that of the chief economist of the World Bank in 1992:

> The increase in population must be limited by it [the means of subsistence]. . . . All children born beyond what would be required to keep up the population to this level must necessarily perish. . . . To act consistently therefore we should facilitate . . . the operation of nature in producing this mortality; and if we dread the too frequent visitation of the horrid form of famine, we should seriously encourage the other forms of destruction. . . . Instead of encouraging cleanliness to the poor, we should encourage the contrary habits. In our towns we should make our streets narrower, crowd more people into the houses and court the return of the plague. In the country we should build our villages near stagnant pools. (Malthus 1976, pp. 179–80)
>
> Shouldn't the World Bank be encouraging more migration of the dirty industries to the LDCs? The measurement of the costs of health-impairing pollution depends on the foregone earnings from increased morbidity and mortality. From this point of view a given amount of health-impairing pollution should be done in the country with the lowest cost, which will be the country with the lowest wages. I think the economic logic behind dumping a load of toxic waste in the lowest wage country is impeccable. . . . I've always thought that under-populated countries in Africa are vastly under-polluted; their air quality is probably vastly inefficiently low compared to Los Angeles. (Summers 1992, p. 66)

The thrust of this chapter has been to point to the historical inadequacy of Malthusian arguments, while simultaneously accepting the reality of the significant demographic changes which have been taking place on the world map. The Malthusian case presented in the 1990s places particular emphasis on environmental quality (both global and regional). In place of food resources, environmental quality has come to be perceived as the key finite resource which cannot tolerate the demands placed upon it by rapid population growth. No one would wish to minimize the seriousness of the environmental challenge facing the world's growing population. However, there seems every reason to expect that mortality patterns associated with future environmental problems will relate primarily to how populations are put at risk through the structures which determine the boundaries on entitlement to access to environments offering a good quality of life, and the extent to which vulnerable, rapidly growing populations are empowered to protect and enhance the physical environment. Technological advances, economic structures and social positions seem likely to protect the interests of wealthy and powerful individuals and groups and regions from environmental catastrophe, just as they have from famine and other Malthusian threats. Meanwhile, "the Malthusian specter" will continue to be paraded by these same groups as a self-justifying explanation for why the most vulnerable populations of the globe suffer from the circumstances of severe environmental degradation.

Despite the general rejection of Malthusian arguments reached by this chapter, this is not a reason to ignore population issues. While the local consequences of unjust global economic systems may be population crises of various kinds (whether in terms of mortality peaks, or surges of mass out-migration), intervention in demographic regimes has a role to play in responding to particular socioeconomic circumstances. Although economic development may tritely be cited as the best contraceptive, promotion of family-planning policies should not be dismissed. The absolute pace of population growth in some developing countries has accentuated the scale of the human tragedy and misery faced by the most vulnerable elements of society. Consider of example the influence of population growth on the labor markets of developing countries. Mexico, with a rate of natural increase of "only" 2.2 percent per annum, must generate approximately one million jobs per year throughout the 1990s simply to sustain its current labor-market position, and to avoid an increase in the numbers of those without work or with inadequate means of supporting themselves and their families. This target (which is half that which the USA has to create in the same time period) has to be achieved by an economy which is only one-thirtieth the size of that of the USA. Rapid population growth in situations of weak economic development can therefore be generally taken to contribute to the plight of those marginalized by the economic system.

There is not scope within this chapter to fully evaluate the benefits which might accrue to vulnerable populations in developing countries from seeking to accelerate fertility decline. The danger is that while many of the states of the developing world have had structural adjustment programs imposed upon them which have led to serious reductions in the standards of living of their populations, at the same time they have had family-planning programs recommended to them by similar international agencies. Fortunately this has not led to an automatic rejection of family planning policies in developing countries, but merely to a questioning of who should pay for them (Lande and Geller 1991). This author at least would concur with Shrestha and Patterson (1990) in rejecting the Malthusian view that rapid population growth is the cause of poverty, while endorsing the perspective which contends that it is social formations which determine the nature and extent of poverty. From this latter perspective, reduced population growth in less developed countries should be encouraged as a labor empowerment strategy which the poorer populations of the world can and should implement in their own interests.

Acknowledgments

I am grateful to Tanis Waugh, editor of the IAG Development Studies Newsletter, who first drew my attention to the parallels between Malthusian texts and recent statements from the World Bank.

11

Global Migration and Ethnicity: Contemporary case-studies

Nurit Kliot

Introduction

Migration is defined as a change of residence across an administrative boundary and, although this change can be intercontinental, international, inter-regional, rural to urban, or some other, in this chapter only international and intercontinental migration are discussed. There are various ways of classifying migration which relate to the extent of movement or the change in culture involved in such movement. Some researchers, however, make a distinction between: *voluntary movement* – when people willingly leave their home in order to improve their livelihood; and *forced*, or *involuntary, movement* – when people are forced to flee their homes because of war or political persecution. We also differentiate between *permanent movements*, when migrants do not intend to return to their country of origin, and *temporary movements*, such as is the case of migrant laborers who return to their countries of origin when their employment in a certain host country has ended. Migration is often analyzed as an outcome of interchange between "push" and "pull" forces, where both "push" and "pull" forces exert pressure on the potential migrants to leave their homes.

Among the "push" forces we include environmental, demographic, economic, and political pressures. The most frequent environmental migratory forces are droughts, floods and earthquakes. The major economic pressures to leave one's country of origin are poverty, lack of employment, and crop failures. Among the political pressures to leave home, are wars, revolutions and coups d'état, persecution by totalitarian governments, expulsion by the state, and so on.

The "pull" forces which attract migrants to a certain destination are very often the result of forces opposite to the "push" forces mentioned above: the economic opportunity to improve living conditions; the chance for personal safety, liberty and freedom; and social opportunities such as educational and cultural opportunities.

Migration flows are shaped by many factors, the most important being the distance between the country of origin and the destination but, for many migrants (especially the poor), the cost of travel will determine how far they will move. The type of relations existing between the source country and the host country has a great influence upon migrant flows; thus, for example, colonial and cultural ties between the two and legal agreements on labor migration and/or quotas of permanent migrants will increase or limit migration flows between certain pairs of states. Finally, migration flows have consequences for both countries of origin and receiving countries since migration may relieve population pressure in the country of origin, while the eviction of certain ethnic groups may either benefit or harm both the emigrant and immigrant countries.

One of the concepts most useful for analyzing the consequences of migration waves is the term *diaspora*. Traditionally, the diaspora was defined as the dispersal and settling of Jews outside Palestine after the Babylonian exile. However, the use of the concept has been expanded to embrace the migration movements and settling of ethnic minority groups in general. Modern diasporas are groups of migrants which reside and function in host countries but maintain strong sentimental and material links with their countries of origin – their homelands (Sheffer 1986). The links between migrants and their countries of origin include regular visits, remittances, and even lobbying and organizing special frameworks to maintain cultural and political ties with their homeland. Diaspora is an appropriate concept to describe migrants who, for various reasons connected to both their country of origin and country of residence, do not fully integrate within their host societies. Ethnic diasporas are classified as either *ancient* (Jews, Greeks, Armenians), *old* (South-Asians, overseas Chinese), or *emerging* (Turks in Germany, Maghrebis in France). Diasporas are going through a process of expansion and are increasing both in heterogeneity and in the role they play in international relations, thus they will receive special focus here.

This chapter will present current migration flows and diasporas according to the following topics:

1. The magnitude of contemporary migrations;
2. The reasons for people migrating, with specific attention being focused on economic and political migrants; and
3. The emergence and role of diasporas in contemporary migration.

The Magnitude of Contemporary Migration

Five categories of migrants can be classified according to their administrative legal status in the host country:

a. *Legally admitted immigrants who are expected to settle permanently in the host country.* In recent years, there has been an average of about 1 million permanent migrants each year to the traditional receivers: the United States, Canada, Australia and New Zealand. A large proportion of these permanent migrants are from developing countries, mainly in Asia.
b. *Legally admitted temporary migrants.* This category encompasses seasonal migrants, non-seasonal contract workers, temporary migrants whose contracts are renewed in the host countries (such as guest workers in Western Europe), and family members who are permitted to join some of these migrants (Rogers 1992). The number of foreign workers legally employed throughout the world is estimated at 20 million (1992).
c. *Illegal (clandestine, undocumented) migrants.* Estimated numbers of undocumented migrants diverge widely. In Western Europe their number for 1991 was estimated as ranging between 2.6 million and 3.0 million. Accurate estimates of the number of illegal migrants are hard to come by but some sources believe that their number could be in the range of 30–40 million worldwide.

d. *Asylum seekers*. These are the close to 1 million people who seek refugee status in a foreign country. In industrialized countries decisions on granting asylum are based on the definition of refugee contained in the 1951 UN Convention.
e. *Refugees*. Refugees, as defined by the 1951 UN Convention, are: "persons who are outside their country because of a well-founded fear of persecution for reasons of race, religion, nationality, membership of a particular social group or political opinion." This definition was primarily devised to apply to refugees in Europe during and after the Second World War. There are currently 17 million registered refugees throughout the world; the vast majority are in the poorest countries like Ethiopia, Malawi, Pakistan, Somalia, and the Sudan. The distribution of refugees in the various regions is as follows: Africa, 5.6 million; the Middle East, 5.5 million; Europe and North America, 3.4 million; South and Central Asia, 2.3 million; East Asia and the Pacific, 398,000; and Latin America and the Caribbean, 107,000 (United States Committee for Refugees 1993).

To sum up, for the years 1991 and 1992 some 70–80 million people were residing outside their countries of origin, either legally or illegally (Appleyard 1992). It is impossible, however, to estimate the length of time these people remain in their host countries, since some refugee groups live for more than 20 years outside their countries of origin and many illegal migrant laborers leave their countries for limited periods of only a year or two. The number of migrants comprises no more than 1.3 percent of the population on earth but they are often a source of political and social unrest since in certain regions, such as Western Europe or East and South Africa, they are concentrated in large numbers (this issue will be dealt with later).

The Reasons for People Migrating

There are two major groups of reasons why people move: the demographic–economic and the political–legal. These factors will be discussed separately and illuminated by case-studies.

Demographic and economic forces upon migration

Population size is among the most important factors responsible for the migration flows from Asia and Latin America to the more developed countries. The world's population has increased very rapidly since the 1960s, and is likely to grow from 6 billion in 1991 to an expected 8 billion in 2022. Over 85 percent of people are expected to be resident in the less developed countries by the middle of the twenty-first century. This process is taking place because most of the countries of the South are in the third phase of the demographic transition, which is characterized by fertility rates which are twice those in most of the countries of the North. The North is in the post-transitional phase and growth is low in most European countries (except in Mediterranean Europe and France). These demographic imbalances created shortages in labor in the developed world during the 1960s and at the beginning of the 1970s, which the countries of the South were ready and willing to fill.

A central issue for developing countries is whether jobs will be available for their fast-growing populations. The ILO (International Labour Organization) estimated that between 1970 and 1990 the economically active population of the less developed world of the South increased by 59 percent, or 658 million. Most observers hold the view that developing countries will not be able to supply jobs to all those who join the labor force. Emigration from the South is likely to be high and more and more people will move to the countries of the North looking for jobs and a better future. The difference in living standards between South and North and the opportunities available also help to explain the increasing pressure for emigration from South to North. The gap in the distribution of wealth between the economic systems of North and South can be summarized by indicators such as per capita GDP, which in 1988 was $17,684 for the North and $775 for the South. By the late 1980s, per capita income in developing countries had slipped back to the level of ten years earlier, while per capita GDP grew in the North in the 1980s, thus widening the gap.

Another factor which leads to economic migration from Eastern Europe and the former Soviet Union is the financial restructuring and economic recession there. The ILO estimates that in 1992 alone, at least 15 million people were unemployed and an additional 30 million were chronically underemployed in the former Soviet Union. Two case studies will demonstrate the above demographic and economic features of migration: Mexican migration to the USA and Eastern European migration to Germany and the West.

South to North migration: Mexican migration to the USA In contrast to many countries of the South, Mexico's GNP per capita of $5,980 (1990) locates that country among the more advanced developing countries. With an average annual population growth of 2.0 percent and a population of 90 million, however, Mexico's resources simply cannot provide enough jobs for the one million people who are now entering its labor force each year. The economy of the United States, which is almost thirty times larger than Mexico's, with a GNP per capita of $21,360 (in 1990), is naturally very attractive to Mexican migrants but the migration tendency is also strengthened by other transactions by the two countries. Mexico is the USA's third largest trading partner, sending 70 percent of its exports there, while almost two-thirds of all foreign investment in Mexico is of US origin. Mexicans have been emigrating to the USA for many generations and the Hispanic communities of the Southwest originated in Mexico. The USA and Mexico have signed a series of agreements concerning temporary workers in American agriculture, the most important being the Bracero Program, and when this agreement ended, in 1964, illegal or undocumented migration to the USA began. Mexican migrants keep close ties with their relatives at home by sending remittances and investing in the purchase of houses and businesses in their country of origin. Often, when they are well established, they will bring their relatives to the USA.

Today, the number of Mexican migrants in the USA is estimated at 2 million, of whom 1.2 million are illegal. In 1986 a new Immigration and Control Act was passed whose purpose was to reduce illegal immigration from Mexico by granting legal status to illegal immigrants already in the country, and the status of 2 million undocumented Mexican immigrants was thus legalized by the Act. The 1986 Act, however, failed to stem the influx of other clandestine migrants from Mexico. The 3,000 km long border between the USA and Mexico is easy to cross and difficult to patrol. As a result more than a million Mexicans without immigration papers are arrested annually and sent back across the border.

Many of the Mexicans who cross the border to the USA, about half of whom come from white-collar and urban skilled jobs, are illegal temporary migrants. Many of the legal migrants to the USA are professionals and technicians, and are part of the "brain-drain" – people who leave their country in order to improve their standard of living and satisfy their expectations and aspirations. This is a type of migration which harms the country of origin.

Migration from Eastern Europe to the West In the past, migration from Eastern Europe (the former Soviet Union, Poland, Hungary, Czechoslovakia, Romania, and Bulgaria) to the West was motivated mainly by the need for the citizens of these countries to flee political repression by Communist regimes. Since the mid-1980s, however, the causes for the increased migration flows to the West have been: (i) political liberalization in all the Eastern European countries (except the Soviet Union), which have adopted legislation and policies allowing their citizens to leave the country unthreatened by possible political consequences; (ii) freedom of information, which has increased knowledge about incomes and job opportunities in other countries; and (iii) ease of travel, sometimes involving short distances and reinforced by family ties with relatives and friends who have already established themselves in the receiving country.

The migration between East and West Europe is economically motivated. The economic reform and the structural adjustment of the countries of Eastern Europe to free market economies has led to a rapid increase in unemployment in both Central and Eastern Europe which, for 1991, was conservatively estimated at 6–10 percent, but which is likely to rise. The gap in incomes and prosperity between Eastern Europe and the West is also a factor shaping migration flows. The migration from Eastern Europe is *not* motivated by population pressure, as in the case of Mexico, since population increases in Central and Eastern Europe are by no means large enough to trigger pressure on migration.

Between 1984 and 1988 migration from Eastern Europe to

Germany, the most popular destination, reached 1,012,985 people (one-third of whom were ethnic Germans), while in 1990, about 2 million people migrated from the Soviet Union and Eastern Europe. This migration is definitely assisted by the proximity and ease of travel between the countries of origin and the receiving country. Thus, migrants from Eastern Europe will first choose to go to neighboring countries such as Germany, Austria, and Italy, and to countries with long standing communities of people whose ethnic origins are the same as those of the migrants. A large number of East Europeans is employed illegally in the West, facilitated by the fact that the borders are easy to cross. Germany allows Czech, Slovak, and Polish workers to cross for daily work in Germany in the districts along the common borders (Hönkopp 1993). Germany has signed guest-workers agreements with Hungary (1989), Poland (1990), and with the Czech and Slovak Republics (1991), which regulate the movement of guest workers in limited quotas of the labor market of Germany.

Most of the Eastern European migrants are between 25 and 50 years old, and their migration is usually short term but, for some, Germany is just a transit station to other Western countries, particularly the USA. Some Eastern Europeans are employed under special contracts which permit foreign and German companies to import specialized personnel into Germany to work in certain categories of jobs such as nursing and even in agriculture and forestry.

The forecast is that until the early twenty-first century emigration from Eastern Europe, mostly for short-term employment, will continue and accelerate as long as the economies of these countries continue to experience the powerful effects of economic restructuring.

Why people move: political and legal causes for migration

Politically motivated migration which creates refugees and asylum seekers is accelerating, although it is becoming more difficult to classify a "genuine" refugee. The increasing convergence of refugee and economically motivated migration has tended to blur the classical distinction made between those who require humanitarian assistance from the international community and those who use the label "refugee" in order to emigrate for other reasons (Widgren 1993).

Flight by refugees might be necessitated because of force exerted by authorities for racial, ethnic, religious, political, or other reasons which produce a life-endangering situation; by the violation of human rights; by civil war or the outbreak of violent conflict between ethnic groups. Some 12 million to 18 million people are currently internally displaced within their countries of origin because of political upheaval and some 17 million have found refuge in neighboring countries. In such a context, it is important to define the concept of asylum. It is the provision of protection to vulnerable human beings who are in grave and urgent need of safety. The main problem for the North is the tremendous increase in the number of people seeking asylum who do not have a genuine claim to the status of refugee. The more the claim for asylum is used as a mechanism for immigration the greater the risk that public, humanitarian support for genuine refugees (such as exiles and refugees from Iran and Sri Lanka who are truly in danger) will erode.

With the end of superpower involvement in many parts of the Third World, there is now in some regions a greater chance for peace and the resolution of long standing problematic refugee situations such as those in Cambodia, Ethiopia, and Afghanistan. Refugees in Central America are returning home, but the bitter conflicts in Rwanda, Bosnia and Somalia and the war in Iraq have led to new waves of refugees and the subsequent international intervention by the United Nations High Commission for Refugees (UNHCR), since the major powers have not been able to prevent or solve the refugee situations.

The two case-studies chosen for discussion are Vietnam and Mozambique.

Vietnam

The Vietnam War ended in 1975 with the unification of the state and refugees had to leave their country of origin because they belonged to the defeated side, the wrong ethnic group, or the wrong political or religious persuasion. Most of the refugees were eventually settled in the West – not in the countries of first asylum to which they initially fled. The acceptance by the West of all Vietnamese asylum seekers for resettlement as refugees largely resulted from the USA's acceptance of its responsibilities for its defeated allies. The movement began in a small way with the initial US evacuation of 130,000 people mainly from Vietnam. As a result of pressures experienced by Vietnam's neighboring first-asylum countries (Thailand, Hong Kong, Malaysia, and Indonesia) a UN-sponsored meeting was convened in 1979 and resulted in offers being made to settle some 250,000 Indo-Chinese refugees in industrialized countries. Between 1975 and 1983, some 775,000 Vietnamese refugees (mostly ethnic Chinese who felt persecuted and oppressed by the new regime) landed in neighboring countries. Eventually, about two-thirds of these refugees were settled in third countries – mostly in the USA, France and Australia, while another 270,000 ethnic Chinese were permanently settled in China. In the succeeding years, however, a consensus developed whereby the majority of Vietnamese, especially the "boat people", were more correctly defined as economic migrants who did not warrant the granting and protection of refugee status. This resulted in the shrinking of resettlement quotas and left some 101,000 Vietnamese refugees stranded in detention camps in Hong Kong (45,000), Indonesia (15,000), Malaysia (10,632), Thailand (12,600), and other countries in Southeast Asia. Efforts to deter Vietnamese from migrating became more determined, particularly in Thailand, Malaysia and Hong Kong, where boats have not been allowed to land and from where Vietnamese are forcefully repatriated. As a result, very few Vietnamese left Vietnam in 1992.

The movement of the Vietnamese to the West has been a long-term permanent movement which has included whole families. The Vietnamese refugees have no intention to return although there is evidence that Vietnamese send remittances to Vietnam and even come to visit their former homeland (Lam 1993). The Vietnamese are well integrated in the economies of the absorbing countries (USA, France, Canada, UK, etc.), and are generally fully employed and socially very mobile. Altogether 2 million Indo-Chinese refugees have integrated very successfully into Western countries.

Mozambique

Mozambique is a typical example of the mass waves of refugees in Africa. There is a total of 1.5 million Mozambican refugees who have found asylum in neighboring countries: 905,000 in Malawi, 264,000 in Zimbabwe, 250,000 in South Africa, 72,000 in Tanzania, and 45,000 in Swaziland. In addition, some 3.5 million have been internally displaced by Mozambique's brutal 16-year civil war, which came to an end in late 1992, due in large measure to a devastating drought that left combatants exhausted and civilians starving. This resulted in the Mozambican refugees and displaced people being totally dependent on food distribution centers, mostly donated by Western countries. The refugee population is slowly being repatriated to Mozambique but, because traveling is not safe and there is no law and order or operational services, many refugees who returned to Mozambique did not remain, and have ultimately returned to their countries of asylum. Among the returning refugees are an estimated quarter of a million children and a high number of physically handicapped people who need special attention. In addition, thousands of land mines left behind by the civil war remain a threat to refugees returning to settle the unoccupied land. To alleviate some of the misery in 1993, international donors contributed $320 million for the reintegration of refugees.

Almost all the Mozambican refugees are "rural to rural" migrants and their influx into host countries raises deep-rooted political and economic issues related to the obligations and the capabilities of poor asylum countries such as Malawi. Refugees represent 12 percent of the population of Malawi and their presence here has had a positive economic impact; for example, by the creation of employment for Malawi citizens in the variety of organizations which assist refugees. In addition, many of the services which have been built for refugees, such as schools, boreholes, health clinics and so on, also serve the local population, and local commerce has been stimulated. The negative impact of the refugees is, however, also enormous with severe environmental damage to land, forests, and water resources, and pressure on the meager Malawi economy and government services. Since it began hosting refugees, Malawi has had increasingly poor relations with Mozambique, which has accused Malawi of discouraging Mozambican refugees from returning home in order to enjoy the benefits of international aid (Dzimbiri 1993). Malawi has also experienced decreased security since both criminal and military elements have entered the country along with the refugees.

African refugees are, perhaps, the most disadvantaged people on earth and the Mozambican case-study does not differ from other major African refugee countries such as Angola, Ethiopia, Sudan and Somalia. All have had social, economic, political, and environmental effects upon the poor countries of asylum.

Diaspora

Almost all countries in the world contain ethnic minority groups who have emigrated voluntarily or involuntarily and preserve their distinct cultural and ethnic identities. Of special interest are the network organizations which diasporas develop in order to sustain relationships with other members of their diaspora and with their country of origin. The two case-studies selected to exemplify the complex effects of diaspora are the "old dìaspora" of the Indian (South Asian) migrants and the new (incipient) diaspora of the Maghrebis in France.

The Indian (South Asian) diaspora

The South Asian (better known as Indian) diaspora was mainly formed during the last 150 years and many of the Indian communities evolved before India and Pakistan received their independence. South Asian communities are widespread throughout 57 countries in all the continents and numbered 13 million in 1990. The largest Indian communities are found in the UK, Mauritius, South Africa, Nepal, Sri Lanka, Fiji, the Persian Gulf States, Guyana, Suriname, and the USA.

South Asian migration has always been, and still is, mostly economically motivated. The first Indian migrants were hired as indentured laborers to work in the plantations of the various colonies of the British Empire – mainly in Natal (South Africa), Mauritius, Fiji, East Africa, Suriname, and in the Caribbean islands of Trinidad, Jamaica, Réunion, and others. South Asians still constitute a majority of the population in Mauritius, about half of the population in Fiji, Guyana, and Trinidad, and form large minorities in Suriname, Malaysia, Nepal and Burma. More recently some 2.5 million immigrant laborers from the Indian sub-continent found temporary employment in the Middle East, particularly in the Gulf countries.

The 1.3 million strong South Asian community in the UK was formed by a combination of forces. Many were attracted because of economic considerations while others used their past colonial ties to emigrate to the metropolis. Some 30,000 Indians who were expelled from Uganda in 1972 emigrated to the UK, 20,000 emigrated to Canada and the USA, and some even "returned" (after having lived all their lives in Uganda) to India. Presently, the Indian community in Fiji, discriminated against by the native Fijian population, is leaving Fiji in growing numbers.

Population growth and poverty in the Indian sub-continent have served as "push" factors and migration from India, Pakistan and Bangladesh (presently mostly a short-term migration as migrant labor) has helped to alleviate unemployment. Remittances from the Gulf play a very important role in the economies of these states and in the mid-1980s amounted to $2.2 billion annually for India and $2.5 billion for Pakistan.

South Asian communities are found at all levels of host societies and are distributed among all the occupations and trades, although they are mainly concentrated in commerce and trade. Politically they are under-represented and discriminated against in South Africa, Malaysia, and Fiji. In most of their host societies the South Asians live in cities and towns, but in Suriname, Guyana and Mauritius they are rural (Clark, Peach, and Vertovec 1990).

The Indian communities in the diaspora have succeeded in maintaining their separate ethnic identity as well as their ties and networks with the homeland and with other Indian diaspora communities. They have succeeded in this despite the fact that the communities are split along religious (Hindu vs Muslims) and linguistic lines, as well as by caste and social class. Separateness and distinctiveness strengthened by spatial isolation, the development of "Indian" political parties, the development of special community facilities to cater for cultural needs, and the maintenance of constant ties with the Indian continent have assisted in maintaining ethnic and cultural uniqueness. A final important note is that the Indian diaspora is still in a process of change and formation – with some communities decreasing in size while others are growing; but there are no signs that that diaspora will be fully assimilated within host societies and simply vanish.

The Maghrebi diaspora in France

Morocco and Tunisia were colonies of France until the mid-1950s, as was Algeria until 1962. Emigration from these colonies began in the 1930s and 1940s, with the common French language and a "Francophone" culture facilitating the process of migration to France, where the migrants were mainly employed in manual jobs in agriculture and manufacturing. After the Second World War, the low natural population increase prevalent in France compelled that country to import labor, and the natural choice was the Maghreb. The proximity and ease of travel increased the number of migrants and their influx even expanded after Algeria gained its independence. France regulated immigration from the Maghreb through annual quotas but in 1974 it stopped admitting migrant workers, mainly because of the economic recession. Since then migration flows to France have continued, mainly because of the bilateral agreements which promote economic exchanges between France and its former African colonies. These agreements, concerning trade, military, and technical assistance, as well as the flow of remittances to the Maghreb, have weakened France's ability to limit immigration (Garson 1992). As a result, 3 million immigrants of Maghreb origin currently live in France. The Maghrebi communities in France preserve their separate ethnic identity by developing communal cultural, educational, and religious institutions and by maintaining strong ties with their homelands. Newcomers are assisted by the existing diaspora and are easily absorbed into France through the mediation of their community. The Maghrebi community in France does not tend to assimilate into normative French society and, as a result, many of the French resent the seclusion and separate identity of the North African ghettos of Paris and other large cities. Resentment is "translated" from time to time into sporadic violence directed at North Africans. The presence of a Maghrebi diaspora in France, like the Turkish community in Germany and other emerging incipient diasporas, exacerbates the problematic complexity of the ethnic composition of European, North American, and Pacific states; but the process does not usually lead to greater tolerance or more liberal migration policies. The opposite is more likely to be true.

Conclusions

The six case-studies presented in this chapter are extremely varied. Some of the migrations have been forced (Mozambique, Vietnam, the Indian expulsion from Uganda) while others have been voluntary (the South Asian, the Eastern European, and the Maghrebi diaspora). Some of the migrants have had to travel large distances in their migration, like the Vietnamese who migrated to the USA and Canada or the Indian diaspora which has spread worldwide. Many migrants simply cross a nearby border in order to reach a safe haven or to find employment, such as the Mozambican refugees, and Polish migrant labor in Germany. The volume of migration is impressive and, in Africa, the magnitude of the refugee problem is overwhelming. Since data on migration and refugees are hard to come by and since countries greatly differ in the way they classify refugees, there are good reasons to believe that the number of migrants, particularly illegal migrants, is greater than the frequently quoted 80 million. Illegal migration is a way of life in many African and Asian states where migrants simply cross the border to work and return to their homes. Only when Asian and African states impose entry regulations will thousands of migrants unwillingly return home.

The "push–pull" matrix has not changed over the years. The fast-growing populations in India, Pakistan and Bangladesh, in Mexico and Latin America or in the Maghreb countries, has been accompanied by the inability of these countries to provide jobs for all those who enter the market place; and this is one of the major push forces. The prosperity of the Western and Northern economies of Europe, North America, and Australia has attracted millions of migrants from the South and from Eastern Europe who look for jobs and the opportunity to improve their place in the world.

Political pressures to move have not diminished in recent years although many regional conflicts ignited by superpower rivalry have been resolved. But new ethnic and tribal conflicts such as those in Afghanistan, Somalia, Angola and former Yugoslavia, and struggles for regional hegemony like the Gulf War, have pushed millions of refugees outside their homelands.

Migration flows, according to most specialists, will accelerate in the future as all the push and pull forces increase their effects. The policies which the North adopted for regulating migration have not succeeded in the past and are probably not going to succeed in the future as long as ambitious people strive for a better life and refugees strive for life itself.

12

Changing Women's Status in a Global Economy

Susan Christopherson

Introduction

The question of women's status relative to that of men is both fascinating and elusive because it takes in so many realms of human experience. Although this examination of relative status will focus on economic and political variables, it should not be taken to imply that women's status in society, in any particular society, can be explained solely with reference to political and economic measures. Even within the economic sphere (as we shall see) there is considerable controversy regarding what it is we are measuring when examining women's status. Behind every measure lie value-laden assumptions. What is equality? What is work? Even, what is woman?

Just as important, status is not static and women's status has to be constantly re-evaluated with reference to changing conditions in time and space, including those shaping the contemporary global economy. The context and conditions within which relative status is measured are in constant flux, affected by technological advances, state–market relations, education, and that infinitely slippery variable, attitude. Although it is possible to sketch substantive trends, the central purpose of this chapter is to argue that while women continue to be disadvantaged relative to men with respect to wages, total work hours, and access to societal resources, they are playing central, critical roles in the processes structuring a global economy. Their continued invisibility within these processes is a result of the measures we use to gauge economic and social contributions; of ideologies that marginalize women as either peripheral to, or victims of, economic change; and of the personal nature of many women's responses to exploitation.

Problems in Measuring Women's Status

Although women's status is typically measured in economic terms, other types of indicators are equally important in providing us with a picture of how women are valued in comparison with men. For example, there is now considerable evidence that gender discrimination has resulted in excess female mortality, especially in the developing world. The number of "missing women" who would be alive were it not for gender discrimination has been estimated at between 60 and 100 million women worldwide (Coale 1991; Sen 1990b). These women have been the victims of pre-natal choices to abort females, of biased health conditions, of unequal nutritional provision, and of infanticide. The sex ratio, defined as females per thousand males, has declined almost continuously since 1900. Although the ratio increased marginally (from 930 to 934) between 1971 and 1981, the sex ratio at birth declined again in the 1980s. The continued preference for boy children reflected in these statistics is a strong indication of the prospects for women throughout their lives. The problem of the "missing women" also suggests that in determining women's status we need to pay attention to what is missing (such as the ability to move freely in the city and the world; and educational opportunity) as well as to what is present.

One attempt to evaluate women's status holistically is represented by the human development indices (HDIs) in the Human Development Report of the United Nations Development Program (1992). The UNDP has developed a gender-sensitive human development index which calculates male and female HDIs and an overall gender-sensitive HDI using figures on life expectancy, adult literacy, number of years of schooling, employment levels and wage rates. The use of a gender-sensitive HDI changes the HDI ranking of many countries. Canada, which is ranked first in overall HDI rankings, slips to 8th place in a gender-sensitive HDI ranking because Canadian women have lower employment and

Table 12.1 Gender-sensitive Human Development Indices (HDI)

	Gender-sensitive HDI	*Female HDI as % of male HDI*
Sweden	0.938	96.16
Norway	0.914	93.48
Finland	0.900	94.47
France	0.899	92.72
Denmark	0.879	92.20
Australia	0.879	90.48
New Zealand	0.851	89.95
Canada	0.842	85.73
USA	0.842	86.26
Netherlands	0.835	86.26
Belgium	0.822	86.57
Austria	0.822	86.47
United Kingdom	0.819	85.09
Czechoslovakia	0.810	90.25
Germany	0.796	83.32
Switzerland	0.790	80.92
Italy	0.772	83.82
Japan	0.761	77.56
Portugal	0.708	83.36
Luxembourg	0.695	74.88
Ireland	0.689	74.89
Greece	0.686	76.10
Cyprus	0.659	72.32
Hong Kong	0.649	71.10
Singapore	0.601	70.87
Costa Rica	0.595	70.61
Korea, Rep. of	0.571	65.53
Paraguay	0.566	88.82
Sri Lanka	0.518	79.59
Philippines	0.472	78.67
Swaziland	0.315	68.74
Myanmar	0.285	74.07
Kenya	0.215	58.60

Source: UNDP Human Development Report, 1992.

wage rates than men. Sweden moves to first place because of the greater equality of status in Sweden (table 12.1, figure 12.1) (UNDP 1992).

When women's status is evaluated specifically in terms of their economic contributions, we again encounter problems with what

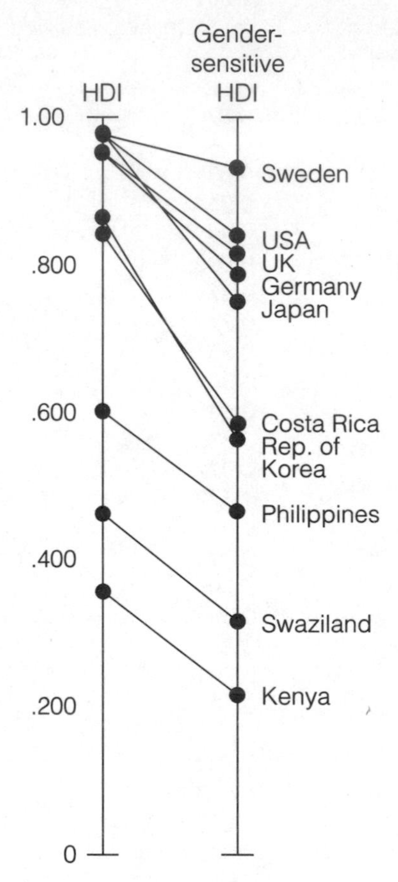

Figure 12.1 Adjusting the Human Development Indices (HDI) for male-female disparities, 1990

is and what is not counted. Since the 1970s, feminist economic researchers have focused on this problem and on the re-evaluation of the concept of work. Much of their empirical research "on work" has been intended to deconstruct taken-for-granted notions of what constitutes socially and economically necessary work and who are the workers (Waring 1988). A rich variety of empirical studies has documented the significance of female labor in subsistence production, domestic production, volunteer activities, and in the informal sector. Although these studies have raised questions about, and in some cases changed, accepted definitions of labor and of the labor force, they have had almost no effect on national accounting of work. As a consequence, there is very little systematic information available on a national level that reflects all that we have learned about the work that women do. Even in cases where there has been some attempt across a number of countries to measure work better in, for example, subsistence agriculture, problems have arisen over the reliability of the data. "Once market criteria did not apply, what was considered an economic activity became arbitrary and differences developed among countries regarding . . . what activities were included in national accounts" (Beneria 1992).

Even in the sectors of the economy where women's work is reported, there are serious problems in understanding the extent and nature of their role. Take the case of part-time work, for example. Countries define part-time work and part-time workers in very different ways. In Japan, part-time workers are defined by their status in the firm – they do not have "regular" employment contracts and so may be laid off at any time. While at work, however, they may be employed for 40 hours per week or more. In some countries, part-time work describes any work that is less than full-time. It may be regularly 39 hours per week. In the United States, part-time work is divided into two categories, voluntary part-time work, which describes situations in which workers prefer part-time jobs because they are going to school or have family responsibilities, and involuntary part-time work, which describes situations in which people are working fewer hours than they would prefer because full-time work is not available to them. Despite the significant differences in the definition of part-time work from one country to another, the prevalence of women as part-time workers is taken to mean that they have a loose attachment to the labor market. In some countries, such as Japan, when part-time workers are laid off they are not counted in unemployment statistics. Because of the problems with definitions such as part-time work, feminists have advocated new approaches to understanding how people are included in the labor market. They suggest, for example, that instead of examining part-time work we might more usefully look at the trends toward de-standardization of employment contracts relative to national norms. Increased variability has been particularly pronounced in those countries in which service employment dominates (such as the UK, Canada, and the US). Women constitute the majority of workers in non-standard contracts, particularly in part-time jobs but also in other forms of non-standard work, such as temporary work on fixed-term employment contracts, and seasonal work. The explanation for women's positioning at the periphery of the labor market is related to their assigned familial role as care givers to both children and the elderly. It is also attributable to their concentration in industrial sectors and occupations which are not protected by collective bargaining agreements such as those which protect the wages and working conditions of many male workers.

There are indications of progress in accounting for women's work in some sectors and countries. It is widely recognized, for example, that a low rate of female labor-force participation may simply disguise the fact that most women are productively employed in family enterprises. This is true in the developed world, in countries such as Italy, Portugal, Spain, and Ireland, as well as in developing countries. The Dominican Rural Women Study, based on a survey of over 2,100 households, resulted in an estimated labor-force participation rate for rural women of 84 percent compared with a 21 percent rate in the 1981 national census. The difference resulted from a broader definition of women's productive work activities, such as garden cultivation, animal care, and cooking for fieldhands (Beneria 1992). Special studies, such as this one, have demonstrated the disparities between official statistics and actual work effort. As yet, the information they provide has still to be incorporated systematically into national data-gathering efforts in such a way that the full range of women's productive activities becomes visible.

The inadequacy in national income accounts and labor-force statistics has implications beyond the recognition of women's contribution. Inadequate measures of women's work in the wage labor force as well as in more informal economic activities have hampered our ability to understand and interpret how women have affected and been affected by recent transformative processes in the global economy. Women are playing a central role in these transformative processes, including the globalization of manufacturing and the enhanced circulation of capital and information. In some regions and countries that role is visible – women are the majority of workers in the new international manufacturing sector. In other regions and countries, particularly in Latin America and Afica, women's role is concentrated in the informal sector but no less significant. Women in much of the developing world are faced with the primary burden of keeping households afloat in a world economy in which trade policies have produced economic instability and high levels of unemployment. Across the world, changes in international markets have interacted with sexually divided patterns of activity to produce differential effects by sector and by region.

Women's Central Role in the Globalization of Manufacturing and Business Services

Between 1953 and 1985, the developed market economies' share of world manufacturing output declined from 72 to 64 percent while that of the developing market economies increased to 11.3 percent. (The share of manufacturing in the centrally planned economies remained stable during this period.) There were also shifts in the share of world manufacturing within the older industrialized countries, with Japan experiencing very high employment growth rates and considerably outpacing the United Kingdom and the US, both of which experienced significant employment losses (Dicken 1992). Women's employment in manufacturing was dramatically affected by these shifts. In those countries losing manufacturing jobs, women lost jobs at a much higher rate than men. In those countries gaining jobs, women constituted the majority of the new workforce. In the United States, for example, between 1973 and 1987 (a period of significant manufacturing job loss) the ratio of goods employment to total employment declined 17 percent for men but nearly 23 percent for women (Bluestone 1990). The high female job losses are explained by women's concentration in manufacturing industries, such as textiles, apparel, and shoes, that were most subject to international competition and thus to shut-downs and lay-offs. Studies in Canada reinforce the finding that women have been more affected by manufacturing job losses than men. An Ontario study of manufacturing job losses in the 1980s shows that women manufacturing workers who lost jobs had a much more difficult time finding another position. While 62 percent of the men were able to find another job, only 38 percent of the women were able to do so and the time it took to find another job was much longer for women than for men. And while the majority of laid-off men in this study improved their wages in the new job, almost half the women reported lower wages in the new job. The result was an increase in the wage gap. Before the lay-offs, the women earned 72 percent of what men earned, after re-employment, the women earned 63 percent of male wages (Cohen 1987).

The overall pattern is even more complex than these figures suggest, however. In the US, the female portion of total manufacturing employment has actually risen in the past 20 years because of the growth of female-dominant manufacturing sectors such as electronics, and increased productivity (and low employment) in male-dominant sectors such as machine tools. So, the picture is not one of wholesale female job loss in manufacturing but rather a shift in employment from some regions to others and from some labor-intensive industries to others.

In the developing countries and emerging high-technology industries, we see a mirror pattern. Both in the new industries, such as electronics, and in the industries such as apparel and shoes, in which labor costs have driven the development of production for export, women constitute the majority of the new workforce. Approximately 80 percent of the total export-industry workforce in developing countries is composed of women (Joekes 1987). These women are frequently in culturally conflicted positions, moving between home, where they are expected to play traditional roles, and the public world of the workplace (Ong 1987).

Women are also a central component in the workforce that has increased with the internationalization of finance and business services. Within the industrialized countries there has been a reordering of regional labor markets in response to the global reorganization of production, distribution, and finance. In those countries and cities which have become the corporate command centers, headquarters functions and business services have expanded dramatically. This has produced a substantial expansion of labor demand in sectors, such as banking, which have historically employed large numbers of women. The expansion of corporate and business service activities has been associated with an increase in the use of the external labor market to provide specialized service inputs. For these workers, including many women, status and mobility are tied less to seniority in the firm than to educational credentials and performance. Workers in these services possess skills that can be applied across industries (computer programming, copy writing, graphic design) rather than firm-specific or even industry-specific skills. Although the number of women in mid-range managerial and professional positions has increased, women are still excluded from upper management positions. Only 4 percent of upper management positions in European and North American international businesses are filled by women. In part, this reflects a belief that businessmen from more gender-conservative countries, such as Japan, find it difficult and uncomfortable to negotiate with women, and that female executives will therefore harm business prospects.

International Flows of People and Information: Women's Roles

The internationalization of services is a direct reflection of the central process associated with globalization – the increased international flow of people, commodities, and information. Women are central to these flows, too, as migrants, and as information providers and processors.

The largest number of skilled migrants from less developed to more developed economies is in the health services. The reasons for this migration are basically economic, but there are some important differences among skilled migrants. Physicians from less developed economies (almost exclusively male) are frequently trained outside their own countries and have become accustomed to higher levels of treatment and diagnostic facilities than are available to them at home. Fewer professional opportunities are available to them because developing countries need general medical practitioners rather than specialists. Migration is thus an avenue to professional development as well as higher remuneration. For nurses, almost all of whom are female, the situation is quite different. Nurses are trained in their home countries and acquire general rather than specialized skills. In several countries, most notably the Philippines but also Korea and Pakistan, the training of nurses reflects a recognition that nursing graduates are being prepared for service in a world market rather than a national one (Ball 1990). While the nursing migration stream to the US comes primarily from Asia as well as from Ireland and Haiti, a parallel migration stream is developing from Eastern European countries into Western Europe.

The increased demand for this category of skilled female labor is the result of a complex set of developments. As employment opportunities have opened for women in non-traditional occupations, nursing has become less attractive as a professional choice. This unattractiveness has been exacerbated by an extremely flat career hierarchy and generally low pay. In addition, the restructuring of the health sector in some countries has considerably worsened working conditions for nurses, causing many of them to leave the profession. Given this shortage, there is considerable pressure to look to foreign sources for nursing graduates to fill available positions.

The case of nurses suggests some ways in which female migrant labor may become more important in industrialized economies. As women in industrialized countries become more qualified and more specialized, occupational mobility may depend on a willingness to spend periods of time outside their native country. Secondly, in economies where the birth rate is low (such as Germany) and where women have moved into more qualified positions, immigrant women may be used to fill the gap for routine service work. In Western Europe and North America, female immigrant labor is increasing in response to a demand for personal services, such as child-care or housekeeping, to replace services previously provided by women who are now employed. The hiring of immigrant women may also be used as a strategy to reduce the cost of providing routine services in industries such as health-care. Finally, on the supply side, there is evidence from data on legal and illegal migration from Mexico to the US, that women form an increasing proportion of the migrant workforce, a workforce that until recently was almost exclusively male. Many of these women identify their occupation as "manufacturing operative" and have entered the workforce as manufacturing workers in industrial zones. Having obtained basic job skills and confidence, they decide to migrate to higher-wage manufacturing sites in the US. Thus the globalization process comes full circle, with workers as well as commodities entering the markets of industrialized countries.

Another area where women are a prominent and increasing component of economic globalization is in information processing. Although services have conventionally been thought of as market driven, some services such as data preparation and processing are aspects of complex production and distribution processes in both manufacturing and service industries. The location of these activities is influenced by the absence of product transport costs (products are transmitted through telecommunications networks), by labor cost, and skill. Barbados, Jamaica, Ireland, the Philippines, Taiwan, Sri Lanka, and China are major locations for data entry. For US companies the high literacy rates in some countries, particularly Barbados, are strong attractions (Howland 1993). The development of information processing industries in these countries demonstrates that even in developing countries the increased educational attainment of women is directly related to employment growth.

Advances in information processing technologies, particularly the replacement of manual data entry with optical scanners, will change the nature of work in this industry and affect thousands of women workers in both industrialized and developing countries who are currently employed as data-entry operators. As was the case with technological change affecting other female-dominant occupations – librarians, nurses, and computer programmers among others – the expansion of necessary skills and specialization may not translate into better jobs or higher wages for women if they are denied access to opportunities for skill acquisition. Instead, the transformation of the work process and its attendant occupations may produce a workforce in which the majority of the more technically skilled jobs are held by men.

New Divisions in the Global Labor Force

In many conventional respects the prospects for women in the global economy look positive. Female labor-force participation is increasing, women's educational attainment has improved

Table 12.2 Female–male gaps

	Females as a percentage of males (see note)										
	Life expectancy 1990	*Popu-lation 1990*	*Mean years of schooling 1990*	*Secondary enrolment 1988–9*	*Tertiary enrolment 1988–9*	*Third-level science students 1987–8*	*Labor force*		*Unemploy-ment 1989*	*Wages 1986*	*Parliament 1990*
							1960	*1988–90*			
Industrial	110	106	96	104	99	65	..	78	..	..	13
Developing	104	96	54	73	53	..	..	52	..	..	14
World	106	99	63	79	65	..	..	58	..	..	14
OECD	109	105	99	104	95	42	44	73	..	66	12
Eastern Europe and USSR	112	109	90	105	105	82	..	89	..	..	15
European Community	109	105	96	103	94	43	..	66	..	76	15
Nordic	109	103	100	105	104	54	..	88	..	84	58
Southern Europe	108	104	90	103	97	38	..	60	..	..	15
Non-Europe	109	104	101	104	96	..	47	78	..	58	6
North America	110	105	102	104	105	..	42	82	..	59	8

Note: All figures are expressed in relation to the male average, which is indexed to equal 100. The smaller the figure the bigger the gap, the closer the figure to 100 the smaller the gap, and a figure above 100 indicates that the female average is higher than the male.
Source: UNDP Human Development Report, 1992.

significantly, women's work is, to a greater degree, recognized, and in some countries the wage gap has decreased. Using a standard in which these indicators explain women's status, we would have to say that women's prospects in the global economy are improving. As was noted at the beginning of this chapter, however, status is not static; and economic, political, and technological developments have occurred which put these achievements in a different light. Both technological and organizational innovations in industries employing women have tended to construct a workforce composed of a small number of higher-skilled jobs and significant reductions in the lower-skilled routine jobs held by women. Although some women are able to move into the highly compensated, high-skilled jobs, women's mobility may be blocked in various ways. Some of these barriers are structured in the production organization. In those countries where labor markets are strongly developed within firms, for example, women are hampered by requirements for continuous employment and geographic mobility. A study of the Japanese information technology industry demonstrates that the career path from programmer to manager is reflected in age cohorts in the various "steps." Of the 20 percent of the software engineers who are female, approximately 40 percent are programmers, 6 percent are systems engineers and 1 percent are managers. This categorization reflects a dual internal labor market with men in the management track and women in the routine service track. So, one way of responding to the increased demand for skilled workers is to routinize the lowest-level job – that of programmer, where most women are employed – in order to induce labor-saving effects and increase the male labor supply for the more skilled jobs (Christopherson 1991).

In countries in which firms rely more on the external labor market, such as the US and UK, another type of dual tracking has evolved. Service firms paying for performance reward those workers able to expend the greatest work effort on behalf of the firm. The result is a "mommy track" for those women who are unable to devote all their waking attention to their jobs.

In both these cases, women's increased educational attainment, which in many industrialized countries now exceeds that of men (table 12.2), has not translated into occupational mobility.

Instead, new organizational mechanisms are being introduced to use a more generally skilled female labor force more effectively without rewarding them with higher compensation or enhanced job responsibilities. These innovations are most notable in the US, where the narrowness of labor regulation allows firms to design new compensation systems without any input from workers. One such innovation is "broad-banding" – a compensation system that groups different types of jobs in broad bands in order to separate skill acquisition, workload, and increased responsibility from the expectation of higher compensation. This system is designed to make effective use of the external labor market by requiring that new employees enter at the lowest wage level of the band without regard for their skill or experience. Broad-banding is now being introduced in large public and private service organizations predominantly employing women. It may constitute a new form of labor segmentation, confining women to broadly defined job "families" and limiting their opportunities for occupational mobility and increased compensation. The case of broad-banding demonstrates that despite women's increased educational attainment, jobs skills, and work experience, increased equality in the workforce is not assured. Just as women master the rules of the game, the rules may change.

Conclusion

In summarizing the effects of globalization on women, one conclusion we could draw with some certainty is that, in any national context, the situation of the "average" woman is less descriptively accurate than it may have been in the 1960s and 1970s. Possibly the signal characteristic of the period of the 1980s was differentiation. With globalization of trade in commodities and increased interaction derived from capital and information flows, women are differentiated from each other in new (as well as old) ways. In addition to dramatic changes in the organization and location of economic activities described above, women's status is affected by the general economic situation in each region. East Asian women, for example, have shared in the increased prosperity of their countries as members of households and because of increased paid work available to them. At the same time, in Africa and Latin America, women's employment prospects have deteriorated because of the structural adjustment measures imposed with the debt crisis. This has been particularly problematic because of the higher incidence in these two regions (than in Asia) of households headed by women, and the greater dependence of children on women's earnings. All these factors suggest that we need to look at women's status as relative but also as constantly evolving.

13

Disease Implications of Global Change

Andrew Cliff and Peter Haggett

Introduction

That the twentieth century has seen a major demographic transition in the human population is well known. In four decades, the world's population more than doubled, from 2.5 billion in 1950 to over 5 billion in 1988. On the United Nations' "medium-growth" assumptions, this total is expected to reach 6.25 billion by the end of the century and 8.5 billion by the year 2025. Although the world's population is continuing to grow, that rate of growth is now decelerating. The global rate increased during the 1950s and 1960s to reach a peak of 2.1 percent per annum in the five years from 1965 to 1970. Since then it has fallen and it stands now at around 1.7 percent annually. If the global fertility rate continues its projected decline, the population growth rate will fall further over the next two decades.

Associated with these familiar demographic changes, there has also been an epidemiological transition in disease patterns. This is partly due to age changes: declining mortality has led to an ageing population, with a relative decrease in the number of childhood deaths and a relative rise in the degenerative diseases of old age. Geographical redistribution also plays a role. For example, it is expected that some 94 percent of population growth until the year 2015 will occur in the developing countries. This will shift the balance of world population still more toward the tropics and low altitudes (e.g., the average temperature of areas inhabited by the global population will rise by around 1°C, from

17°C to 18°C [Haggett 1991, table 3]), thus exposing a greater share of the world's population to conventional tropical diseases.

The disease transitions relating to global population growth are well known and will not be pursued further here. Rather, this chapter will look at three less familiar themes which may result in significant changes in global patterns of disease in the future. These are, first, the epidemiological impact of increased long-distance travel; second, the geographical implications for disease of global warming; and third, the factors behind the emergence of new – or apparently new – diseases. Although we illustrate them with specific local examples (many drawn from our own country, the United Kingdom), we try to emphasize throughout, the worldwide implications which arise from them.

Increases in Long-distance Travel

The ways in which travel patterns have changed over recent generations has been shown in an interesting way by the distinguished epidemiologist D. J. Bradley (1988). Bradley compares the travel patterns of his great-grandfather, his grandfather, his father and himself: figure 13.1 shows the results. The life-time travel track of his great-grandfather around a village in Northamptonshire could be contained within a square with sides of only 40 km. His grandfather's map was still limited to southern England, but it now ranged as far as London and could easily be contained within a square with sides of 400 km. If we compare these maps with that of Bradley's father (who traveled widely in Europe) and with Bradley's own sphere of travel, which is worldwide, then the enclosing square has to be widened to sides of 4,000 km and 40,000 km respectively. In broad terms, the spatial range of travel has increased tenfold in each generation so that Bradley's own range is one thousand times wider than that of his great-grandfather.

Against this individual cameo, we can set some broader statistical trends from recent years. Figure 13.2 plots the trend in world tourism arrivals over thirty years and compares it with global population increases over the same period. The precise rates of flux or travel of population both within and between countries are difficult to catch in official statistics. But most available

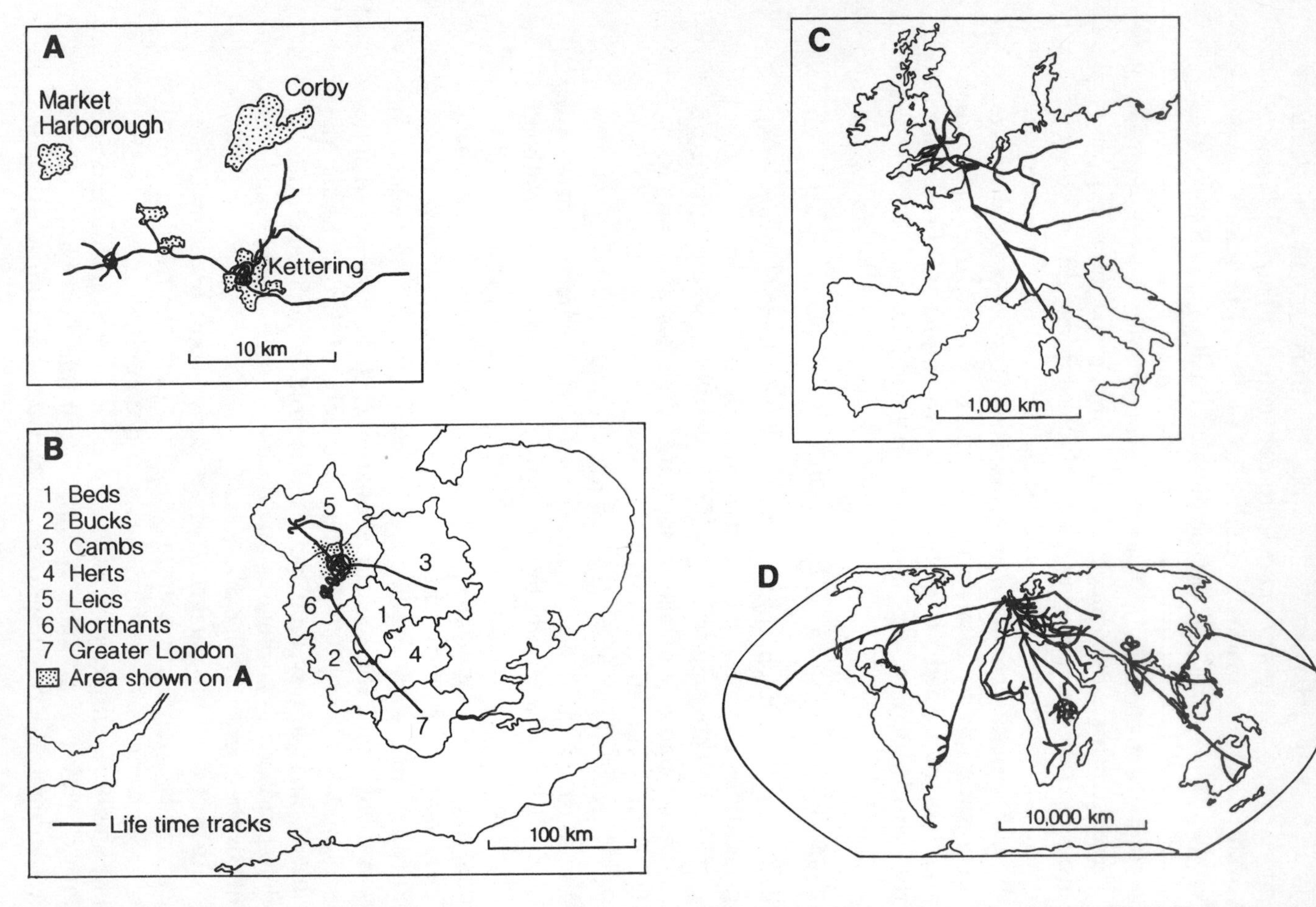

Figure 13.1 Increasing travel over four male generations of the same family. (A) Great-grandfather. (B) Grandfather. (C) Father. (D) Son. Each map shows in a simplified manner the "life-time tracks" in a widening spatial context, with the linear

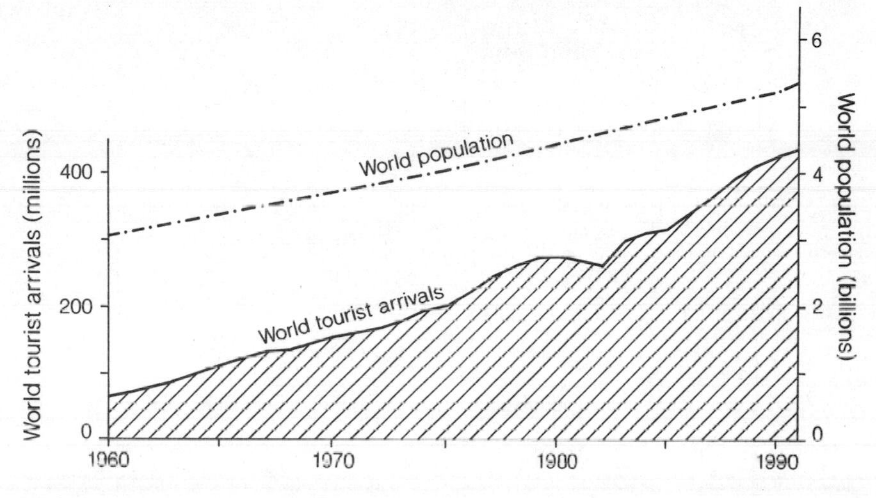

Figure 13.2 Trends in the growth of population and one indicator of its spatial flux. Estimated world population and estimated world tourism arrivals, 1960–2. Source: United Nations *Demographic Yearbooks* and *World Travel and Tourism Review.*

evidence suggests that the flux over the last few decades has increased at an accelerating rate. While world population growth rate since the middle of the twentieth century has been running at between 1.5 and 2.5 percent per annum, the growth in international movements of passengers across national boundaries has been between 7.5 and 10 percent per annum. One striking example is provided by Australia: since the 1950s, its resident population has doubled, while the movement of people across its international boundaries (that is, into and out of Australia) has increased nearly one hundredfold.

The implications of increased travel are twofold. First, there are possible long-term genetic effects. With more travel and longer-range migration, there is an enhanced probability of partnerships being formed and reproduction arising from unions between individuals from formerly distant populations. As Khlat and Khoury (1991) have shown, this can bring advantages from the viewpoint of some diseases. For example, the probability of occurrence of multifactorial conditions such as cystic fibrosis or spinal muscular atrophy is reduced; the risk of these conditions is somewhat higher in children of consanguineous unions. Conversely, inherited

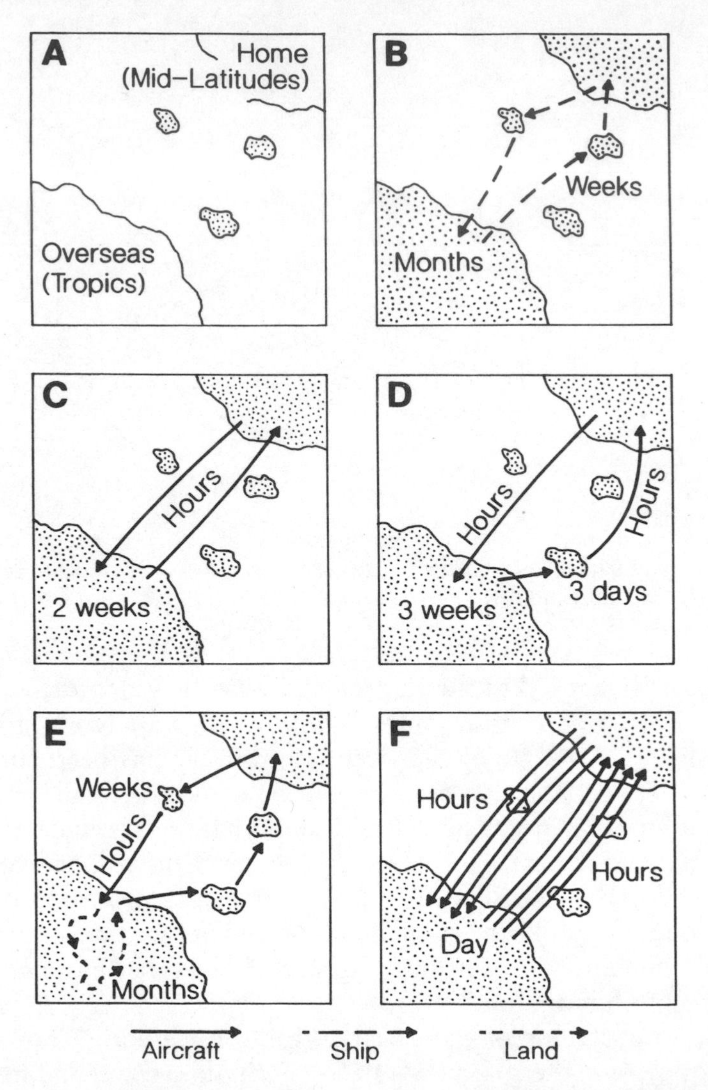

Figure 13.3 Schematic diagram of typical overseas journeys. (A) Travel environment. Basic context with homeland (upper right), ocean with stopover sites (stippled), and tropical overseas area (lower left). (B) Pre-aircraft travel. Typical journey to the tropics using sea transport. (C) Package tour. Typical package tourist journey using air transport. (D) Stopover package tour. Typical package tourist journey with stopover on return flight. (E) Student travel. Typical student overlander journey with long internal travel within the tropics and multiple stopovers. (F) Business travel. Typical business trips with multiple short visits within a short period. Source: redrawn from Bradley (1988, figures 5–10, pp. 6–7).

disorders such as sickle cell anaemia might become more widely dispersed.

Another and more immediately important effect is the exposure of the traveling public to a range of diseases not encountered in their home country. Again, Bradley has vividly illustrated the kinds of journeys that might be made (see figure 13.3), although visits to a tropical area (figure 13.3A) may not always be necessary for diseases transfer to occur.

One way in which international aircraft from the tropics can cause the spread of disease to a non-indigenous area is seen in the occasional outbreaks of tropical diseases around mid-latitude airports. Typical are the malaria cases that appeared within 2 km of a Swiss airport, Geneva-Cointrin, in the summer of 1989 (Bouvier et al. 1990). Cases occurred in late summer when high temperatures allowed the in-flight survival of infected *anopheles* mosquitos that had been inadvertently introduced into the aircraft while at an airport in a malarious area. The infected mosquitos escaped when the aircaft landed at Geneva, to cause malaria cases among several local residents, none of whom had visited a malarious country.

The last few decades of air travel have increased the potential for disease spread through two factors (Westwood 1980). We can illustrate these for the United Kingdom. First, travel times to and from London have decreased so sharply that if a passenger is in the infectious stage of a virus fever before overt and recognizable symptoms appear, fellow passengers are all potentially exposed but may disperse without being aware of this. Thus the speed and range of aircraft mean that people will be arriving in UK airports (and dispersing throughout the population) long before clinical symptoms for many diseases have had time to develop. Less than one-tenth of the world's major cities are now more than a day's flying time from London. To illustrate this point, table 13.1 gives some examples of passenger flows into London (Heathrow and Gatwick combined) on 1988 scheduled air services.

A second factor with modern aircraft is their increasing size. Bradley (1988, p. 4) postulates a hypothetical situation in which the chance of one person in the traveling population having a given communicable disease in the infectious stage is 1 in 10,000. With a 200-seat aircraft, the probability of having an infected

Table 13.1 Direct (non-stop) passenger flows on scheduled aircraft into London (Heathrow + Gatwick) from the 25 largest world cities.

City (Country)	*Estimated population (millions)*	*Number of scheduled flights*	*Passengers (thousands)*	*Projected national population changes, 1990–2010*
Mexico City (Mexico)*	19.4	Indirect flight		+ 1.7%
New York (USA)	18.0	5,333	1,304.4	+ 0.6%
Los Angeles (USA)	13.5	1,413	360.4	+ 0.6%
Cairo (Egypt)*	13.0	688	115.0	+ 2.6%
Shanghai (China)*	12.5	Indirect flight		+ 1.1%
Peking (China)*	10.7	Indirect flight		+ 1.1%
Seoul (South Korea)	9.7	29	4.3	+ 0.8%
Calcutta (India)*	9.2	Indirect flight		+ 1.8%
Moscow (USSR)	8.8	549	65.6	+ 0.6%
Paris (France)	8.7	10,491	1,418.7	+ 0.3%
Sao Paulo (Brazil)*	8.6	21	4.3	+ 1.6%
Tokyo (Japan)	8.3	488	144.3	+ 0.3%
Bombay (India)*	8.2	800	207.8	+ 1.8%
Chicago (USA)	8.1	764	167.1	+ 0.6%
Jakarta (Indonesia)*	6.5	Indirect flight		+ 1.3%
Lagos (Nigeria)*	6.0	265	47.0	+ 3.3%
San Francisco (USA)	6.0	587	159.3	+ 0.6%
Manila (Philippines)*	5.9	Indirect flight		+ 2.0%
Istanbul (Turkey)	5.9	573	50.1	+ 1.6%
Tehran (Iran)*	5.8	38	9.6	+ 3.2%
Hong Kong	5.7	296	85.9	+ 0.7%
Delhi (India)*	5.6	512	156.5	+ 1.8%
Bankok (Thailand)*	5.3	391	115.3	+ 1.3%
Rio de Janeiro (Brazil)*	5.2	244	48.6	+ 1.6%
Karachi (Pakistan)*	5.1	18	4.3	+ 2.6%

Source: ICAO Statistics: Traffic by Flight Stage (1988). Note that figures for city sizes may vary depending on the precise urban definitions used. Figures for passenger flow refer to direct (i.e. non-stop) flights only. Figures for projected national changes are annual rates for the country as a whole and not for the individual city. Cities located in developing countries are marked by an asterisk.

passenger on board (x) is 0.02 and the number of potential contacts (y) is 199. If we assume homogeneous mixing, this gives a combined risk factor (xy) of 3.98. If we double the aircraft size to 400 passengers, then the corresponding figures are $x = 0.04$, $y = 399$, and $xy = 15.96$. In other words, doubling the aircraft

Table 13.2 Indicators of international exposure of resident populations.

Exposure in relation to resident population	*Global*	*UK*	*Developed countries*	*Developing countries*
Tourist arrivals/1,000 residents	69	270	223	20
Aircraft movements/10,000 residents	50	201	167	12

Source: 1989 ICAO estimates.

size increases the risk from the flight fourfold. Thus the new generation of wide-bodied jets presents fresh possibilities for disease spread, not only through their flying range and their speed, but also from their size. Table 13.2 shows two very crude indices of the exposure of the UK population to overseas contacts.

Global Warming

Of the many global scenarios for disease and the environment in the early part of the twenty-first century, it is the health implications of global warming that have caught the attention of governments and press worldwide. There have already been major studies of its potential health implications in at least three countries: the United States (Smith and Tirpak 1989), Australia (Ewan, Bryant, and Calvert 1990), and the United Kingdom (Bannister et al. 1991). The World Health Organization also has a committee looking at this issue.

A number of health effects have been postulated as following from a worldwide increase in average temperature from global warming and these are summarized in table 13.3. Few attempts have yet been made to compute the relative burden of morbidity and mortality that would be yielded by these effects. Any such calculation would also need to offset losses against gains that might accrue (for example, reductions in hypothermia against increases in hyperthermia).

The magnitude and spatial manifestations of global warming are still speculative. One of the main conclusions of the report of the Intergovernmental Panel on Climate Change (IPCC) in 1990 was how far research still had to go before reliable estimates of

Table 13.3 Possible health effects potentially attributable to global warming.

Nature of the effect	*Source*
Hyperthermia due to summertime heat-related mortality	Kalkstein et al. (1989)
Infectious illness due to changes in geographical range of pathogens, vectors and reservoirs	Leaf (1989) Dobson (1989)
Respiratory disease due to photo-chemical air pollutants	Lucier and Hook (1989)
Skin cancer, melanomas, cataracts, and immune suppression linked to increase in ambient UV	Illig (1987), Hersey et al. (1983), Giannini (1990)
Malnutrition and starvation due to changes in location of optimum food-producing areas	Brown et al. (1990)
Refugee problems related to enforced migration from areas of ecological disasters	Schneider (1989)

Source: McCally and Cassel (1990, pp. 469–70). Full references are given there.

global warming could be identified. But some rough orders of magnitude can be computed from the estimates of the different models that have been used. In global terms, warming appears to range from "a predicted rise from 1990 to the year 2030 of 0.7°C to 1.5°C with a best estimate of 1.1°C" (Intergovernmental Panel of Climatic Change 1990, p. 178).

Equivalent estimates for the United Kingdom itself are not given, but we can obtain some idea of the implications of the predicted global shift for local mean temperatures. For the UK's major cities, the current differences between the coldest (Aberdeen, latitude 57.10N) and warmest (Portsmouth, latitude 50.48N) is 2.4°C; this is well beyond the postulated IPCC warming effect by the year 2030. Climate is a much more complex matter than average temperature, but – if the global warming models carry over to the UK – then, by 2030, Edinburgh might have temperatures something like those of the English Midlands, and London something like those of the Loire valley in central France. If we accept the much higher estimate of +4.8°C warming over 80 years, this brings London into the temperature bands of southern France and northern Spain. Provided that these projections are sensible, something might be gained by comparative studies of disease incidence within the UK and adjacent EC countries and disease incidence in warmer climates that match those predicted for the UK.

The biological diversity of viruses and bacteria is partly temperature dependent, and it is much greater in low than in high latitudes. Conditions of higher temperature would favor the expansion of malarious areas, not just for the more adaptable *Plasmodium vivax* but also for *P. falciparum*. Rising temperatures might also allow the expansion of the endemic areas of other diseases of human importance: these include, for example, leishmaniasis and arboviral infections such as dengue and yellow fever. Higher temperatures also favor the rapid replication of food-poisoning organisms. Warmer climates might also encourage the number of people going barefoot in poorer countries, thereby increasing exposure to hookworms, schistosomes and Guinea worms. But not all effects would be negative. Warmer external air temperatures might reduce the degree of indoor crowding and lower the transmission of influenza, pneumonias, and "winter" colds.

While modest rises in average temperatures are the central and most probable of any greenhouse effects, they are likely to be accompanied by three other main changes: (a) sea-level rises of up to a meter; (b) increased seasonality in rainfall, thus reducing the level of water available for summer use; and (c) storm frequency increases (Henderson-Sellars and Blong 1989). Since these can be measured with even less precision than temperature changes, they are not considered further here.

Emergence of New Diseases

The last quarter-century has seen the global eradication of one major human disease (smallpox, by 1979), but the emergence of many new ones. In addition to AIDS, recognized around the early 1980s, we can add Legionnaires' disease, Lyme disease, and toxic shock syndrome. We have also the mysterious outbreaks of African tropical diseases that occasionally erupt into middle-latitude consciousness, caused by the Lassa, Marburg and Ebola viruses. Figure 13.4 shows the routes of green monkey shipments to European cities and the resulting pattern of cases and deaths in August and September 1967.

One indicator of the expanding number of diseases now recognized is provided by examining past medical literature. In 1917,

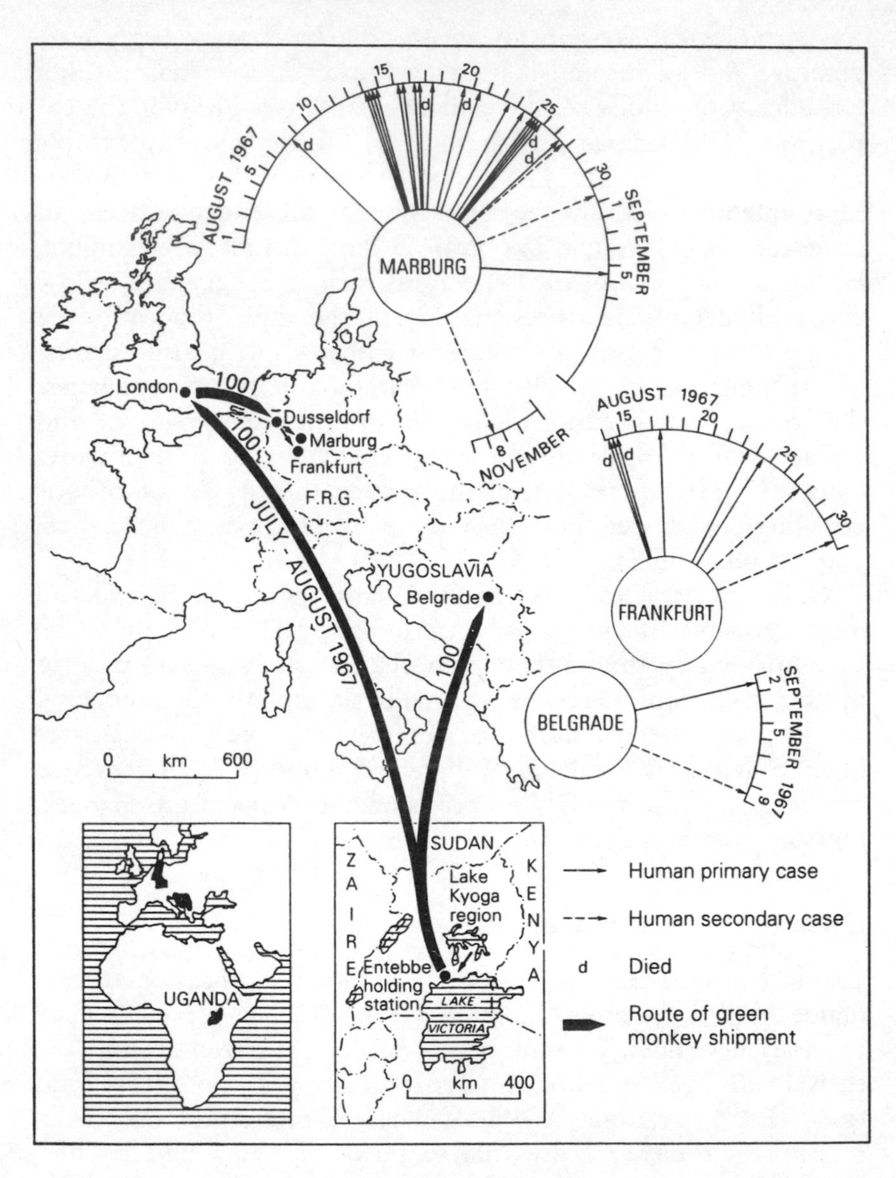

Figure 13.4 History of the 1967 Marburg fever outbreak in Europe. The vectors show the routes of green monkey shipments from Uganda. The three city diagrams show the timing of primary and secondary cases with deaths (d) indicated. Source: Smallman-Raynor, Cliff and Haggett (1992, figure 3.4A, p. 133).

the American Public Health Association published the first edition of its pioneer handbook on *Control of Communicable Diseases in Man* (Benenson 1990). It listed control measures for 38 communicable diseases, all those then officially reported in the United States. Since 1917, the number listed has expanded steadily so that the most recent edition, the fifteenth, now lists some 280 diseases.

The discovery of apparently new diseases raises afresh the question of where and when diseases originate, whether they will spread around the world, and whether new diseases will continue to be added to our existing list. Such questions have been given increased significance by the recent emergence of AIDS as a major human disease. This has been accompanied by an unprecedented torrent of literature, some of it of geographical interest (see Smallman-Raynor et al. 1992). Notwithstanding this literature growth, one still unsolved question is when and where the epidemic of AIDS began. Conventional wisdom, based upon a variety of circumstantial evidence, reviewed in Smallman-Raynor et al. (1992, chapters 3 and 4), is that the causative agent, the human immunodeficiency virus (HIV), jumped the species barrier from monkeys in Africa – possibly from the chimpanzee in Central Africa for HIV-1 and from the sooty mangabey in West Africa for HIV-2. This is speculative, however, and the independent emergence of two strains of the same virus in two widely separated geographical areas has a low probability, a point to which we return below. Further, while we know that the epidemic spread of the disease in the United States began in 1981, there is increasing evidence that AIDS has been internationally present in a non-epidemic form for many years before the first epidemic cases. In the United States, Huminer and colleagues have identified a score of cases between 1953 and 1981 that meet the strict US Centers for Disease Control criteria for AIDS (Huminer et al. 1987). Other scholars have argued for a much older origin.

AIDS is not the only epidemic of a previously unrecognized disease to have occurred since the 1970s. As we noted above, the list includes Lyme disease, Legionnaires' disease and toxic shock syndrome, all first identified in the United States. Such modern examples of "new" infectious diseases share a number of common features both with each other and with earlier historical manifestations: (a) the onset of the new diseases appears to be sudden and unprecedented; (b) once the disease is recognized, isolated cases that occurred well before the outbreak are retrospectively identified; and (c) previously unknown pathogens or toxins account for many of the new infections.

Ampel (1991) suggests four factors which may explain these observations:

1. The infection was present all along, but was previously unrecognized and unrecorded in the International Classification of Diseases (ICD) list. This may have been true for progenitors of HIV, but it is unlikely that toxic shock syndrome, Lyme disease, or AIDS existed in their present forms in a large number of patients long before their recognition. We need to look further than a simple "unrecorded" hypothesis.
2. Pathogens responsible for these new diseases existed in the past but in a less virulent form. Some event, such as a genetic mutation, then converted the organism to its virulent form. Rosqvist and co-workers (1987) showed that double-point mutations in the bacterium *Yersinia pseudotuberculosis* resulted in a marked increase in the virulence of this organism *in vitro*. Carmichael and Silverstein (1987) postulated that the marked increase in mortality associated with smallpox in sixteenth-century Europe could have been due to mutations in the causative virus, *variola*. Such events could also have occurred with regard to HIV-1. Smith et al. (1988) examined the genomic sequences of conserved areas of HIV-1 and compared these with human immunodeficiency virus type 2 (HIV-2), the related and potentially less virulent strain, and with the simian immunodeficiency virus, which is virulent for non-human primates but not for humans. They concluded from these data that HIV-1 diverged from HIV-2 in 1951, plus or minus three years, and that "the onset of conspicuous diversification of these viruses correlates strongly with the known historical rise of the pandemic" (Smith et al. 1988). This contrasts with the "independent origins" hypothesis outlined

earlier. Finally, we note that the classic case of genetic change in an already existing virus was the emergence of the Spanish influenza virus in 1918 (Cliff, Haggett, and Ord 1986).

3. Environmental and behavioral changes provide a new environment in which the disease-causing organisms may flourish. Legionnaires' disease is related to the increased use of cooling towers and evaporative condensers from the 1960s; Lyme disease, to the growth of deer population in woodlots that grew up on the abandoned fields of New England. Toxic shock disease is related to a behavioral change, the increasing use of tampons by menstruating women. Further back in time, the kind of changes which accompanied agriculture may well have brought shifts in disease patterns. One hypothesis needing study is that malaria began to attack humans 10,000 years ago when Africans shifted from hunting on the savannah to farming in the forests.
4. A new epidemic arises from the introduction of a virulent organism into a non-immune population (a so-called *virgin soil* epidemic) or from the arrival of new settlers in a previously unsettled area. This is a theme with rich historical parallels: one historically-important example is the epidemic of smallpox that arrived in Mexico in 1520 with the Spanish *conquistadores*, decimating the native Aztec population; Cliff and Haggett (1985) have described the first introduction of measles into Fiji in 1875, in which one-quarter of the islands' population died within a two-month period. Most recently, the spread of HIV provides striking examples of tropical to middle-latitude disease transmissions. Figure 13.5 shows the spread of HIV-1 from tropical African countries into the then Soviet Union; figure 13.6, the documented sources of HIV-2 transmission from West African countries into Western Europe.

New settlers following the logging roads into Amazonia encountered heavy malarial infections, partly because the land-use changes had increased the forest-edge environments suitable for certain mosquito species. Likewise, in 1977, the building of

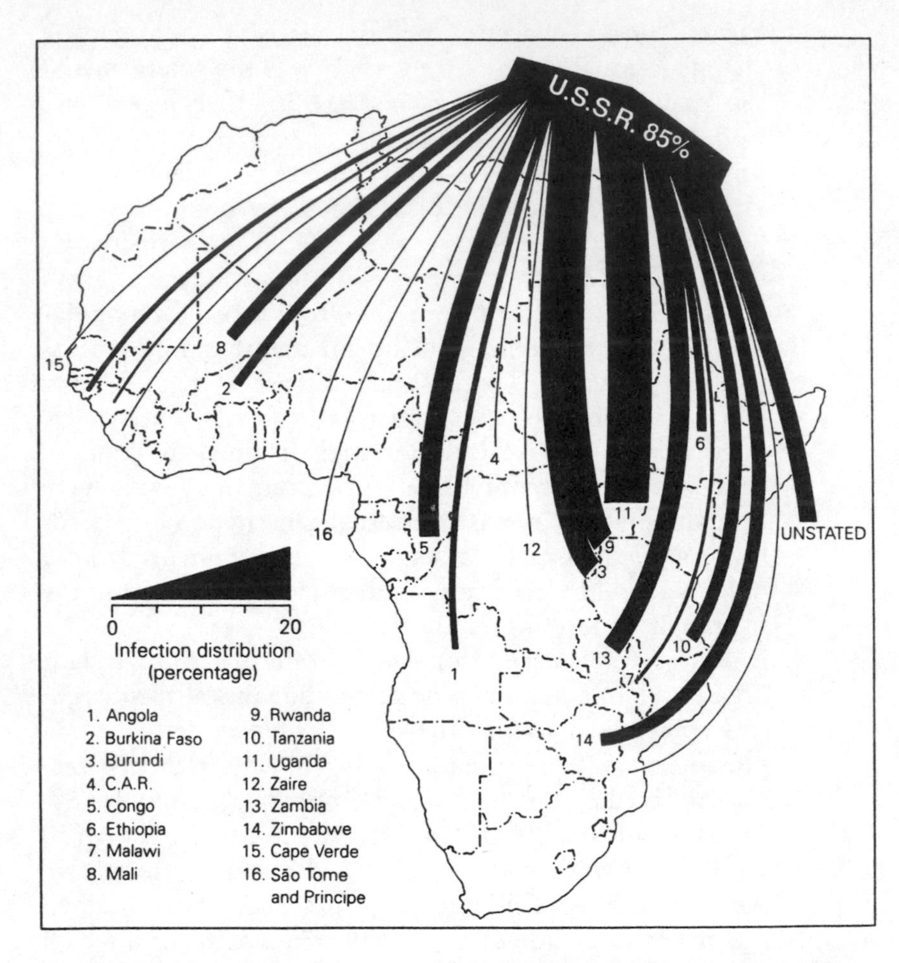

Figure 13.5 Documented sources of HIV-1 transmission from tropical African countries to the Soviet Union. Country of origin of Soviet HIV-1 infections based on the first Soviet national screening programme in 1987. Source: Smallman-Raynor, Cliff and Haggett (1992, figure 4.7B, p. 171).

the Aswan dam created just the right breeding grounds for virus-carrying mosquitos, allowing the northward spread of Rift Valley Fever, previously confined to sub-Saharan Africa.

The emergence and persistence of new diseases need very special conditions: there are probably many cases where diseases failed to emerge from contact with a disease-bearing organism. For virus diseases, McKeown (1988, pp. 4–5) has noted that when an infection comes into contact with a strong human host, three things may occur: the virus may fail to multiply and the

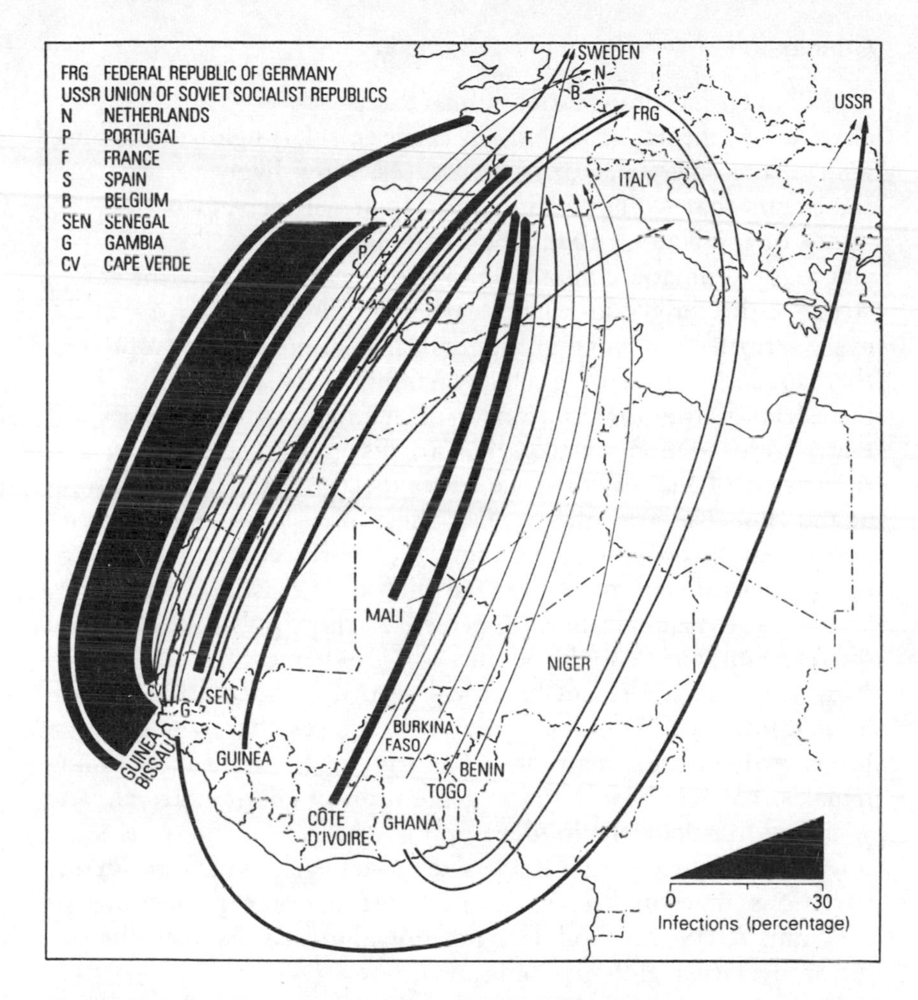

Figure 13.6 Documented sources of HIV-2 transmission from tropical West African countries to Western Europe. Vectors are based on 235 record infections. Source: Smallman-Raynor, Cliff and Haggett (1992, figure 4.8B, p. 177)

encounter pass unnoticed; the virus may multiply rapidly and kill the host without being transmitted to another host; or virus and host populations (after a period of adaptation) settle down into a prolonged relationship which we associate with sustainable diseases. The relative probabilities we can attach to the three outcomes is unknown, but the fragmentary history of puzzling and unsustained disease outbreaks suggests that the third option is the rarest.

Conclusion

This chapter has drawn attention to three aspects of future disease changes: (a) the impact of long-distance travel, (b) global warming, and (c) the emergence of new diseases. How likely to occur are the events that we have outlined? Much will depend on the certainty of predictions about the controlling factors. For example, claims of dramatic climatic change are not new and there is a string of dire predictions from earlier prophets that have – in the event – turned out to be unfounded. It is important to emphasize that *absolute* proof of global warming – given the other non-terrestrial factors that can affect the planet's climate – does not exist. Climate modelers continue to disagree on the timing and the extent of the warming we can expect. What is clear is that, at the time we write this report, there is a large consensus (67 percent of those polled) amongst the world community of climatic scientists that moderate warming of at least 2°C will occur, and an overwhelming view (90 percent of those polled) that carbon dioxide emissions should be curbed (Holden 1990).

By contrast, the recent historical record of population growth is unambiguous. The demographic processes that will generate future growth are already in place: the larger numbers of young females that will bear the next generation of children are already with us (particularly in developing countries) and it is only questions of projected family size which give much room for variations in estimates. Even the most extreme projections of HIV and AIDS will modulate but not alter in substance the significantly larger global population.

Increased spatial mobility from long-distance travel lies somewhere on the probability scale between the climatic and demographic estimates. Economic growth and economic integration will foster greater short-term exchanges of people; regional disparities in growth will tend to foster greater illegal immigration. It is much more difficult to foresee forces – other than another worldwide conflict – that will reduce the number of residents from Western countries traveling abroad.

The public implications of these predicted global changes in disease incidence are hard to judge. Psychological research shows that public concern (and thus political pressure on government bodies) over particular disease hazards rarely keeps exactly in tune with the scientific estimates of the size of those hazards (Douglas and Wildavsky 1983). Slowly developing health catastrophes may generate far less public fear than those that hit the press headlines. There is no obvious alternative way forward, between complacency and hysteria, other than that presently used – to continue to monitor and map disease trends very carefully, and periodically to re-evaluate global hazards as the scientific evidence hardens or weakens, as it is certain to do, over the next few decades. The situation is one which will bear very careful watching.

PART IV

Geocultural Change

McDonald's

Introduction to Part IV: Modernity, Identity, and Machineries of Meaning

I am prepared to take off my ski mask if Mexican society will take off its own mask. Both would show their faces but the great difference will be that Marcos always knew what his face was really like, and civil society will wake from a long and hazy dream that "modernity" has imposed on it at the cost of everything and everyone. (Subcomandante Marcos, the masked rebel leader of the Mexican Chiapas rebellion) (Zapatist National Liberation Army), January 1994).

On January 1, 1994, the day the North American Free Trade Agreement (NAFTA) sent trade barriers tumbling between Mexico, Canada and the US, a guerrilla group was seizing half a dozen towns in Chiapas, southern Mexico. Mr Salinas's economic restructuring policies had widened already huge economic disparities; 40 percent of the Mexican population is said to be poor or extremely poor. According to the Zapatist guerrilla leader, Subcomandante Marcos, NAFTA represented a "death sentence" for the poor.

On the other side of the Atlantic, Radoslav Unkovic, director of the Institute for the Protection of Cultural, Historical and Natural Heritage in the Bosnian Serb Government, is in charge of the current linguistic purge. Even the name "Bosnia" is a target for linguistic cleansing since it has become, in the course of the break-up of Yugoslavia, a symbol of Turkish and Austro-Hungarian oppression. "By eliminating this term," says Unkovic, "we are eliminating the memory and the consequences that stem from it" (*San Francisco Chronicle*, May 2, 1993, p. B-10). Among the casualties, he continues, are those names which "in an ethnic-linguistic sense do not correspond to Serb traditions." Names disappear from maps, street signs, dictionaries, travel directories, and encyclopaedias.

This erasure – the annihilation of meaning on the basis of a putative tradition – presumably extends to Radoslav Unkovic himself since his own name derives from the Turkish word "un" meaning forward.

The horrors of the Balkan wars and the chimerical presence of Subcomandante Marcos pose quite sharply the paradox of the late twentieth century. To wit, the collapse of many forms of existing socialism and the radical shift toward market integration under the auspices of free trade and heavy-handed interventionism by global regulatory institutions such as the World Bank and the IMF. What has been called market triumphalism has not produced the End of History, as Francis Fukuyama (1989) would have us believe – the final triumph, in other words, of liberal democracy. Rather, we have seen the proliferation of militarized ethno-nationalisms and other forms of illiberal politics whether radical Hindus in India, Zulu secessionist movements in South Africa, or the Islamic Front for Salvation in Algeria – what Ben Anderson has properly called the New World Disorder (Anderson 1992). Against the backdrop of the instabilities of the nation state, and the erosion of national economies and identities in the face of unprecedented globalization, the post-Cold War period is both a complex and a dangerous moment (Hall 1991a). In the vacuum created by the demise of Stalinism and the undermining of state-led modernization, new sorts of territorial place-based identities spring up which, when conflated with racial, ethnic, religious, or class-based identifications, can be "one of the most pervasive bases for both progressive political mobilization and reactionary exclusionary politics" (Harvey 1993, p. 4).

Imagined communities of various stripes – Iranian Muslims in Houston, Punjabi separatists in India, Palestinians in Gaza, fundamentalist Christians in Waco, Texas – all lie at the intersection of, and in some way are products of, local and global forces. Put differently, the late twentieth century has witnessed not only radical economic restructuring at the hands of globalized markets and flexible accumulation, but also a globalization of culture – that is to say, publicly meaningful forms – and new patterns of cultural complexity, diffusion, innovation, and polycentralism (Hannerz 1991). Mapping this cultural cosmopolitanism is key to understanding the contours of the New World Disorder and the variety of new social movements and postmodern alternatives to development and modernity.

The specifically "cultural" content of global culture refers to both meaningful social forms and the means by which they are rendered meaningful, recognizing that both forms and means are in some fundamental sense global. Ideas and modes of thought, the forms in which meaning is made accessible, and the way in which the cultural inventory is distributed, have globalized aspects and features (ibid., pp. 6–7). The

Muslim diaspora, for example, is worldwide, interconnected through media and global networks (including the hadj) that link disparate communities in Houston, Oxford, and Kuala Lumpur; some of the most active Punjabi nationalists are resident in Canada, an instance of what Ben Anderson calls "fax nationalism." All states participate in the creation of a public culture in which the cultural materials and cultural capitals are drawn from a global system of signification. Adeline Masquelier (1992) illustrates this point in her description of how a Dunlop tyre advertisement featuring a European woman was converted into a road siren – a peculiar conflation of local spirit traditions and Western advertising imagery – which spoke to the contradictions experienced by rural people in postcolonial Niger. As people make history they do so within a historically shaped imagination and often with what Taussig (1993) calls mute things that are given human significance. These mute things, however, now have a global reach and distribution, facilitated by the machinery of global media and transnational capital. All countries create their own modernity, as Paz (1992) says, but draw upon local and global meanings to fashion it, while specific constituencies participate in and contest its creation.

The machineries of meaning, and the global media in particular, are instrumental in the distribution and externalization of cultural meanings, and this is a process which is aided and abetted by the extraordinary mobility of people. Cities, and certain world cities such as London, São Paulo, and New York, emerge as centers of extra-ordinary cultural diversity, as loci in the world system of signs, and as sites of what one might call a critical cultural cosmopolitanism. Even Tokyo, long distinguished by its cultural closure, has working-class concentrations of Chinese, Bangladeshis, Koreans, and Filipinos. But the fluidity of people and culture signals not so much the Frankfurt School nightmare of cultural homogenization – a sort of erasure of difference on the grounds that everyone watches Dallas, drinks Pepsi and worries about Michael Jackson's personal life – as much as it signals a destabilization of sites and places, in which hybridity and the refashioning of new identities seems to proliferate:

> Not only had the border [between the West and the Rest] been punctured porous by the global market [and capital and migration] . . . but the border has dissolved and expanded to cover the lands it once separated such that all land is borderland. (Taussig 1993, pp. 248–9)

A world of flows and hypermobile capital does not so much produce powerless places and unremitting deterritorialization, but rather, results in the endless production of difference with a complex and unstable globality. Cultural complexity stands less as a monument to the final annihilation of space than as a testament to the complex mappings between force fields of local and global provenance.

All of this returns us to territorial and other sociocultural identifications. The efforts to erase place, community, and memory in Bosnia by ethnic and linguistic cleansing does not reflect irreducible ethnic hatreds; indeed the history of killings really only began in 1928. For war to occur in the context of the collapse of the former Soviet Union, nationalists speaking the language of self-determination had to convince neighbors and friends that they had been killing one another for millennia. The events of the 1980s "turned the narcissisms of minor differences into the monstrous fable that people on either side were genocidal killers" (Ignatieff 1993, p. 3). It is this rewriting of history, the reinvention and reimagining of cultural traditions, that runs through the Balkans, not eternally fixed identities from some essentialized past. It is the continual play of history, culture and power in an increasingly globalized world capitalist system.

If the border and the hybrid are the metaphors for late capitalist culture, this is no less the case for the knowledges by which capitalist modernity is interpreted and understood. The challenging of meta-narratives and of totalizing histories has come from many quarters (Gregory 1993). But it is perhaps entirely appropriate that the crisis of representation has above all emerged from those voices who in some way stand at the margins – feminisms of various strands have drawn self-consciously from this marginality – and more often than not at the borderland of First and Third Worlds. It is from the vantage point of the postcolonial subject, and from cultural studies more generally, that so much of the ferocious attack on the West has emerged. The West is itself a difficult and ambiguous category in this regard since its very existence is rooted in the creation and history of the Rest. Whether the celebration of fragmentation, difference, and marginalization that accompanies much of this critique, and its abandonment of a sense of totality or system, is to be uncritically welcomed is a matter for debate, of course, but the impact of the subaltern critique from within the belly of "big science" and "grand theory" has been unmistakable. Moreover, to take the case of the critical rejection of development as a trope for those parts of the world that are poor and peripheral, the postmodern assault reflects a critical engagement between the ideas of intellectuals and the political practice of new social movements. For some, these popular movements, often emerging from the ashes of Draconian structural adjustment during the 1970s and 1980s, are held aloft as the bearers of new sorts of politics, new sorts of subjectivities, and new sorts of democratic practice, all socially embedded and expressed through local cultural traditions (Escobar 1992a). Whether grassroots movements around dam construction in Brazil, in the barrios of Rio, or in the name of eco-feminism in India do actually contain the potential for feasible alternatives to development, any more than Derridean deconstruction represents the means to refigure human geography, strikes to the heart of both contemporary debate over theory and, one should add, *real politique* outside the academy. In all of this, culture has returned to center stage. No longer an epiphenomenon or a residue of political economy, the realm of meanings – their construction, contestation and interpretation – stands at the very epicenter of the reworking of *fin de siècle* modernity (Pred and Watts 1992).

14

World Cities and the Organization of Global Space

Paul L. Knox

Introduction

World cities are both cause and effect of economic and cultural globalization. They must be seen as the product of the combination of a new international division of labor, of the internationalization of finance, and of the global strategies of networks of transnational corporations – all of which have been facilitated by new modes of regulation and by revolutionary process technologies and circulation technologies. At the same time, world cities must be seen as the places and settings through which large regions of the world are articulated into the space of global capital accumulation: centers of economic, cultural, and political authority that give shape and direction to the interdependent forces of economic and cultural globalization. The result is a "smaller" world, in which our lives are lived and shaped through the global metropolitanism of "larger" cities. These world cities – no more than a couple of dozen of them altogether – are not necessarily the largest in terms of population but they are the most capacious in terms of economic and cultural capital and innovation. Embedded within them are the nodal points of a "fast world" of flexible production systems and sophisticated consumption patterns. This fast world currently extends from the world's triadic economic core to its dependent theaters of accumulation in the megacities of semi-peripheral and peripheral world regions. Its corollary is the "slow world" of catatonic rural settings, declining manufacturing regions, and disadvantaged slums, all of which are increasingly disengaged from the culture and lifestyles of world cities. Yet these slow worlds are not altogether separate from the global metropolitanism of world cities. Both the internal and the external proletariat of world cities contribute capital (economic, human, and cultural) to the cause of global metropolitanism, and in so doing they are unavoidably inscribed into the economic and cultural landscapes of world cities.

Globalization and Urbanization

As we have seen in earlier chapters, globalization is by no means a new phenomenon; nor is it novel that a few major cities play a key role in the capital-accumulation circuits of the world-system. A globalized infrastructure of unitary nation states, international agencies and institutions, global forms of communication, a standardized system of global time, international competitions and prizes, and shared notions of citizenship and human rights had all been established by the mid-nineteenth century, with roots stretching back to the sixteenth century (Robertson 1990). Meanwhile, the emerging international division of labor was organized around a series of major cities – Venice, Madrid, Rome, Amsterdam, Vienna, London, Manchester, Berlin, Paris, New York, Chicago – that were themselves nested within national and regional urban systems (Braudel 1984; Chase-Dunn 1985).

The recent acceleration of globalization and the emergence of a distinctive iteration of world cities was grounded in the twentieth-century development of this legacy. What is distinctive about the globalization of the late twentieth century is, first, that there has been a decisive shift in the proportion of the world's economic activity that is transnational in scope. At the same time, there has been a decisive shift in the nature and organization of transnational economic activity, with international trade in raw materials and manufactured goods being eclipsed by flows of goods, capital, and information that take place within and between transnational conglomerate corporations. A third distinctive feature, interdependent with the first two, is the articulation of new world views and cultural sensibilities: notably the ecological concern with global resources and environments and the postmodern condition of pluralistic, multicultural, non-hierarchical, and de-centered world society. All this adds up to an intensification of global connectedness and the constitution of the world as one place – at least for that portion of the world's population that is in fact tied in to global systems of production and exchange and to global networks of communication and knowledge. For us – you and me and another 800 million or so in the fast world – there has been a profound redefinition of our roles as producers and consumers, and an equally profound reordering of time and space in social life.

Much of this change has been transacted and mediated through world cities, the nodal points of the multiplicity of linkages and interconnections that sustain the contemporary world economy. Yet world cities themselves have to be understood not only as the legacy of past phases of globalization and urbanization but also as the product of enabling technologies, of the strategies of transnational corporations, and of the responses of local, national, and supranational governments and institutions. In this context, we can draw on several aspects of the geoeconomic and geopolitical change described in Parts I and II of this volume:

1. The "New International Division of Labor" (see chapter 3), which has resulted in a locational hierarchy, the top of which is constituted by the concentration of high-level management functions, mostly in major cities of the world's core economies. It has been the expanded management, planning, and control operations of transnational corporations that have formed the nucleus of contemporary world-city formation.
2. New production technologies and advances in telematics that have made for a more variable geometry of economic activity, a faster-paced economic and social environment, and a compression of space and time around the world. The development of these technologies, however, has followed patterns of initial economic advantage, so that a "tripod skeleton" of three main circuits (Europe/North America, Europe/Far East, and Pacific Rim) of traffic has come to dominate the "space of flows" in the "informational economy" (Castells 1989; Sklair

1991). Cities with major concentrations of transnational corporate headquarters, of international news, information and entertainment services, and of international business services – world cities – have naturally succeeded to nodal positions in this tripod skeleton.

3. The trend toward global neo-Fordism (chapter 3), which has required more sophisticated, internationalized financial and business services. This in turn has resulted in the renewed importance of major cities as sites not only for management and coordination but also for servicing, marketing, innovation, the raising and consolidation of investment capital, and the formation of an international property market (Sassen 1991; King 1990). These trends have been reinforced by the corporate strategies of transnational companies: joint ventures, alliances, and global networks, together with "global insider" marketing that concentrates on simultaneous penetration of the increasingly integrated "mass-consumer technopoles" of the world's core economies (Ohmae 1990).
4. A new mode of regulation that features deregulation, public–private cooperation, selective trade reforms, less restrictive labor laws, and heavy subsidies for telematics, for high-tech infrastructure, and for science and technology with commercial potential. The ideology of competitiveness attached to this new mode of regulation has resulted in a distinctive geopolitics of "techno-nationalism" (Petrella 1991) whereby state policies in the leading countries have generally protected the interests of those involved in commercial innovation and corporate control, which has in turn fostered the development of world cities within the core of the world-system.
5. A proliferation of transnational, non-government organizations (NGOs) – partly as a result of global geopolitics (see chapter 7), and partly in response to economic globalization. Between 1973 and 1980, the total number of transnational NGOs doubled, from around 2,000 to more than 4,000. With this proliferation there has also been a consolidation and a localization of transnational NGOs in centers of international politics and mediation:

London, Paris, New York, Brussels, Strasbourg, Geneva, Vienna, and Helsinki.

It must be acknowledged that these same changes have made for a great deal of economic decentralization: from core economies to semi-peripheral ones, from rustbelts to sunbelts, from metropolitan areas to smaller towns and cities, and from downtowns to suburbs and edge cities. World cities are not so much an exception to this decentralization as they are a consequence and shaper of it. As Amin and Thrift (1992) point out, centeredness is essential within a globalized world economy. First, centers – world cities – are needed for their *authority*: their knowledge structures and their ability to generate and disseminate discourses and collective beliefs relating to economic strategies and business climate. Secondly, they are needed for their ability to sustain settings of *sociability*, in which key actors can gather information, establish and maintain coalitions, and monitor implicit contracts. Thirdly, they are needed for their ability to foster *innovation*: places where there are sufficient numbers with the specialized knowledge to identify gaps in markets, develop new uses for technologies, and produce innovations; where there is sufficient mass in the early states of innovation; and where social networks provide rapid reactions within a sophisticated market.

World Cities

World cities, then, are nodal points that function as control centers for the interdependent skein of material, financial, and cultural flows which, together, support and sustain globalization. They also provide an interface between the global and the local, containing economic, sociocultural and institutional settings that facilitate the articulation of regional and metropolitan resources and impulses into globalizing processes while, conversely, mediating the impulses of globalization to local political economies. As such, there are several functional components of world cities:

- They are the sites of most of the leading global markets for commodities, commodity futures, investment capital, foreign exchange, equities, and bonds.

- They are the sites of clusters of specialized, high-order business services, especially those which are international in scope and which are attached to finance, accounting, advertising, property development, and law.
- They are the sites of concentrations of corporate headquarters – not just of transnational corporations but also of major national firms and of large foreign firms.
- They are the sites of concentrations of national and international headquarters of trade and professional associations.
- They are the sites of most of the leading NGOs (non-governmental organizations) and IGOs (inter-governmental organizations) that are international in scope (e.g. the World Health Organization, UNESCO, ILO (International Labour Organization), the Commonwealth Lawyers' Association, the International Federation of Agricultural Producers).
- They are the sites of the most powerful and internationally influential media organizations (including newspapers, magazines, book publishing, satellite television), news and information services (including newswires and on-line information services), and culture industries (including art and design, fashion, film, and television).

There is a great deal of synergy in these various functional components. A city like New York, for example, attracts transnational corporations because it is a center of culture and communications. It attracts specialized business services because it is a center of corporate headquarters and of global markets; and so on. At the same time, different cities fulfill different functions within the world-system, making for different emphases and combinations of functional attributes (i.e. differences in the *nature* of world-city-ness), as well as for differences in their absolute and relative localization (i.e. differences in the *degree* of world-city-ness).

Table 14.1 illustrates this with reference to four (rather crude) criteria: the headquarters offices of major service corporations (banking, insurance, etc.), the headquarters offices of major industrial corporations, the headquarters offices of transnational

Table 14.1 Some world city functions.

City	*Fortune Global Service 500*[a] *Headquarters*	*Fortune Global 500*[b] *Headquarters*	*Index of Primacy*	*NGOs and IGOs*
Amsterdam	4	1	1.20	45
Atlanta	5	3	0.30	5
Brussels	6	4	2.00	485
Chicago	7	10	0.68	21
Dallas	4	4	0.29	3
Düsseldorf	4	3	0.17	17
Frankfurt	8	3	0.19	37
Hamburg	5	1	0.48	20
Houston	4	5	0.37	8
London	28	35	6.50	283
Los Angeles	7	10	1.04	15
Madrid	9	4	1.79	40
Milan	4	1	0.55	28
Montreal	7	3	1.64	26
Munich	6	3	0.36	27
New York	25	12	0.96	77
Osaka	20	21	0.32	1
Paris	28	23	2.69	475
Philadelphia	4	4	0.55	12
Rome	6	2	1.94	52
San Francisco	6	7	0.18	9
Seoul	3	10	2.79	4
Stockholm	7	8	1.56	89
Sydney	4	5	1.17	17
Tokyo	86	83	2.54	59
Toronto	9	3	0.61	25
Washington	6	5	0.43	76
Zürich	7	3	2.01	62

[a] includes banking, financial, savings, insurance, retailing, transport, and utilities services

[b] includes all industrial companies

NGOs and IGOs, and cultural centrality (as reflected by the index of primacy relative to the city's national urban system). Based on these criteria, only London, New York, Paris, and Tokyo have both a full range of world-city functions *and* a high degree of localization of these functions. A number of cities – Chicago, Düsseldorf, Frankfurt, Los Angeles, Madrid, Montreal, Munich,

Rome, Toronto, Washington, and Zürich – are "rounded" world cities but with a significantly smaller share of each function. Osaka is a significant world city in business terms, but not in terms of cultural centrality or international affairs; Brussels and Stockholm, on the other hand, are distinctive for their importance as centers for international agencies and organizations. It is also worth noting that several metropoli that are often cited as world cities – including Hong Kong, Mexico City, Miami, São Paulo, Singapore, and Vancouver – are entirely absent from table 14.1. This is simply because their scores on all four criteria *used here* are low.

These variations notwithstanding, world cities have come to be regarded as settings with distinctive attributes. John Friedmann (1986), writing largely in the context of the New International Division of Labor, hypothesized that world-city formation would result in metropolitan restructuring to accommodate not only the physical settings for concentrations of international activities and their supporting infrastructure, but also the new class fractions and the spatial and class polarization that is consequent upon evolving local labor and housing markets. King (1990), Fujita (1991), Machimura (1992), and Sassen (1991), among others, have explored these restructuring processes and elaborated the attributes and characteristics of world-city labor markets, housing markets, and property markets.

The linkages between world cities, along with their relationships to processes of globalization, have been subject to rather less attention. World-system theory tends to portray world cities as the "cotter pins" that hold together the global hierarchy of core, semi-periphery, and periphery (Rodriguez and Feagin 1986). This fits comfortably with the widespread notion of a global hierarchy of world cities, a hierarchy dominated by London, New York, and Tokyo, with a second tier of cities of regional transnational importance (e.g. Amsterdam, Frankfurt, Los Angeles), a third tier of important international cities (e.g. Madrid, Seoul, Sydney, Zürich), and a fourth tier of cities of national importance and with some transnational functions (e.g. Houston, Milan, Munich, Osaka, and San Francisco – see table 14.2, and Friedmann 1994). One might add a fifth tier that includes the likes of Atlanta, Georgia, Rochester NY, Columbus, Ohio, and Charlotte, NC, and the 19 Japanese "technopolis" new towns: places where an

Table 14.2 A hierarchy of world cities.

1. Global financial articulations:
   # London*A (also national articulation)
   # New York A
   # Tokyo*A (also multinational articulation: S. E. Asia)
2. Multinational articulations:
   # Miami C (Caribbean, Latin America)
   # Los Angeles A (Pacific Rim)
   # Frankfurt C (Western Europe)
   # Amsterdam C or Randstadt B
   Singapore*C (Southeast Asia)
3. Important national articulations: (1989 GDP>$200 billion)
   # Paris*B
   # Zürich C
   Madrid*C
   Mexico City*A
   São Paulo A
   Seoul*A
   # Sydney B
4. Subnational/regional articulations:
   Osaka–Kobe (Kansai region) B
   # San Francisco C
   # Seattle C
   # Houston C
   # Chicago B
   # Boston C
   # Vancouver C
   # Toronto C
   Montreal C
   Hong Kong (Pearl River delta) B
   # Milano C
   Lyon C
   Barcelona C
   # Munich C
   # Düsseldorf–Cologne–Essen–Dortmund (Rhein–Ruhr region) B

KEY:

A	10–20 million inhabitants
B	5–10 million inhabitants
C	1–5 million inhabitants
*	national capital
#	major immigration target

Source: Friedmarn (1994), table 2.1.

imaginative and aggressive leadership has sought to carve out distinctive niches in the global market place. Columbus, for example, with a substantial "informational" infrastructure that includes CompuServe, Sterling Software/Ordernet, Chemical Abstracts, the Online Computer Library Center, and the Ohio Supercomputer Center, has managed to have itself designated as an "Infoport" by the UN Conference on Trade and Development, which is seeking to facilitate international trade through computer networks and electronic data interchange. In the Japanese "technopolis" program, collaboration between government, business, and the academic community is aimed at establishing new urban forms using fibre-optic systems, integrated digital network services, international teleports, local-area computer networks, and so on, as platforms for the next phase of neo-Fordist economic globalization (Gibson et al. 1992; Rimmer 1993).

World Cities and Metrocentric Global Cultures

Throughout the urban system represented by this hierarchy, the "transnational practices" of the "transnational producer–service class" necessary to globalization have begun to generate new cultural structures and processes which echo and reverberate through the daily practices and spatial organization of the rest of the "fast" world. The global metropolitanism resulting from these transnational practices is not merely a state of interconnectedness and a shared, materialistic culture-ideology of consumerism (Sklair 1991). It involves, at various levels, not only cultural homogenization and cultural synchronization but also cultural proliferation and cultural fragmentation. It involves both the universalization of particularism (i.e. dissolving the traditional boundaries of space and time, the relativism of postmodernity continuously propagating and redefining uniqueness, difference, and otherness) and the particularization of universalism (i.e. crystallizing transnational practices around specific regional, class, gender, and ethnic groups) (Robertson 1991).

This global metropolitanism is closely tied to the compression of the world and the speeding up of production and consumption, of politics, and development. Yet globalization involves much more than the speeding up and spreading out of people's activities. While traditional links to family, neighborhood, region, and nationality are subverted by high-tech, high-speed networks and devices, *quantitative* changes – more decisions, more choices, more mobility, more interaction, more objects, more images – become *qualitative* – new lifestyles, new world views. In short, the global metropolitanism of world cities facilitates new processes affecting the construction of identity, new forms of meaning, and new movements that can take on material representations (Hannerz 1992).

Global metropolitanism is, of course, closely tied to the material culture promoted by transnational capitalism: designer products, services, and images targeted at transnational market niches, promoted through international advertising agencies, the motion picture industry, and television series. One common interpretation of this is that it represents the homogenization, universalization and "Americanization" of global culture through the economic and political hegemony of the United States and of US-based transnational corporations (Mattelart 1979). But though the world may "dream itself" to be American, this means different things to different people, depending on the ways in which American imagery is appropriated and, more often than not, subverted by the mere fact of becoming iconic (Olalquiaga 1992). Furthermore, although American-based transnational capitalism and media may be the indisputable locus of pop-culture mythologies (Blonsky 1992), it is clear that America has difficulty competing with Japanese cars, cameras, and hi-fi systems, Italian design, German engineering, French theory, and British TV comedy shows. Meanwhile, it is now clear that the recourse to Orientalism – the Western world view that developed as a repository for all the exotic differences and otherness repressed or cast out by the West as it sought to construct a coherent identity (Said 1978) – has had to yield to the plural histories, diverse modernities, and alternative moral orders uncovered by globalization.

Rather than suggesting cultural homogenization, then, global metropolitanism invokes a differential and contingent reach, through world cities, that embodies tensions and oppositions rather than convergence and uniformity. McGrew (1992) characterizes these oppositions as follows:

- *Universalism v. Particularism* – although globalization tends to universalize many spheres of social life (e.g. the iconography of materialistic consumerism, the idea of citizenship, the ideology of the nation state), it also provokes a sociospatial dialectic in which the social construction of difference and uniqueness results in particularism (e.g. the resurgence of regional and ethnic identities).

- *Homogenization v. Differentiation* – just as globalization fosters similarity in material culture, institutions, and lifestyles, the "differential of contemporaneity" means that "global" tendencies are articulated and imprinted differentially in response to varying local circumstances.
- *Integration v. Fragmentation* – the functional integration of labor markets, consumer markets, political institutions and economic organizations that unites people across traditional political boundaries also gives rise to new cleavages. Labor, for example, has become fragmented along lines of race, gender, age, and region.
- *Centralization v. Decentralization* – another aspect of the sociospatial dialectic provoked by globalization involves new movements (e.g. localized environmental movements and the "postmodernism of resistance" that seeks to deconstruct Modernism [Foster, 1985]) in opposition to concentrations of power, information, and knowledge.
- *Juxtaposition v. Syncretization* – Whereas time–space compression and global economic interdependence tend to juxtapose civilizations, lifestyles, and social practices, it can also fuel and reinforce sociocultural prejudices and sharpen sociospatial boundaries.

The global metropolitanism mediated and reproduced by world cities is thus complex, dynamic, and multi-dimensional. The most direct contribution of world cities to global metropolitanism stems from the critical mass of what Sklair (1991) calls the transnational producer–service class, with its "transnational practices" of work and consumption. These are the people who hold international conference calls, who send and receive faxes and e-mail, who make decisions and transact investments that are transnational in scope, who edit the news, design and market the international products, and travel the world for business and pleasure. World cities not only represent their workplaces but are the proscenia for their materialistic, cosmopolitan lifestyles, the crucibles of their narratives, myths, and transnational sensibilities. These new sensibilities are, in turn, adopted by the mass-market consumers of the "fast" world. The lingua franca of this populist dimension of global metropolitanism is the patois of soap operas and comedy series; its dress code and world view are taken from MTV and the sports page, its politics from cyberpunk magazines, and its lifestyle from promotional spots for Budweiser, Carlsberg, Levis, Pepsi, Reebok, Sony, and Volvo.

Of course, the more this global pop culture draws from the hedonistic materialism of the transnational élite, the more the latter is driven toward innovative distinctiveness in its attitudes and material ensembles. The more self-consciously stylish the transnational bourgeoisie, the more tongue-in-*chic* the wannabees and the cyberpunks. As this dialectic has unfolded (via global networks in television and advertising), more people have come to see their lives through the prisms of others' lives, as presented by mass media. Consequently, *fantasy* has become a social practice characteristic of global metropolitanism. But these fantasies, too, become caught in the sociospatial dialectic of the fast world, the result being the further confusion of spatial and temporal boundaries and the collapse of many of the conventions that formerly distinguished fantasy from reality. The cognitive space of world cities, emptied of traditional referential signifiers, thus comes to be filled with *simulations*: iconographies borrowed from other times, other peoples, and other places (Olalquiaga 1992). Nowhere is this more apparent than in the large, set-piece developments that have come to characterize the built environment of world cities (and aspirant world cities): the "variations on a theme park" (Sorkin 1992) that constitute the landscapes of transnational power.

These, however, are but one dimension of the reorganization of space within and between world cities. The "space of flows" of the "informational economy" is made manifest through a series of sociocultural flows that both reflect and reproduce global metropolitanism. Though they are by no means isomorphic in structure, these flows may be conceptualized in categorical terms. Appadurai (1990) suggests that there are five principal categories:

- *technoscapes* (produced by flows of technology, software, and machinery disseminated by transnational corporations, supranational organizations, and government agencies);
- *finanscapes* (produced by rapid flows of capital, currency,

and securities, and made visible not only through teleports and concentrations of financial service workers but also through the rapidly-changing geography of investment and disinvestment);

- *ethnoscapes* (produced by flows of business personnel, guestworkers, tourists, immigrants, refugees, etc.);
- *mediascapes* (produced by flows of images and information through print media, television, and film); and
- *ideoscapes* (produced by the diffusion of ideological constructs, mostly derived from Western world views – e.g. democracy, sovereignty, citizenship, welfare rights).

To these I would add a sixth category – "commodityscapes" in Appadurai's terminology – produced by flows of high-end consumer products and services: the ensembles of clothes, interior design, food, and personal and household objects that are the signifiers of taste and distinction within the culture-ideology of consumption propagated by the transnational producer–service class.

These flows are just as important to the organization of global space and to core–periphery patterns as were the flows of raw materials and manufactured products to earlier phases of capital accumulation. They are, in this regard, positive forces mainly for the "haves"; for the "have-nots," they are more often constraining. Doreen Massey (1991, p. 26), writing about flows of what I have called commodityscapes, illustrates this point with the example of the residents of the *favelas* of Rio de Janeiro, "who know global football like the back of their hand, and have produced some of its players; who have contributed massively to global music, who gave us the samba and produced the lambada that everyone was dancing to last year in the clubs of Paris and London; and who have never, or hardly ever, been to downtown Rio. At one level they have been tremendous contributors to what we call time–space compression; and at another level they are imprisoned in it."

World Cities as Places

The geographical overlay of the economic, built, and social environments that creates a sense of place for both residents and visitors is inevitably ruptured by the need to restructure both metropolitan form and metropolitan labor markets in response to the imperatives of global capital accumulation. Just as inevitably, the ubiquity of transnational architectural styles, retail chains, fast-food chains, clothing styles, and music, together with the ubiquitous presence of transcontinental immigrants, business visitors and tourists, tends to propagate a sense of placelessness and dislocation. Ethnographers have often stressed this in terms of *deterritorialization*: "As groups migrate, regroup in new locations, reconstruct their histories, and reconfigure their ethnic "projects" the *ethno* in ethnography takes on a slippery, nonlocalized quality. Groups are no longer tightly territorialized, spatially bounded, historically unselfconscious, or culturally homogeneous" (Appadurai 1991, p. 191).

Yet the common experiences engendered by globalization are still mediated by local reactions. The structures and flows of global metropolitanism are variously embraced, resisted, subverted, and exploited as they make contact with specific political economies and sociospatial settings. In the process, places are reconstructed rather than effaced. Often (and perhaps unexpectedly) it involves *reterritorialization* and a revaluation of place. Strassoldo puts it this way (1992, pp. 46–7):

> Post-modern man/woman, just because he/she is so deeply embedded in global information flows, may feel the need to revive small enclaves of familiarity, intimacy, security, intelligibility, organic–sensory interaction in which to mirror him/herself. . . . The possibility of being exposed, through modern communication technology, to a near infinity of places, persons, things, ideas, makes it all the more necessary to have a center in which to cultivate one's self. The easy access of the whole world, with just a little time and money, now gives new meaning to a need for a subjective center – a home, a community, a locale – from which to move and to which to return and rest.

World cities have also come to be special kinds of cultural spaces, sites for the construction of new cultural and political identities, for new discourses, texts, and metaphors through which the struggle for place is enacted. One way of interpreting the outcomes of these processes is in terms of *loyalty shifts* (DiMuccio and Rosenau 1992). In terms of the organization of global space, the most significant of these are *outward* shifts, whereby loyalties are redirected toward entities (e.g. transnational employers), classes (e.g. Sklair's transnational producer–service class), supranational organizations (e.g. the European Community), or movements (e.g. global ecology, human rights).

But in terms of the organization of metropolitan space and the social construction of place the most significant loyalty shifts are *inward*. At one level, we can see these shifts in people's apparent need for stability, identity, and centeredness within the infinite relativism of postmodernity. This impulse has been articulated through housing markets (most strikingly through gentrification and through private master-planned communities), and commodified through neo-traditional urban design and the merchandising of local histories. Among the affluent within the fast world, reterritorialization thus results from colonizing or invading spaces that can be given both social meaning and spatial identity. But perhaps the most dramatic examples of reterritorialization are those deriving from the lived experience of low-income transnational migrants, exiles, and refugees. Within world cities, such groups are able to establish new networks and new cultural practices that define new spaces for daily life. In addition to the transformation and adaptation of old neighborhoods and obsolete sociocultural spaces within consolidating ethnic enclaves, this involves the emergence of otherwise marginalized voices and alternate representations. These voices and representations can be carried over into the "host" society and carried back into the "homeland." The latter has the potential, at least, for a kind of "transnational grassroots politics" (Smith 1994). The former not only contributes to the cosmopolitanism of world cities but also has the potential (realized only in a few world cities – London, New York, San Francisco, and Los Angeles) to foster distinctive and innovative multicultural spaces – the latest phase in the sociospatial dialectic of global metropolitanism.

15

The New Spaces of Global Media

Kevin Robins

Introduction

Though they have received surprisingly little attention within the discipline, the mass media provide a particularly rich focus for geographical analysis. There is the dimension of industrial structure and the organization of media economies – production, distribution, and consumption – within places and across spaces. There is the role of the media in the constitution of political cultures, and in the functioning of democratic life within particular territories. And, then, there is the question of cultural geographies, concerned with how media products relate to collective and territorial identities. In the 1990s, we are seeing dramatic transformations in media industries and media cultures. In geographical terms, these transformations may be seen in terms of the shift from national to global media systems. Taking the example of broadcasting and audio-visual media, I want to consider the nature and the significance of those new spaces of global media. Particularly, I shall be concerned with the consequences of the changing economic geography of the media for political and cultural geographies.

From National to Global Media

Until very recently, what has prevailed in Britain, as elsewhere in Europe, has been the system of public service broadcasting, involving the provision of mixed programming – with strict controls on the amount of foreign material shown – on national channels available to all. The principle that governed the regulation of broadcasting was that of "public interest." Broadcasting should contribute to the public and political life of the nation; in the words of the BBC's first Director General, John Reith, it should serve as "the integrator of democracy" (quoted in Cardiff and Scannell 1987, p. 159). Broadcasting should also help to construct a sense of national unity. In the earliest days of the BBC, the medium of radio was consciously employed "to forge a link between the dispersed and disparate listeners and the symbolic heartland of national life" (ibid., p. 157). In the post-war years, it was television that became the central mechanism for constructing this collective life and culture of the nation. In succession, radio and television have "brought into being a culture in common to whole populations and a shared public life of a quite new kind" (Scannell 1989, p. 138). Historically, then, broadcasting has assumed a dual role, serving both as the political public sphere of the nation state, and as the focus for national cultural identification. (Even in the very different context of the United States, where commercial broadcasting was the norm from the beginning, national concerns were paramount; the "national networks" of CBS, NBC and ABC served as the focus for national life, interests, and activities.) We can say that (on both sides of the Atlantic) broadcasting has been one of the key institutions through which listeners and viewers have come to imagine themselves as members of the national community.

Now, however, things are changing, and changing decisively. During the 1980s, as a consequence of the complex interplay of regulatory, economic and technological change, dramatic upheavals took place in the media industries, laying the basis for what must be seen as a new media order. What was most significant was the decisive shift in regulatory principles: from regulation in the public interest to a new regulatory regime – erroneously described as "deregulation" – driven by economic and entrepreneurial imperatives. Within this changed context, viewers are no longer addressed in political terms, that is to say as the citizens of a national community, but, rather, as economic entities, as parts of a consumer market (Robins and Webster 1990). The political and social concerns of the public service era – with democracy and public life, with national culture and identity – have come to be regarded as factors inhibiting the development of new media markets. In the new media order, the overriding objective is to dismantle such "barriers to trade." No longer constrained by, or responsible to, a public philosophy, media corporations and businesses are now simply required to respond to consumer demand and to maximize consumer choice.

What concern us particularly, of course, are the geographical implications of these developments. What of space and place in this new media order? Driven now by the logic of profit and competition, the overriding objective of the new media corporations is to get their product to the largest number of consumers. There is, then, an expansionist tendency at work, pushing all the time toward the construction of enlarged audio-visual spaces and markets. The imperative is to break down the old boundaries and frontiers of national communities, which now present themselves as arbitrary and irrational obstacles to this reorganization of business strategies. Audio-visual geographies are thus becoming detached from the symbolic spaces of national culture, and realigned on the basis of the more "universal" principles of international consumer culture. The free and unimpeded circulation of programs – television without frontiers – is the great ideal in the new order. It is an ideal whose logic is driving ultimately toward the creation of global programming and global markets – and already we are seeing the rise to power of global corporations intent on turning ideal into reality. The new media order is set to become a global order.

What does this mean? Is it a good or bad thing? In considering these questions, let us begin with what I would call the mythology of global media. A fine example of this is the "Worldview Address" delivered to the 1990 Edinburgh International Television Festival by the late Steven Ross (1990), then head of the world's largest media corporation, Time Warner. According to Ross, Time Warner stands for "complete freedom of information," that is to say, for the "free flow of ideas, products and technologies in the spirit of fair competition." National frontiers he sees as a relic of the past: "the new reality of international media is driven more by market opportunity than by national identity." We are, says

Ross, "on the path to a truly free and open competition that will be dictated by consumers' tastes and desires." It is a world order in which the consumer is truly sovereign.

The world that Time Warner is anxious to construct will be "a better world" (for Time Warner and for the consumers of its products). "The competitive market place of ideas and experience can only bring the world closer together," Ross maintains. "With new technologies, we can bring services and ideas that will help draw even the most remote areas of the world into the international media community." A world closer together will be a more democratic world. Thinking of recent events in Eastern Europe, Ross associates the free flow of communications with the overthrow of totalitarian societies. "Who," he asks, "could have imagined the satellite, the fax machines, CNN, television – and even records and movies – as tools of democratic revolution?" Media corporations are now at the cutting edge of the new world order. "It is up to us," says Ross, "the producers and distributors of ideas, to facilitate this movement and to participate in it with an acute awareness of our responsibility as citizens of one world. . . . We can help to see to it that all peoples of all races, religions, and nationalities have equality and respect."

In the 1960s, the Canadian writer Marshall McLuhan put forward the argument that electronic media and communications were turning the world into a "global village." For Ross, the global media corporations of the 1990s are now finally and truly realizing what McLuhan predicted. In his eyes communication is a good thing and the more freely it flows, the better it is; experiences shared on a global scale through the new communications media will help us to transcend the differences between different cultures and societies and to work toward "genuine mutual trust and understanding." The message is simple and uplifting: it is the story of progress toward freedom and democracy on a world scale, and of the responsibility of companies like Time Warner to "actively lead the way toward making the world a better place."

Steven Ross offers us one way of interpreting the significance of globalization in the media industries. And, indeed, there is something attractive about his "world view," with its appeal to international democracy and the "interconnection of cultures." Nonetheless, it is a world view that we should regard with

Table 15.1 The world's top 20 audio-visual companies by turnover.

rank			*home country*	*total turnover*		*turnover from audio-visual activities*		*audio-visual as % of total turnover*	
1991	*1990*			*1990 $m*	*1991 $m*	*1990 $m*	*1991 $m*	*1990 %*	*1991 %*
1	1	Time Warner	USA	11,517.0	12,021.0	7,101.4	7,390.5	61.66	61.48
2	2	Sony Corporation	Japan	24,977.7	28,372.1	5,055.5	5,702.8	20.24	20.10
3	13	Matsushita Electric Industrial	Japan	45,578.4	55,303.2	2,538.7	4,651.0	5.57	8.41
4	3	Capital Cities/ABC	USA	5,386.0	5,382.0	4,360.0	4,329.8	80.95	80.45
5	4	NHK	Japan	3,743.4	4,029.0	3,743.4	4,029.0	100.00	100.00
6	5	ARD	Germany	3,557.8	3,580.2	3,557.8	3,580.2	100.00	100.00
7	10	Philips (Polygram)	Netherlands	30,624.0	30,478.7	2,884.8	3,380.1	9.42	11.09
8	8	Fininvest	Italy	6,310.8	8,138.0	3,171.2	3,330.9	50.25	40.93
9	9	Fujisankei*	Japan	5,127.0	6,000.0	3,122.9	3,259.2	60.91	54.32
10	11	Bertelsmann	Germany	8,972.6	9,613.7	2,810.2	3,187.9	31.32	33.16
11	7	General Electric/NBC	USA	58,414.0	60,236.0	3,236.1	3,120.2	5.54	5.18
12	6	CBS	USA	3,260.0	3,035.0	3,260.0	3,035.0	100.00	100.00
13	15	News Corporation	Australia	8,248.0	8,590.6	2,292.9	2,750.7	27.80	32.02
14	14	RAI	Italy	2,502.6	2,732.6	2,324.7	2,727.1	92.89	99.80
15	16	Walt Disney Company	USA	5,843.7	6,182.4	2,250.4	2,594.1	38.51	41.96
16	12	BBC	UK	2,687.4	2,743.3	2,565.4	2,576.5	95.46	93.92
17	18	Thorn EMI	UK	6,352.5	6,343.0	1,813.6	2,455.4	28.55	38.71
18	17	Paramount	USA	3,869.0	3,895.4	2,054.1	2,380.1	53.09	61.10
19	19	Nintendo	Japan	3,114.7	3,767.4	1,638.3	1,702.9	52.60	45.20
20	23	Tokyo Broadcasting System	Japan	1,343.4	1,586.4	1,343.4	1,586.4	100.00	100.00

Source: IDATE.
* Estimate

considerable skepticism. In whose interests will it work? It is not difficult to see how Ross's global cultural order might actually serve the interests and priorities of the global corporate order. It is all too easy to see how the "free circulation" of media products might, in reality, be about corporate power and profits, rather more than it is about a "better world." What we are being asked to buy is very much an idealized image of a new media order. It is an image which, as I shall go on to argue, bears little relation to the real order that is presently taking shape around us.

The New Media Order

Herbert Schiller's interpretation of the new media order differs considerably from that offered by Steven Ross. "The actual sources of what is being called globalisation are not to be found in a newly achieved harmony of interests in the international arena," Schiller argues. What he sees is "transnational corporate cultural domination"; a world in which "private giant economic enterprises pursue – sometimes competitively, sometimes co-operatively – historical capitalist objectives of profit making and capital accumulation in continuously changing market and geopolitical conditions" (Schiller 1991, pp. 20–1). What is emphasized here is the historical continuity, and consistency, in corporate motivations. What is recognized and acknowledged is that, in the 1990s, the context of this drive for market and competitive position has been significantly transformed. The struggle for power and profits is now being waged at the global scale (Aksoy and Robins 1992).

What we are seeing is the construction of the media order through the entrepreneurial devices of a comparatively small number of global players, the likes of Time Warner, Sony, Matsushita, Rupert Murdoch's News Corporation and the Walt Disney Company (see table 15.1). For viewers, the new media order has become apparent through the emergence of new commercial channels such as BSkyB, CNN, MTV, or the Cartoon Network. What we are seeing is the development of a new media market characterized by new services, new delivery systems (satellite and cable), and new forms of payment (figure 15.1). In place of the mixed-programming channels of the "traditional"

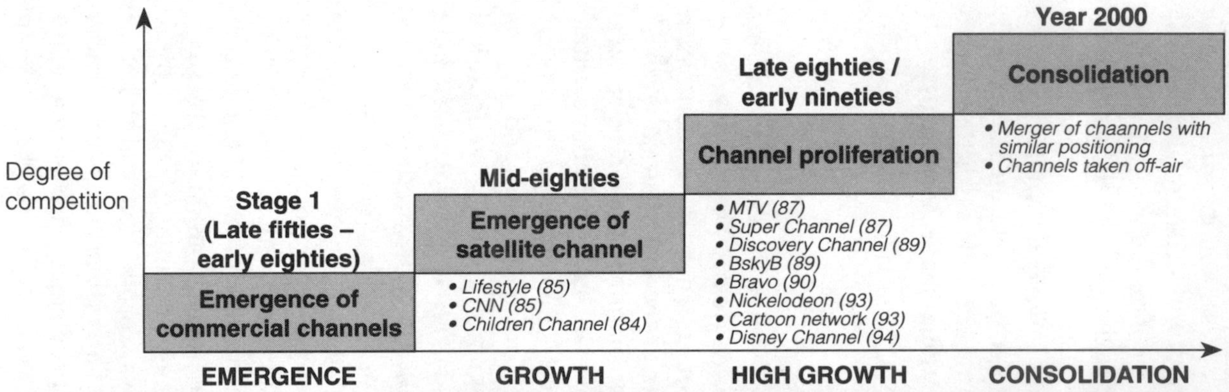

Degree of competition

Stage 1 (Late fifties – early eighties)
Emergence of commercial channels

Mid-eighties
Emergence of satellite channel
- *Lifestyle (85)*
- *CNN (85)*
- *Children Channel (84)*

Late eighties / early nineties
Channel proliferation
- *MTV (87)*
- *Super Channel (87)*
- *Discovery Channel (89)*
- *BskyB (89)*
- *Bravo (90)*
- *Nickelodeon (93)*
- *Cartoon network (93)*
- *Disney Channel (94)*

Year 2000
Consolidation
- *Merger of chaannels with similar positioning*
- *Channels taken off-air*

	EMERGENCE	GROWTH	HIGH GROWTH	CONSOLIDATION
	STAGE 1	**STAGE 2**	**STAGE 3**	**STAGE 4**
Introduction of competition	Foundation of commercial television	Satellite channel emerging	Channel proliferation	Shake-out and further proliferation of satellite channels
Programme offering	Broad based entertainment	Thematic channel Direct to home satellite	Thematic channel	Further segmentation Thematic channels
Transmission	Terrestrial	Satellite to cable Direct to home satellite	Satellite to cable Direct to home satellite	Increasingly through cable
Funding	Advertising	Advertising	Advertising and subscription	Increasingly also subscription funded/ Minipay packages

Figure 15.1 Development of a new media market.
Source: *Booz Allen & Hamilton.*

broadcasters, we now have the proliferation of generic channels (sport, news, music, movies). It is estimated that the 59 channels licensed to operate in the UK in 1992 will increase to around 130 by 2002 (Booz·Allen & Hamilton 1993, p. 9). In the United States, there are likely to be more than two hundred. It is, of course, the global media players that are investing in these channels (and the UK is only one small part in their global jigsaw).

Global corporations are presently maneuvering for world supremacy. There are three basic options open to media corporations: "The first is to be a studio and produce products. The second is to be a wholesale distributor of products, as MTV, CBS, and HBO are. The third is to be a hardware delivery system, whether that hardware is a cable wire or a Walkman" (Auletta 1993b, p. 81). The objective for the real global players is to operate across two or even all three of these activities. It is this ambition that has motivated the recent takeovers of Hollywood studios (Universal by Matsushita, Columbia/TriStar by Sony, Fox by Rupert Murdoch, and currently Paramount, which is attracting rival bids from both the US cable giant Viacom, and the home-shopping cable channel, QVC). As Steven Ross (1990) observes, "mass is critical, if it is combined with vertical integration and the resulting combination is intelligently managed." The issue for media corporations now is to decide what scale of integration they need to achieve, and are capable of managing, in order to build globally.

But there is more to it than just integration within the media sector. What we are beginning to see is a much more fundamental process of transformation in which entertainment and information businesses are converging with the telecommunications industry. A sign of things to come was the attempted merger, in October 1993, of the telecommunications company Bell with the largest US cable company, TCI (which was also a major backer of QVC in its attempt to acquire Paramount). The deal was not completed because of regulatory difficulties: the result would have been the world's largest media corporation. It had been described, by Bell Atlantic's chairman, as "a perfect information age marriage" and "a model for communications in the next century" (Dickson 1993). The new "multi-media" giant would have provided not only conventional cable television, but also telecommunications services, computer games and software, home banking and shopping, video on demand, and other interactive services. The aim was to develop information and communications "super-highways" to move us beyond the era of mass media and into that of personalized media and individual choice. Despite this initial setback, such grand mergers remain a corporate priority.

In describing the development of a global cultural order, I have emphasized that it will be a global corporate order. Global corporations are securing control over programming (production, archives), over distribution, and over transmission systems. The developments described above indicate that the flow of images and products is both more intensive and more extensive than in the past. What should also be emphasized is how much American cultural domination remains a fundamental part of this new order, though now American or American-style output is also the staple fare of non-US interests too (Schiller 1991). As a writer in the *Financial Times* recently observed, "soon hardly anywhere on earth will be entirely safe from at least the potential of tuning in to cheerful American voices revealing the latest news or introducing the oldest films" (Snoddy 1993).

What corporate maneuvers and machinations are seeking to bring into existence is a global media space and market. In the mid-1980s, the global advertising company Saatchi and Saatchi was talking about "world cultural convergence," and arguing that "convergences in demography, behaviour and shared cultural elements are creating a more favourable climate for acceptance of a single product and positioning across a wide range of geography." Television programs such as *Dallas*, or films such as *Star Wars* and *E.T.*, were seen to "have crossed many national boundaries to achieve world awareness for their plots, characters, etc." (Winram 1984, p. 21). Theodore Levitt, whose influential book *The Marketing Imagination* helped to shape the Saatchi outlook, was, at the same time, pointing to the increasing standardization and homogenization of markets across the world. "The global corporation," he argued, "looks to the nations of the world not for how they are different but for how they are alike . . . it seeks constantly in every way to standardise everything into a common global mode" (Levitt 1983, p. 28). Of course, if it is profitable to do so, global companies will respond to the demands of particular segments of the market. In so doing, however, "they will search for opportunities to sell to similar segments throughout the globe to achieve the scale economies that keep their costs competitive" (ibid., p. 26). The strategy is to "treat these market segments as global, not local, markets" (Winram 1984, p. 19).

This would still seem to be the logic at work in the 1990s. American movies – now it is *The Flintstones* and *Jurassic Park* – are still breaking box-office records across the world (hence the keen struggle to acquire Hollywood studios and archives). Satellite and cable channels are also making headway in marketing a standardized product worldwide. MTV, recently invited into Lithuania to help promote democracy, and CNN, now on twelve satellites beaming "global village" news the world over, seem to have come close to finding the answer to global marketing. The new "super-highways," still in their early stages of development, seem set to push processes of standardization further. But they are also likely to add more complexity, delivering "personalized" and "individualized" services to specialized and niche markets. Such strategies, it should be emphasized, "are not denials or contradictions of global homogenization, but rather its confirmation ... globalization does not mean the end of segments. It means, instead, their expansion to worldwide proportions" (Levitt 1983, pp. 30–1).

So much for the logic of corporate ambition. The question that we must now consider is how this logic unfolds as it encounters and negotiates the real world, the world of already existing and established markets and cultures. Let us first take the example of CNN, which might be seen as the very model of a global operator. Launched in 1980 by the American entrepreneur Ted Turner, its phenomenal success has been achieved through the distribution worldwide of a single, English-language news service. Increasingly, however, the channel is confronting the accusation that it is too American in its corporate identity. CNN's global presence is interpreted as an expression of American cultural domination, and this clearly raises problems as to its credibility as a global news provider. Back at company headquarters, this also translates into a fundamental dilemma over market strategy and position. CNN's present news service has been successful in reaching the world's business and political élites, but it has not significantly penetrated mass markets, where local affiliations and attachments are far stronger. To reach such viewers, "CNN would have to dramatically change its vision of a single, English-speaking global network," and "to effect that change Turner would need to seek partners and would need to localise" (Auletta 1993a, p. 30). What CNN is having to recognize is that the pursuit of further success will entail the production of different editions, in different languages, in different parts of the world. To this end, collaboration with local partners will be essential. In the context of growing competition – from, among others, Sky News, BBC World Service Television, and Reuters – CNN must learn to reconcile global ambitions with local complexities.

The case of Star TV provides another good example of the necessary accommodation between global and local dynamics. As part of their strategy for global hegemony, media corporations have sectioned the world into large geoeconomic regions. Star, a Hong Kong-based company which began broadcasting in August 1991, has effectively constructed the Asia region; stretching from Turkey to Japan, from Mongolia to Indonesia, it encompasses 38 countries (though only 13 receive Star signals at present). The station combined pan-Asian programs and advertising with a certain amount of material targeted at "spot markets," such as India or Taiwan. It also sought to balance Asian programming (Indian or Chinese pop music and films, for example) with "Western" channels (MTV, BBC World Service Television, Prime Sports), many of which are highly popular and welcomed as forces of internationalization and "modernization" (Poole 1993). Acknowledgment of its success across this vast region came in July 1993 when Rupert Murdoch's News Corporation paid $525 million for a controlling share of Star TV. For Murdoch, the Asian region was part of his "global dream," and he will clearly seek to market his Sky channels there. But he also recognizes the enormous cultural and linguistic differences within the region, and is planning to create separate services for India and China, and possibly also for Indonesia (Snoddy 1993). Given the diversity and complexity of this market, and given the enormous political (China, Indonesia) and religious (Malaysia) sensitivities, Murdoch's "local" partners are crucial to the future success of

Star. Success will depend on finding the right balance between market integration and market diversity.

The new media order involves, then, the articulation of both global and local factors (Robins 1990). In his "Worldview Address," Steven Ross (1990) acknowledged that global media must "be sensitive to the cultural environment and needs of every locale in which we operate." Anxious to avert charges of cultural homogenization and domination, global corporations are concerned to develop local credentials and credibility (though in this context, of course, "local" may amount to a multi-national region).

Culture and Politics in the New Media Order

This chapter will finally consider some cultural and political issues associated with globalization. Here it will take as its focus developments in Europe, another geoeconomic region in the new media order. During the 1980s, we saw heroic efforts by the European Community, on behalf of European media corporations, to construct a "European audio-visual area." In so doing, it inevitably raised questions about the cultural life and identity of the continent, and about the nature of Europe as a community (Robins 1989; Morley and Robins 1989). In the context of Europe, we can identify some of the tensions and problems arising out of the globalization process.

The clear objective of the European Community has been to bring into existence the European equivalents of Sony and Time Warner. It has sought to make the painful transition from the old public service era, in which broadcasters provided a diverse and balanced range of programs for citizen-viewers, to a new regime in which the imperative is to maximize the competitive position of European media business committed to satisfying the needs of consumers in global markets. It is the logic of industrial concentration and integration, working toward the creation of a few media giants. It is the logic of globalization, pushing toward the greater standardization and homogenization of output, and detaching media cultures from the particularities of place and context.

And yet there is also another, and contrary, force at work, challenging the imperatives of globalization. As an antidote to the internationalizaton of programming, and as compensation for the standardization and loss of identity that is associated with global networks, we have seen a resurgent interest in regionalism within Europe, appealing to the kind of situated meaning and emotional belonging that seem to have been eroded by the logic of globalization. This new regionalism puts value on the diversity and difference of identities in Europe, and seeks to sustain and conserve the variety of cultural heritages, regional and national. Broadcasting has been seen as a major resource in the pursuit of this objective, and in the 1980s we saw a growing interest in promoting media industries and activities within the regions and small nations of Europe. In most cases, principles of local public interest have been mobilized against the interests of transnational market forces. In lobbying for support from the European Community, the argument has been put that "in the particular case of regional TV programming in the European vernacular languages, the criteria should not be based on audience ratings and percentages of the language-speaking population, nor on strict, economic cost-effectiveness" (Garitaonandia 1993, p. 291). Since the late 1980s, a certain level of support has been elicited from the EC, particularly through its MEDIA program, which provides loans and support for small producers across the continent. Within the Community there has been increasing sensitivity toward cultural difference and commitment to the preservation of cultural identities in Europe.

Here, again, we have an example of the global–local nexus. "Local" in this case, however, means something quite different from what it means in the corporate lexicon. In this context, it relates to the distinctive identities and interests of local and regional communities. In these global times, there are those who desire to "reterritorialize" the media, that is to re-establish a relationship between media and territory. They are determined that the media should contribute to sustaining both the distinctiveness and the integrity of local and regional cultures against the threatening forces of "deterritorialization" and homogenization. "Local" in this sense constitutes a challenge to the strategies of global corporate interests.

If the processes of globalization provoke fear and resentment, these tend, for the most part, to become attached to the perceived threat of American culture and "Americanization." American mass culture has, for a very long time, been seen as a force that is eroding and dissolving European culture and tradition. The cultural domination of Hollywood has seemed to jeopardize the very survival of Europe's cultural industries. The culture of the continent is seen to be "in thrall to American money – and ultimately American values"; put simply, "Hollywood is the enemy" (Malcolm 1990). The American share of the European cinema market is now 75 percent (whilst the non-American share of US box-office takings stands at only 2 percent). In consequence, quotas have been imposed on non-European (that is, in effect, American) programming, with the aim of both protecting cultural sovereignty and enhancing the competitive position of domestic producers.

This question of cultural sovereignty was a key issue in the Uruguay round of GATT negotiations. GATT is clearly a fundamental issue across all sectors of the world economy, from agriculture to informatics. It is significant that it was trade in audio-visual products that proved so contentious. For the United States, the audio-visual industry was second only to aeronautics as a source of export revenue. For the European negotiators, particularly the French, what was at issue was the very survival of cultural identity. Whilst the US was calling, in the name of free trade and the free circulation of ideas, for the scrapping of quota restrictions, European interests were resolved to preserve them, in order, as they saw it, to defend the cultural specificity and integrity of European civilization. In France, there has been considerable hostility against Britain for having afforded Ted Turner access to European audiences. According to one critic, "Turner is only the avant-garde of the big US companies who are sitting back to see how Europe reacts. If he gets in, Disney and Time Warner will follow" (Powell 1993). The European stance has been seen as a battle for freedom of expression: "We want the Americans to let us survive. Ours is a struggle for the diversity of European culture so that our children will be able to hear French and German and Italian spoken in films" (ibid.). Again, the emphasis is on particularity and difference, in the face of what seems to threaten their dissolution.

In and across the new spaces of global media, there is a complex interaction between economic and cultural dynamics. But where is politics in all this? At the beginning of this chapter, I made the point that public service broadcasting acted as the focus not only for national culture, but also for political and democratic life. Questions of identity and questions of citizenship were bound together. What is presently happening to public service systems now raises very real questions about the future of political culture. In a discussion of what it might mean to move from a national to a continental political space and public sphere, Stig Hjarvard (1993, p. 90) observes that "at the European level there is no public with the ability to perform a critical function or represent alternative interpretations or definitions of the political agenda." There are very real difficulties in the way of constructing mechanisms for effective publicity and debate across a transnational space. The European Community has so far failed to develop an adequate political culture or a basis for European citizenship. Questions of identity and questions of citizenship have become dissociated, and the very real danger now is that, within the European audio-visual space, the compensations of cultural identification will be made to prevail over the political objectives of public communication and debate. And, of course, this "democratic deficit" is not only an issue of concern in the European zone of the world's news media order.

Acknowledgments

I would like to thank James Cornford at CURDS; David Hancock of IDATE; Booz·Allen & Hamilton; and the ITV Network Association.

16

Resisting and Reshaping the Modern: Social Movements and the Development Process

Paul Routledge

Introduction

In contemporary, Western, popular parlance, development means change but it also implies betterment and advance. The term conveys a sense of optimism and expectation of progress, and of a general improvement in the human condition. Indeed, the fruits of development are potentially manifold, and have included, for example, advances in medicine and education. The project of development and modernization has been particularly directed at countries in the South (i.e. the states of Latin America, Africa [excluding South Africa], and Asia [excluding Japan]) by the advanced capitalist countries. It has also been directed at indigenous peoples within the developed world, such as Native Americans in the United States and Canada. In contrast to the expected improvements that it would engender, the development project has caused widespread environmental damage and disrupted or destroyed the cultures and economies of numerous traditional (subsistence) and tribal communities. In response to this assault, social movements have emerged across the planet to pose challenges and alternatives to the process of development. This chapter will critically examine the nature of the development project and discuss some of those social movements that attempt to resist and reshape the forces of modernization.

Development as Discourse

The word "develop" was first associated with ideas concerning the nature of economic change in the nineteenth century, implying the notion of a society passing through definite evolutionary stages. The term "underdeveloped" began to be used after 1945, referring to lands in which natural resources have been insufficiently developed and exploited, and to economies and societies destined to pass through predictable "stages of development" according to a known model (Williams 1983, pp. 102–4); although, as Peet and Watts (1993b) have noted, these notions were already embedded in the colonial development and welfare acts of the French and British colonies during the prewar period.

The term was used by US President Truman in 1949 to describe those areas of the world that had yet to reach the standard of living experienced by those in the West. Truman went on to advocate an era of development whereby the West would provide assistance to "underdeveloped" areas to enable them – via "scientific advances" and "industrial progress" – to improve and grow (Esteva 1992, p. 6). The division of the world into developed and underdeveloped areas and the articulation of a particular remedy to this situation (the modernizing project of development) formed the basis of the discourse of development, becoming part of the process by which the "colonial world" was reconfigured into the "developing world" (Peet and Watts 1993a, p. 232).

Discourse is a field of strategies (statements, views, theories, concepts, and objects of analysis and their interrelations) that create knowledge about something and create differentiations by posing limits on what can be said and by whom (Foucault 1977b; 1979). The development project has been deployed by many of the Western states through discourses of underdevelopment that include, for example, theories of development, World Bank policies, and the apparatus of development programs. These discourses privilege the West's economic systems, institutions, and policy "experts" at the expense of those of the South, imagining the West as "the transcendental pivot of all analytical reflection" (Slater 1992, p. 312). Indeed, the development discourse serves to manage, control, and create the South politically, economically, ideologically, and culturally.

Ideologically, the development project is based upon a linear theory of progress and evolution (the meaning of development in the biological sciences) that places the industrialized states at the top of the evolutionary scale ("the developed") and Southern countries at the bottom ("the underdeveloped"). This differentiation, therefore, requires the countries of the South to evolve further – to develop – so that they can reach their full potential.

Economically, the project of development is based upon a program of economic growth, modernization, and industrialization that is seen to be essential to the alleviation of poverty. Since 1945, a plethora of development theories has been proposed variously privileging the role of the market (e.g. Rostow's (1960) modernization theory, and Bauer's (1966) theory of neo-classical economic development), civil society (e.g. Hirschmann's (1958) notion of Weberian modernization), and the state (e.g. those proposed by Mao and Nehru in China and India respectively) in the process of development (Peet and Watts 1993b). While a variety of new approaches to development have emerged since 1980 (see Peet and Watts 1993b), recent recommendations by the United Nations' Brandt and Brundtland Commissions continue to argue that economic growth is the necessary condition for development and the subsequent alleviation of poverty (Sachs 1992, Esteva 1992).

Culturally, the development project defines people as poor, because they do not participate in the market economy. It emphasizes Western values (of capitalist production, economic growth, rationality, calculability, etc.) and devalues indigenous and traditional systems of knowledge, economy, and culture. By deeming culturally-specific economic practices such as subsistence agriculture as backward, development discourse has legitimized the intervention of the processes of modernization into traditional economies (Bandyopadhyay and Shiva 1988).

Politically, the development project was initiated as a response by French and British colonial powers to anti-colonial movements in regions such as Africa. Development became a means by which the perils of independence could be negotiated, and colonialism continued (Peet and Watts 1993a, p. 232). More recently the development project has served as a means to encourage capitalist market economies, thereby providing conditions under which Western-style "democracy" could flourish. This purpose of the project emerged in the West at a time when the Soviet Union had emerged as a superpower competitor to the United States. As Rostow's (1960) "non-communist manifesto" implies, the development project was constructed as a strategy to bring Southern states into the geopolitical orbit of the United States and its allies (Sachs 1992).

In development discourse, then, the South does not represent itself, rather it is represented by Western academics, experts, professionals, bankers and government officials. Communication and information technologies such as the mass media, TV, and commercial cinema serve to reinforce the stereotypes by which the South is viewed (e.g. poor, non-modern, undemocratic, etc). The discourse of development serves three purposes. First, it creates problems and abnormalities (such as underdeveloped countries, poverty, etc.) which require treatment and reform. Second, the project of development is professionalized through the creation of development-studies courses, the classification of problems, and the formulation of policies. Third, the development project is institutionalized through the establishment of international agencies, governments, national planning bodies and local-level development agencies, the "professionals" within which determine what is best for developing countries (Escobar 1984). Two types of intervention have co-evolved with these processes: (i) intervention in the South by international corporations and institutions such as the World Bank and the International Monetary Fund (development as dependence); and (ii) intervention in the traditional economies and cultures within particular states by national governments, which are seen as the only legitimate form of political authority.

Development as Dependence

The extent to which the political economies of the Southern countries are influenced by a global economy dominated by the advanced capitalist countries has been analyzed by a variety of scholars known as the "dependency theorists." The analysis of dependency includes a variety of related theories including those formulated by the Economic Commission for Latin America – see Love (1980), Frank (1967), Dos Santos (1970), Cardoso (1972), Emmanuel (1972), Amin (1976), and Cardoso and Faletto (1979).

Dependency theory focuses on the unequal economic and political exchange that takes place between the advanced capitalist countries (the "core") and the countries of the South (the "periphery"). The economies of the periphery are seen as conditioned by, and dependent upon, the development and expansion of economies in the core. The process of development is seen as selective, reinforcing the accumulation of wealth in the core at the expense of the periphery. Dependency is seen as both the relationship between states – an industrialized core and an impoverished periphery – and also the relationship between groups and classes within states. It is important to note that while certain broad trends are identified below, regional and intra-state differences exist.

In adopting the Western-defined development project, the countries of the South are confronted with various problems that place them at a disadvantage in relation to the developed countries. First, the particular economic and political conditions that enabled the West to industrialize – based upon the domination and exploitation of natural and human resources in the West's colonies – are different from those that confront the South. For example, whereas the West was able to support its modernization upon an expanding resource base (in the colonies), the Southern states are confronted with diminishing resources within their political borders. Secondly, the process of colonialism integrated the South into a world division of labor whereby the major function of the colonized economies was the production of raw materials for the European colonizers, and the export of those raw materials for the colonizer's use. This facilitated the industrialization of the economies of the West at the expense of those in the South. It also integrated the economies of the South into a world economy dominated by the industrial centers in the West, placing them in a dependent relationship to the West that has continued, albeit in different forms, since decolonization.

The development project fosters an economic dependence of

the South on developed countries that has several features. First, most of the technology required for development (especially information technology such as satellites, computers, etc.) is produced in the developed countries, who determine the price and availability of such technologies to the South. Much of this technology is also capital-intensive whereas traditional and subsistence economies are labor-intensive. Hence, when adopted as part of the process of modernization, traditional labor is displaced, resulting in unemployment. Secondly, the South relies upon foreign investment to accelerate the development process, to access new technology, and to gain new markets. The markets are often represented by Multinational Corporations (MNCs). This results in the MNCs being in a position to exert considerable economic and political influence over Southern economies. For example, many of the countries in the South have pursued export-led economic policies involving, for example, the production of cash crops and manufactured goods for export to developed countries (e.g. the US and the EC) on whose markets they are dependent (Chandra 1992).

Finally, although development is mostly financed from domestic sources, foreign aid is important to the South as an added source of foreign exchange, human resources, and technical assistance. However, various factors, including the interest payments made by Southern countries on the debts incurred by foreign loans, and price fluctuations in commodities sold by Southern countries to the West, mean that a "reverse transfer" of resources has occurred from the former to the latter. For example, since 1983 the Southern states have transferred over US$ 30 billion in net wealth to the industrialized countries (Franke and Chasin 1992). This process has had enormous ecological, cultural, political, and economic repercussions in the South, since in the past up to half of the loans and credits from international banks have gone to environmentally-sensitive projects such as irrigation, forestry, agriculture, and dam construction. For example, in 1983–4 Brazil borrowed US$ 950 million per year to develop farming for export, which necessitated deforestation and human displacement in the Amazon (Bandyopadhyay and Shiva 1988, p. 1231). Through imposing structural adjustments (such as privatization) upon a country as the condition for awarding loans, such institutions as the World Bank also influence the long-term economic policy and not just single projects.

An example of the development project creating dependency is provided by the green revolution, which emerged in the 1960s as a development strategy designed to improve the yields of food production in developing countries. The project was capital-intensive and relied on the application of expensive new technologies such as pesticides, chemical fertilizers, and high-yield variety seeds (HYVs), which were manufactured by MNCs and had to be imported from the West. The project was based upon the introduction of HYV monocrops such as rice, maize and wheat, whose increase in productivity was achieved by undermining the productivity and availability of other locally important crops such as pulses. The project destabilized traditional farming methods, which further rationalized the use of new technologies from the West, and the displacement of traditional foodstuffs by the HYVs. For example, during 1965–6 HYV sorghum was introduced, under irrigated conditions, in Dharwar district in Karnataka, India. It replaced indigenous sorghum varieties that had been previously cultivated with pulses and oilseeds. Because of its susceptibility to pests, the HYV sorghum required extensive pesticide spraying, which destroyed the pest–predator balance in the area. As a result, new pests appeared which gradually destroyed the traditional varieties of sorghum. Because sorghum is the main food-crop in the region, farmers were compelled to plant HYV sorghum at the expense of indigenous crops so that by 1980–81 no area was sown with traditional crop varieties. Traditional sorghum cultivation involved mixed cropping with other important protein sources such as pulses, and the use of sorghum straw for animal fodder. Its replacement by the HYV monocrop decreased overall foodgrain and fodder yields and availability, and reduced soil fertility (see Shiva 1989, pp. 123–5).

The green revolution increased the South's dependence upon the West in various ways. First, Southern economies were dependent upon foreign banks, which provided the necessary credit to farmers to purchase the expensive inputs. Secondly, the project supplanted traditional agriculture with Western methods, which required the intervention of Western professionals and institutions into the South's agricultural sector. Thirdly, the project made

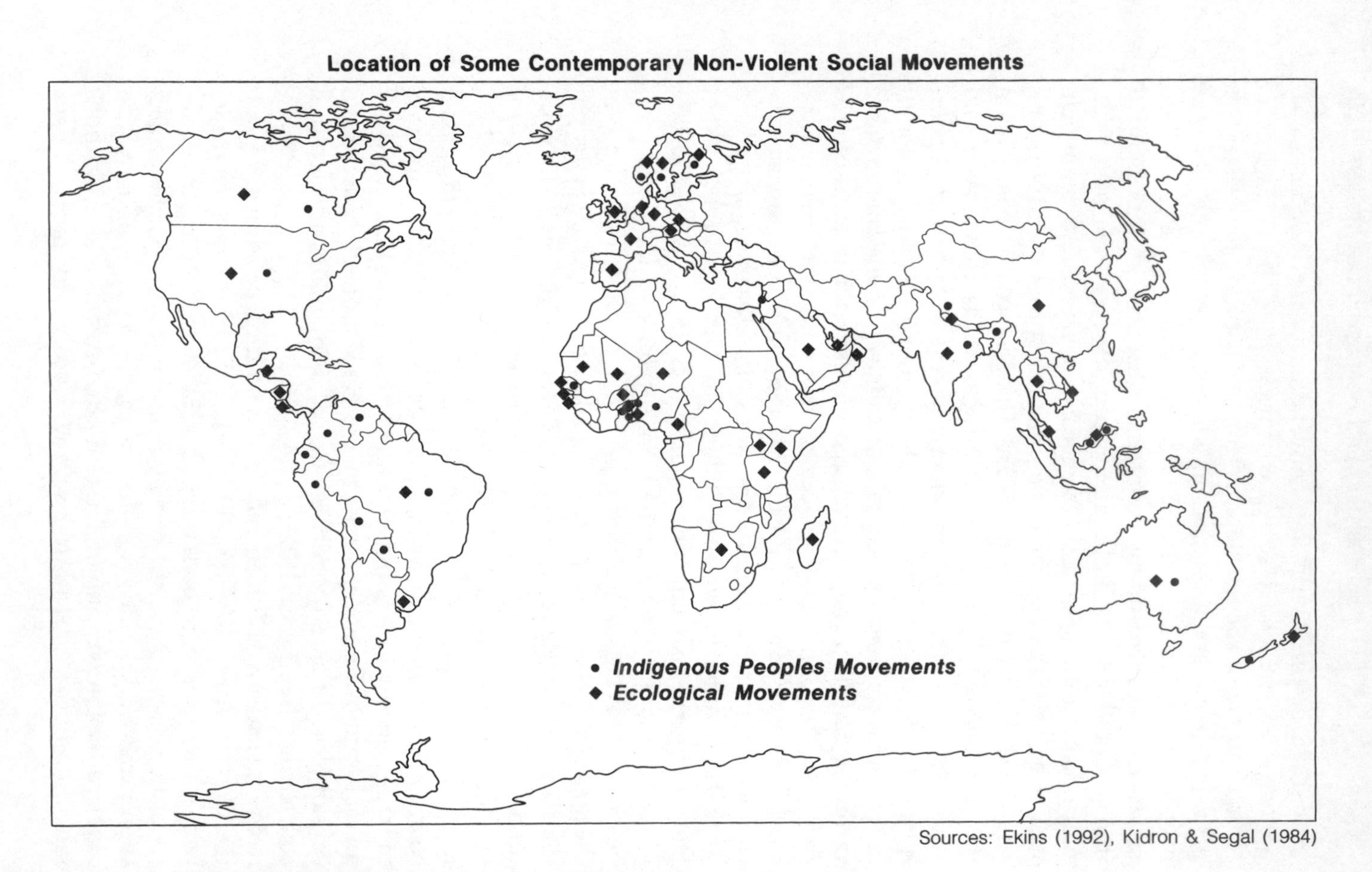

Figure 16.1 Location of some contemporary non-violent social movements.

agriculture dependent upon industrial outputs (e.g. fertilizers, pesticides, expert advice, credit, etc.) thus subordinating agriculture to the requirements of industrial growth (Alvares 1992).

Throughout the entire development project, the majority of people in the South (peasants, women, children, tribal peoples) are viewed as impediments to progress and modernity and excluded from decision-making and participation in a process that has an enormous impact on their lives. State-directed development and state-acquired modern science and technology are seen as being opposed to the "unscientific" and "irrational" lifestyles of traditional (peasant, tribal) cultures. The development project has facilitated the state's securing control over natural and financial resources and consolidated the power of those directing and benefiting from the development apparatus – national ruling élites, and international institutions (Nandy 1984). In the process, traditional subsistence economies and their associated cultures are being destroyed; people face displacement from their homes and lands, losing access to their resources, and become economically marginalized. In response to these processes numerous social movements have arisen in the South (and in the developed North) to challenge the violence of the development project, and it is to this resistance that I will now turn.

Social Movements: Resisting and Reshaping the Modern

In response to the development project, myriad social movements have emerged articulating struggles for cultural, ecological and economic survival. Although this chapter has focused primarily on countries in the South, it is important to acknowledge that the development project continues to be directed against indigenous peoples in the developed world, where it has also been met with resistance (see figure 16.1). Examples of such movements include the Cree struggle to prevent the Hydro-Quebec dam in James Bay, Canada; Aboriginal struggles against mining in Australia; and Yakima Indian struggles for fishing rights in the United States (Moody 1988).

Many of the movements involved in resistance in the South and the North have been termed "new social movements" by a variety of social theorists (e.g. Melucci 1989; Castells 1983; Laclau 1985; Offe 1985; Touraine 1985; Guha 1989). These movements are considered "new" for a number of reasons. First, at the level of the economy, they have articulated conflicts over productive resources and other issues hitherto neglected by political organizations. For example, "traditional" social movements frequently focussed upon class struggles in the factories and the fields, whereas ecological movements articulate a new kind of struggle over natural resources such as forests and water, and women's movements have articulated struggles in the workplace and in the household on sexual grounds. Also, the economic demands of new social movements are not restricted to a more equitable distribution of resources between competing groups in those areas of production largely ignored by earlier movements, but are also involved in the creation of new services such as health and education in rural areas (Guha 1989). Indeed, social movements have emerged in many areas, including civil liberties, women's rights, science and health, that are themselves related to problems caused by the development project.

Secondly, at the level of politics, the new social movements are frequently autonomous of political parties (although some have formed alliances with voluntary organizations and non-government organizations). Their goals frequently articulate alternatives to the political process, political parties, the state, and the capture of state power. By articulating concerns of justice and "quality of life," these movements have enlarged the conception of politics to include issues of gender, ethnicity, and the autonomy and dignity of diverse individuals and groups (Guha 1989). Many of these new social movements are also multi-dimensional, simultaneously addressing, for example, issues of poverty, environment, and culture. This multi-dimensionality is indicative of an alternative politics that seeks to create autonomous spaces of action outside of the state arena (Peet and Watts 1993b). However, because many of these new movements have a limited social base, they sometimes form working relationships with other organizations such as trade unions and political parties while retaining autonomy from them. They can also play complementary roles to class-based movements, by enlarging the purview of such movements to include issues of gender, health, ecology, etc. (Guha 1989).

Thirdly, on the level of culture, new social movement identities and solidarities are formed, for example, around issues of kinship, neighborhood, and the social networks of everyday life, in addition to, or instead of, "traditional" solidarities of class. Movement struggles are frequently cultural struggles over material conditions and needs, and over the practices and meanings of everyday life (Escobar 1992b).

New social movements are by no means homogeneous. A multiplicity of groups including squatter movements, neighborhood groups, human-rights organizations, women's associations, indigenous rights groups, self-help movements amongst the poor and unemployed, youth groups, educational and health associations, and artists' movements are involved in various types of struggle (Corbridge 1991). Many of these struggles take place within the realm of civil society, i.e. those areas of society that are neither part of the processes of material production in the economy nor part of state-funded organizations. For these movements, civil society represents "the domain of struggles, public spaces, and political processes. It comprises the social realm in which the creation of norms, identities, and social relations of domination and resistance and located" (Cohen 1985, p. 700). Although not exclusively, these movements frequently employ non-violent methods of resistance (see Routledge 1993).

While new social movements in the North and South share some of the broad characteristics mentioned above – e.g. they articulate such issues as ecology, gender, and ethnicity that were not addressed by class-based movements – there are also important differences between them. In the North, new social movements have often concentrated upon "quality of life" issues, whereas in the South, movements have often focused upon the access to economic resources. An example is the issues faced by ecological movements. In the North, the ecology movement has taken much of the industrial economy and consumer society for granted, working to preserve nature as an item of "consumption," as a haven from the world of work. In the South, however, those affected by environmental degradation – poor and landless peasants, women, and tribals – are involved in struggles for economic and cultural survival rather than the quality of life (Guha 1989).

As responses to the effects of the development project, new

social movements in the South reflect the crises of ecology (e.g. struggles to prevent deforestation and pollution), economy (e.g. squatter movement struggles to secure urban housing), culture (e.g. struggles to protect the integrity of indigenous people's communities), and politics (e.g. struggles for increased local autonomy). Many of these movements are also place-specific and frequently involved in what I term "constructive resistance." That is, not only do these movements articulate dissent from (and often non-compliance with) central and state government policies, they also actively seek to articulate and implement alternative development practices. Viewing the state-directed development process as inimical to local tradition and livelihood, many social movements actively affirm local identity, culture, and systems of knowledge as an integral part of their resistance. In doing so, these movements articulate localized "terrains of resistance" to the dominating discourse of development, expressing their own counterhegemonic (or antidevelopment) discourses and practices. Hence particular places become contested terrains between different social groups who assign different meanings and values to those places: between local communities, with their traditional lifestyles, and the state (and private corporations) as the implementers of the modernizing development project.

As noted above, these social movements articulate resistances at the economic, ecological, political, and cultural levels of society, all of which can be interrelated, At the level of economy, social movements articulate conflicts over the productive resources in society such as forest and water resources, involving demands for a more equitable distribution of resources, the creation of new services, and the integrity of local, traditional forms of economic practice. For example, during the mid-1980s in the Philippines, farmers, peasants, and scientists joined forces to demand the closure of the green revolution's International Rice Research Institute, which they perceived as being instrumental in the destruction of their traditional farming economy (Alvares 1992).

At the ecological level, social movements struggle to protect remaining environments from further destruction, and to ensure the economic (and cultural) survival of peasant and tribal populations. For example, resistance against the Nam Choan Dam in Thailand during the 1980s arose because the construction of the dam threatened to destroy wildlife sanctuaries and riverine forest environments and to displace up to 6,000 Karen and Hmong tribal people (Eudey 1988).

At the political level, social movements challenge the state-centered and development-biased character of the political process, articulating critiques of development ideology and of the role of the state. For example, in Baliapal, India, during the late 1980s, peasants resisted the construction of a missile base by the government because it would have entailed their forced eviction from their homes and lands. They argued that their lives should take precedence over the national security demands of the state (Routledge 1993).

At the cultural level, social movements frequently affirm and regenerate local (place-specific) identity, knowledge and practices, which at times are expressed in the language and character of the struggles. Local resistance may incorporate local linguistic expressions, such as songs, poems, and dramas that imbue and affirm local experiences, beliefs, and cultural practices. For example, in Colombia during the 1980s the Indigenous Authorities Movement sought to reaffirm the integrity of tribal territory, community, and culture against incursions made by the hacienda system (Findji 1992).

The responses of state authorities to social movements vary, according to the type of resistance movement, and the character of the government involved. When faced with social movement challenges, governmental responses include repression, cooption, cooperation, and accommodation. Repression can range from harassment and physical beatings, to imprisonment, torture, and the killing of activists. For example, in El Salvador during 1983, 73 members of the National Association of Indian People of El Salvador (ANIS), who were involved in struggles for economic rights, were assassinated by the military (The Environmental Project of Central America 1987). In concert with, or instead of, repression, governments may attempt to coopt the leadership of oppositional movements through bribes, the offer of government jobs, etc. If neither of these approaches is effective, governments may be forced to accommodate the demands of the social movement, or at least cooperate partially with the social movement by instituting certain reforms. For example, during the late

1980s, in the Philippines' capital of Manila, the squatter movement, Sama Sama, was able to negotiate a partnership with the government to plan and implement the urban land reforms that the movement had demanded (Parnell 1992).

As already mentioned, social movements can be either defensive or assertive (or both) in character, attempting to resist and reshape the development project. Examples of defensive resistance are widespread and include recent struggles in Brazil and Sarawak. In Brazil during the 1980s, numerous conflicts developed over access, use and exploitation of the Amazon and its forest and mineral resources. For example, between 1985 and 1987 the number of people involved in land conflicts rose form 566,000 to 1,363,729. These included indigenous tribes, gold miners, rubber tappers, land speculators, and mining and timber concerns. One conflict, in the Amazonian state of Acre, pitted land speculators and ranchers against rubber tappers and subsequently indigenous groups. In order to move the rubber tappers off the land, the ranchers and speculators engaged in the wholesale burning of the tappers' houses and crops, and destruction of the rubber trees. In resistance, a rubber tapper, Chico Mendes, organized the rubber tappers to conduct direct action in the form of *empats*, or standoffs. When confronted with the destruction of the rubber trees, men, women, and children of the rubber tapper communities began blocking the tree clearers to prevent them from carrying out their work. During the 1980s, literally millions of acres were saved by *empates*, and the rubber tappers joined forces with various indigenous groups to form the Forest People's Alliance to resist ranching and logging practices in the area (Hecht and Cockburn 1990).

In Sarawak, between 1963 and 1985, over 2.8 million hectares of forest were logged by commercial timber companies to trade tropical hardwoods on the world market. The massive deforestation led to the destruction of forest wildlife, pollution, and massive soil erosion into the river systems, which decreased fish supplies. The forests are the traditional homes of the Orang Ulu peoples, which include the Penan, Kelabit, and Kayan communities. The Orang Ulu depend entirely on the forest and river resources for their subsistence. In 1987, in response to the logging of their homes, the Penan, Kelabit, and Kayan communities formed human barricades across the logging tracks in an attempt to prevent the further destruction of the forests. Within three months, blockade sites had been established along a 150 km length of road in Sarawak. One of the corporations targeted was the Limbang Trading Company, owned by Malaysia's Minister for Environment and Tourism. As a result, the Orang Ulu faced severe police repression. However, their struggle had received support from the Sahabat Alam Malaysia (Friends of the Earth in Malaysia) non-governmental organization, and continues to resist deforestation (Sahabat Alam Malaysia 1987; Ekins 1992).

In their assertive forms of resistance social movements are attempting to reshape the development project by articulating alternative practices at both the local and national levels. For example, the Sarvodaya Shramadana Movement in Sri Lanka is active in 8,000 of Sri Lanka's 23,000 villages (incorporating 3 million of the country's 15 million population). The movement is involved in providing health-care, education, housing, irrigation, and agricultural resources via the creation of Shramadana Camps. These employ donated labor by the villagers for collective projects, each program being structured into separate autonomous organizations. In Burkina Faso, the NAAM Movement has established traditional village organizations of principally young people to establish culturally appropriate agricultural projects adapted to local needs. By 1987 there were over 2,700 NAAM groups in the Yatenga area of the country with approximately 160,000 members (Ekins 1992, p. 113). Meanwhile, in Kenya, the grassroots Green Belt Movement has conducted reforestation programs, encouraged soil conservation practices, and set up nurseries, to counter deforestation, and provided incomes for local women. By the mid-1980s the movement had established about 600 tree-nurseries, had planted about 2,000 green belts of at least 1,000 trees each, and had helped between 2,000 and 3,000 women earn their own incomes (Ekins 1992, p. 151).

Social movements also combine the strategies of resistance and the articulation of alternatives to development. One of the most celebrated examples is the Chipko movement in India. The Chipko Movement is a peasant movement that emerged in 1972 in response to the effects that ecological destruction (especially deforestation) had had upon the local culture and economy in the Garhwal Himalaya region of Uttar Pradesh. Resistance was articulated through a variety of non-violent methods including hugging trees to prevent them from being cut down (the movement took its name from the Hindi word "chipko" which means to hug), demonstrations, and the uprooting of eucalyptus saplings in social forestry plantations. Although led by men, the movement saw the active involvement of women – since their deep involvement in the local subsistence economy was threatened by the effects of deforestation. The movement articulated local religious and ecological beliefs in their resistance, which served to motivate and educate peasants regarding the need to protect the forests. For example, movement activists conducted *padyatras* (foot marches) from village to village to educate and organize local peasants regarding the ecological consequences of deforestation. The *padyatra* evoked the image of the Hindu pilgrimage, an important religious custom, with which local peasants could identify.

As a result of their resistance, the Indian government granted limited concessions to the movement, only partly reducing the process of deforestation in the region. Although certain state forestry practices were reformed, other commercial practices such as dam construction and mining have continued to have an impact on the area. The movement consisted of three factions, two of which have increasingly cooperated with the central government on environmental development schemes, leading to criticisms that the participation of local villagers has declined. Also, few low-caste peasants have been involved in the movement.

On the positive side, the movement has articulated various alternative development practices as part of their resistance, including reforestation schemes, soil conservation schemes, and the conducting of ecodevelopment camps through which peasant communities come together to discuss and plan alternative development programs in their villages. However, the impact of these measures has been limited to particular areas of the Garhwal Himalayas (Routledge 1993).

Conclusion

Despite its claims to bring prosperity and the alleviation of poverty through economic growth, the development project has caused enormous environmental destruction, and the impoverishment, displacement and, at times, cultural ethnocide of poor and landless peasants, urban workers in both the formal and informal sectors, women, and tribal peoples. In resistance to these processes, social movements have emerged throughout the world, attempting to protect their homes, lands, and cultures. While in the past the oppressed often focused their actions and struggles on objectives such as industrial employment, wage increases and access to public services within the existing culture of domination created by the development project, there is a growing realization amongst many social movements that resistance should also articulate alternatives to the dominant culture. Thus many movements are regenerating traditional health practices, re-embedding learning in local culture, regenerating environmentally-sustainable agriculture, and recovering a definition of their own needs and autonomous ways of living that were dismantled and redefined by the development project. While these movements tend to be active in place-specific contexts, the issues that they address – such as ecologically sustainable practices – have both national and international importance. In addition, social movements have made visible the particular ideology of development, its inherent injustice and non-sustainability. In doing so they are both resisting the modernization process and articulating alternatives – economic, ecological, political, and cultural – to it. As such they are attempting to reshape and redefine the development juggernaut.

17

Understanding Diversity: the Problem of/for "Theory"

Linda McDowell

> One day, the philosopher of science Paul Feyerabend, then teaching at the University of California Berkeley, confessed to a female African-American student his reservations about teaching imperial Western philosophy to representatives from different cultures, who had their own ways of knowing and their own values. She replied: "You teach. We choose." (Interview with Feyerabend, *The Higher*, 10 Dec. 1993).

Introduction

This story is the key to this chapter. A set of challenges to the theoretical basis of disciplinary knowledge currently poses a major challenge to a discipline like geography whose *raison d'être* is the explanation of difference and diversity. Recent critiques of Western science, exposing its purportedly universal and objective truths as ethnocentric, masculinist, and historically specific assumptions, seem to have left us marooned in a quagmire of relativism in which different ways of knowing have apparently equal claims to validity. An angry debate, ranging across disciplinary boundaries, within and between the social sciences and the humanities, and reaching into the sciences, has erupted in which adherents of different perspectives exhort each other variously to celebrate difference or to hold on to old verities. The bad-tempered arguments within a related discipline, anthropology, give a flavor of the debate and the passions with which each side holds to its arguments for or against notions of rational science and cultural relativism (see Gellner 1992). Our own discipline, too, has seen fierce exchanges about the nature of theory between, for example, Marxists, feminists, and a number of adherents of various versions of postmodernism.[1]

I want to argue here, however, that abandoning old certainties does not necessarily entail the whole-hearted embrace of relativism. Recognizing different ways of knowing does not mean abrogating responsibility for distinguishing between them. For teachers, as well as students, one of the main purposes of theory is to permit and to justify such choices. As geographers, we need theoretical perspectives that not only permit the elucidation of the main outlines of difference and diversity, of the contradictory patterns of spatial differentiation in an increasingly complex world, where ever-tighter global interconnections coexist with extreme differences between localities; but perspectives that also allow us to say something about the *significance* of these differences. Thus, I argue here that geographers must develop *Principled Positions* (Squires 1993) as a basis for constructing a *politics* of difference.

In this chapter, I shall try to summarize the main features of three sets of theorists of diversity – feminists, poststructuralists, and postcolonial theorists. This is an almost impossible task as the literatures in each area are now huge, both within and without geography. References to other introductory summaries are listed in the Further Reading section. It is an important task, however, as without a flavor of these debates it is impossible to address the question of relativism that arises if we argue that discourses are geographically and socially specific, or the possibility of distinguishing between different ways of knowing.

What is Theory / What is Theory For?

In debates about the status of knowledge, the term theory has a privileged position. A major consequence of the debates is that its definition has changed. A theoretical analysis of a geographic issue consists of a rule-governed activity. These rules establish "what counts as evidence, what constitutes an argument, what appeals are to be made to empirical truth, explanatory adequacy, rhetorical effectiveness, or any other standard of judgement" (Landry and Maclean 1993, p. 136). Conventionally, this has meant a scientific, or objective, analysis: this is usually what is meant by "theory" in the technical sense and is the definition that informs geographic approaches such as spatial science or Marxist analyses. But increasingly, geographers recognize that theories themselves are "texts" or narratives that tell a single and particular story, and not others that might have been told. Thus a second kind of theoretical analysis is becoming common in geography: analyses of theories themselves, revealing the ideological assumptions in the work; the contradictions, unvoiced assumptions, the limits to the texts' logic (see Barnes 1992). Scholars from a range of positions – feminist, poststructural, postmodern, and postcolonial – have raised new questions about whose point of view is represented in geographic writing. These scholars point out the absence of certain voices in geographic narratives – in the main, those of the powerless and dispossessed – arguing against conventional scientific methods that assume a correspondence between "reality" and its representation by geographers, the idea that maps and texts make transparent a singular meaning for the reader. Rather, they argue that knowledge reflects and maintains power relations, that it is partial, contextual, and situated in particular times, places, and circumstances. Representations of these partial truths are produced by authors who are "raced," gendered, and classed beings with a particular way of seeing the world. Further, these socially-produced texts have no necessary or fixed meaning; the reader is implicated in the construction of meaning. A reader's own set of assumptions and sociogeographic location affects how texts are read and interpreted.

These arguments are tremendously powerful and potentially liberating, raising new questions about power and knowledge and, it seems, particularly pertinent questions for geographers about the relationship between place and representations, about the social construction of geographic knowledge and our research methods. They are also unsettling, posing difficult questions about the significance of different views of the world. If there is no singular Truth any more, but diverse truths, how, we might ask Feyerabend's student, does she go about chosing *between* them?

To understand the extent of the challenge posed by what we might term the new theorists of diversity, it is helpful to reflect on the history of theoretical endeavor in the discipline. Surveys of the history of geographical explanation tend to give the impression that the replacement of one approach by another is orderly progression in which increasingly sophisticated theories replace earlier and, by implication, discredited perspectives. But this is a false impression. With the possible exception of spatial science – and even this approach never achieved complete dominance – a more accurate picture is one of competing knowledge claims, of interpretative communities with different views of the world and different methodological approaches. Geography, like most social sciences, has a history of multiple perspectives, competing paradigms, and theoretical diversity. However, these diverse ways of seeing the world were regarded as *alternatives*. The adherents of each approach argued that their view was the correct one. Each held to a notion of a single truth embodied in their own view of the world and uniquely revealed by their approach to research – beliefs that are crucial in arguments that geography is "scientific" but are disrupted by the claims by the multiple "others" who have recently appeared in geographic texts. The voices of women, working class subjects, "Third World" peoples, of the underclass and transgressors, make different claims – that the singular geographies we have been accustomed to are equally partial; they reflect the view of the world of the powerful rather than a rational, scientific, or objective view of reality as claimed. The theoretical challenge of representing the partial perspectives of multiple "others" rather than what Haraway (1991) has termed "the view from nowhere" is subsumed in part of my title: the problem *of* theory. The current problem *for* theory is how to respond to it.

The Problem *of* Theory: the Status of "Truth/truth"

Science, ideology, knowledge, discourse, narrative: these five words indicate alternative positions in the intellectual landscape of contemporary geography. The first two reflect enlightenment views of Truth that underpin the liberal and Marxist perspectives in geography, whereas the latter three are common terms in new theoretical positions. Whereas Marxist analysts in geography are concerned to demonstrate what Marx himself termed "the economics of untruth" in liberal theories of the world, more recently other geographers, especially those influenced by the French philosopher Foucault, have challenged the scientific basis of Marxist theory in their focus on "the politics of truth" (the term is Foucault's). I shall briefly spell out these differences, but interested readers will find a more thorough discussion in Michele Barrett's *The Politics of Truth* (1991).

The first challenge to the scientific basis of positivist approaches in geography came from the work of Marxist scholars and from socialist feminists, who argued that the work of too many theorists bore little relationship to the reality that it claimed to portray. Positivist spatial science, for example, portrayed a landscape without power, poverty, or political struggles: areas to which Marxist and socialist feminist analysts drew particular attention. Although adherents of Marxian perspectives continue to hold to notions of scientific knowledge and Truth, they argue that positivist science is ideological, justification of the privileges of a class society dressed up as truth. Here, we see the idea that knowledge is positional or embedded, a version of the world that embodies a particular justification of the distribution of rewards and privileges. However, it is within that set of knowledges variously labeled postmodernist, poststructuralist, and postcolonialist, and in recent developments in feminist scholarship (McDowell 1992a and 1992b), that resistance to truth claims has been most strenuous. The rest of the chapter discusses these arguments.

Introducing the "Others": Feminism, Poststructuralism and Postcolonialism

What was it that happened to disrupt theoretical confidence in a singular notion of Truth? What were the challenges to geographers' ways of seeing the world? How were spaces opened up in which ideas about plurality, diversity, multiplicity, difference, disruption, and contradiction challenged the singular emancipatory vision of Western science, that central notion of development and progress that sustained both the liberal/scientific and the Marxist/socialist discourses? The answer lies partly in the political movements developing from the late 1960s onwards as well as in theoretical shifts in the academy. Within geography, a group of committed and radical academics were active not only in class struggles but also in feminist, green, and ecosocialist struggles. (The founding of the journal *Antipode* in 1969 gave these radical geographers a disciplinary standpoint.) Their combination of academic and political work raised questions about the assumptions behind the notions of social justice and progress that informed both liberal-humanist notions of individual equality and socialist versions of the future. A sustained attack was launched from several quarters on the supposedly universal "we" of both these discourses. A choir of voices – women's, gay men, colonial subjects, prisoners, and other "deviants" – was raised to protest that the supposed universal "we" of humanist and socialist discourses in fact excluded their specific experiences, based as they were on a particular view of the world: a view through the eyes of the Western, male, bourgeois subject.

Theoretical developments in the critical social sciences coincided with these political movements to necessitate the re-evaluation of the privileged status of existing theoretical frameworks, with their commitment to notions of a single universal truth and a unitary, universal human subject (Barrett 1991). Instead of a search for Truth and universal laws, we have seen a number of geographers, especially those working in the social, historical, and cultural areas of the discipline, turning toward the theorization, and in some cases the celebration, of difference and diversity, in opposition to the totalizing tendencies in what have become known as the "grand narratives" of liberal humanism and Marxism. Instead of searching for general laws these geographers insist on the particularity and plurality of knowledge, on analyses of the specificity of its social construction by certain social groups located, in particular, in space–time frameworks. While debates about the situatedness of knowledge have permeated the social sciences and humanities more generally, for geographers they strike a particular chord as they coincide with our long-held interest in difference, diversity, and place-related specificity.

The key players in the displacement of old certainties have, appropriately, been many and various and their arguments overlap. However, I shall give a flavor of their claims under three headings: feminism, poststructuralism and postcolonialism, indicating the implications for geographical work, although many of the key theoretical texts are outside our discipline.

Feminism

Despite the insistent claims of a parade of grand (and mainly European) male theorists – Michel Foucault, Jacques Derrida, Jacques Lacan, Jean-François Lyotard, Richard Rorty et al. – some of the earliest attempts to "deconstruct" the humanist subject were made by feminist theorists, "as long ago as 1792 by Mary Wollstonecraft in her demand for women to be included in the entitlements claimed by the 'Rights of Man'" (Soper 1990, pp. 229–30). As innumerable feminists have demonstrated since, Western humanism succeeded in writing women out of the theoretical agenda by constructing the category "Woman" in opposition to "Man," in such a way that women were defined as the "Other," as an absence or a lack, compared with the "One," that is an idealized rational, disembodied subject or individual, that, in fact, embodied masculine (and Western) attributes. Through the construction of a set of binaries that equated masculinity with science, rationality, objectivity, and the public and political arenas of life, mapping femininity onto opposite characteristics – emotion, irrationality, subjectivity, the private and domestic sphere – women were relegated to an arena that apparently needed no theoretical investigation but might be taken for granted as "natural." By naming femininity in opposition to the category "Man," which was assumed to be human, woman was situated in a very different category from that of man (Pateman and Gross 1986). Feminism has convincingly demonstrated how "women's silence and exclusion from struggles over representation have been the condition of possibility for humanist thought: the position of woman has indeed been that of an internal exclusion within Western culture" (Martin 1988, p. 13).

The early achievements of feminist scholarship involved establishing women's right to inclusion, based on an appeal to liberal notions of social justice. As women, too, were individuals, to deny them the rights and benefits available to men was patently unjust. By the mid-twentieth century, however, it became clear to feminists that a claim for equality within humanist thought was impossible. As the humanist subject was masculine, women would always be excluded. So, it is argued, humanism as such was inherently flawed and must be replaced by an alternative set of knowledge claims (although based on what claims and in what form is contested and uncertain). It is here, in the *deconstruction (the challenging and taking apart) of knowledge claims*, that recent feminist, poststructuralist, and postmodern theories unite to pose a common challenge to the theoretical basis of conceptions of knowledge, truth, and science: a challenge that is resulting in the "displacement of the problematics of science and ideology, in favour of an analysis of the fundamental *implications* [original emphasis] of power-knowledge and their historical transformations" (Morris 1988, p. 33). By this, Morris means that we must look at whose interests theories support, how they vary across space and how they change over time. It is this emphasis on change that perhaps distinguishes poststructuralism from Marxism but there is also a significant difference in the way in which power relations are theorized, as I shall demonstrate below.

Poststructuralism

The term power/knowledge is a reference to the work of Foucault, who, although resistant to the labeling of his work, is generally regarded as a poststructuralist. What is important here is his rejection of the notion of ideology, and a reconceptualization of power relations as complex and unstable, part of the social construction of self rather than an external system "imposed" on people from above. Let us take these claims in turn.

Foucault rejects the concept of ideology on several grounds: first, that it necessitates an opposite notion, that of truth; secondly, that it stands in a secondary position to the material or economic determination of society; and thirdly, that it rests on the humanist notion of an individual subject. Poststructuralists have developed instead the concept of "discourse," a concept that is dependent upon Foucault's redefinition of power. Rather than seeing power as a social and political system in which a repressed group is dominated for the benefit of the oppressors, Foucault (1970, 1977, 1980) argued that language, texts, discourse, as well as social practices, are part of the maintenance of particular power relations in modern societies. He provided what he called a different "grid" for deciphering history – a new way of looking at and analyzing power which focused on the ways in which discourses, and the pleasures and powers they produce, have been deployed in the service of creating and maintaining hierarchical relations in Western societies (Foucault 1978). Thus power relations are maintained not through the false consciousness of the repressed, who will eventually act as a class for themselves, but rather as part of the very construction of identity. Thus an individual subject is constructed through a grid of discourse and practice. In poststructuralist theory there is no notion of a prior or essential moral self, as in liberal and Marxist theory, and so traditional ideas of "truth" and "self" are rejected. As Foucault argued, "power seeps into the very grain of individuals, reaches right into their bodies, permeates their gestures, their posture, what they say, how they learn to live and work with other people" (quoted in Sheridan 1980, p. 217). Our bodies, identity, and sense of self have no prior existence but are brought into play through language, discursive strategies, and representational practices, and so conventional distinctions between thought and action, between language and practice, are collapsed as all social actions have a discursive aspect.

While all discourses are changeable, open to contestation because they are social practices, Foucault termed the establishment of an interlocking set of successful or dominant discursive formations a *regime of truth*. The term truth, however, is used here in a different sense from conventional notions. It is impossible to distinguish between what is true or false by appealing to the "facts" of the matter, as the "facts" themselves are part of a particular discursive formation. The "truth" of a situation is determined by the outcome of struggles between competing discourse.

Postcolonialism

It is within recent work under this heading, linked to Third World feminists' critiques of the ethnocentric biases of Western feminism, that perhaps the most powerful arguments about contested and incompatible knowledges, as well as multiple subject positions, have been advanced. Within Western theory, "women of color" have criticized the racially exclusionary theory and practices of white, middle-class feminism, problematizing assumed commonalities as women. At the international level, the theoretical focus is on imperialism rather than racism, and here the most influential work has been that of Said in his arguments about the construction of a colonial "Other" by the West, and the work of the Subaltern Studies group, who are rewriting the history of colonial India from the point of view of peasant insurgents. In both cases, despite his profound Eurocentrism, Foucault is a key inspiration, as he has been for a number of feminists who argue that women's exclusion from struggles over representation within Western culture are another demonstration of Foucault's concept of power (Martin 1988).

Said's *Orientalism* (1978) has probably been the most influential text in the development of new ways of thinking about Western imperialism. Drawing on Foucault's notion of discourse, Said argued that "without examining Orientalism as a discourse one cannot possibly understand the enormously systematic discipline by which European culture was able to manage – and even produce – the Orient politically, sociologically, militarily, ideologically, scientifically, and imaginatively during the post-Enlightenment period" (p. 3). Through a range of discursive strategies and social practice, people of the Orient were constructed as the "Other" to the Western "One," in ways such that "European culture gained in strength and identity by setting itself off against the Orient as a sort of surrogate and even underground self" (Said 1978, p. 3).

Whereas Said focuses on the discursive construction of the Orient in Western texts, the Subaltern Studies group is interested in the impact of colonialism on native cultures. Spivak (1988) argues that the imposition of colonial rule in India effectively ruptured and silenced indigenous cultures, a fracturing that she terms, after Foucault, "epistemic violence." The colonial subject is unable to answer back in a language that is untouched by imperial contact, and which may only be glimpsed in the interstices of colonial texts. The truly marginalized, the "subaltern" of the group's title and Spivak's most well-known paper, is unable to speak at all.

These arguments have interesting implications for how geographers undertaking ethnographic and field-based work in postcolonial societies theorize and analyze the narratives collected from "native informants." The implications of Spivak's arguments are that no one falls outside the field of imperialist discourse. Whether engaged in peasant insurgency or postcolonial critique, the native is forced to speak the language of imperialism. However, rather than resulting in capitulation, Bhabha (1990) has argued that what he terms "hybridization" is the result, a sort of double displacement whereby native mimicry of imperialist discourses embodies cultural and political resistance, disrupting authoritative representations of imperialist power. Thus rather than either a nostalgic search for lost origins or a desire to re-establish authentic traditions, the postcolonial critic analyzes the ways in which resistance lies in hybridization. In the context of the West, where the huge migration flows related to slavery, imperialism, and global capitalism have resulted in peoples who are of neither First nor Third World, Bhabha's notion of hybridization also seems relevant. Indeed, the sociologists Gilroy (1993) and Hall (1996) have developed interesting parallel arguments about cultural translation. Gilroy, for example, in his work on African-American peoples, argues that a hybrid Black Atlantic culture has developed. Similarly, "women of color" in the US are exploring the position of women who are the product of international and cultural borders – women of mixed origins, whose identity is constructed out of differences (Anzaldua 1987; Sandoval 1991).

In combination, the work summarized under the above three headings has altered the subject of geographic research, refocused theoretical and methodological assumptions, and challenged the political practice of "radical" geographers. Many have realized that, in the words of Landry and Maclean (1993, pp. 125–6), "the white male heterosexual worker of the First World industrial nations just won't do any longer as the alternative or insurgent subject of history, the locus of resistance to ruling class dominance." Drawing in particular on qualitative and ethnographic methods, with increasing attention to everyday social relations and to discursive strategies, a number of geographers have turned to detailed analyses of the "other" – women in different positions (see my reviews – McDowell 1992a and b), female travelers (Domash 1991), native peoples and ethnic minorities (K. Anderson 1992), the mad (Philo 1989), and the perverse

(Valentine 1993), and of the transgressive landscapes of desire of "alternative" sexualities (Bell 1994; Bell, Binney, Cream, and Valentine 1994). There has also been a "discursive turn" in which the methods of discursive and textual analysis have become influential in the interpretation of landscapes of power and oppression, as well as social relations in localities. As Peet and Watts (1993a) have suggested, "one of the great merits of the turn to discourse, broadly understood, . . . is the demands it makes for nuanced, richly textured empirical work" (p. 248): a type of work admirably illustrated in Watts' book with Pred (Pred and Watts 1993).

Recently, attention also has turned to the discursive strategies of geographers, to textual analyses of our own writing. Geographic language is recognized as being not simply a medium expressing reality "out there" but as playing an active role in creating social and cultural worlds. Interpretative schemes "in part . . . create the reality that they seek to interpret" (Barnes 1992, p. 118). Theories "have the imaginative capacity to represent and reconfigure the world like *this* rather than like *that*" (original emphasis: Gregory 1994, p. 182). In new histories of geography, theoretical perspectives are seen not as alternatives in a struggle for hegemonic status but as alternative, and incompatible, interpretative communities. The purpose of theoretical endeavor in our discipline is thus disrupted. Rather than a struggle to establish universalizing claims, the grand and certain aim of grand theory, we (not this time united by theoretical consensus, but a motley collection of geographers with different views of the world) have in common what Gregory terms a "strategy of supplementarity – *dis*-closing what theory closes off" so that "the certainties of theory can be capsized, its confidence interrupted, and its conditional nature reasserted" (Gregory 1994, pp. 181–2). The singular geographical imagination so eloquently defined by David Harvey in 1990 has become the partial and plural geographical imaginations of Derek Gregory's (1994) latest survey.

Problems *for* Theory

Claims that geographies are multiple, that knowledge is contested and that space is discontinuous and fluid, a discursively constructed set of relationships that discipline, control, or privilege (Lefebvre 1991), a "plane of contest" (Ashley 1987) rather than a set of fixed regions defined by Cartesian coordinates, have led to an exciting set of debates in the "new" geography but also an unease. Fears of relativism, enunciated not only by those critical of claims that knowledge is embedded and situated but also by many who generally welcome the challenge to Enlightenment notions, are perhaps the most serious, although it is important to note Barrett's claim that poststructuralist critiques of truth and knowledge are "not so much relativist as highly politicised" (Barrett 1991, p. 161). Theoretical acceptance of plural and fragmented identities, of multiple and local sites of power, has been enormously empowering for those "others" condemned as handmaidens, helpmeets, "perverse," or slaves of Western "civilisation," whether marginalized by the universalist pretensions of humanism, or left at home/outside by Marxists while the "real" struggles take place elsewhere. But questions remain. How, for example, might we judge between competing knowledge claims? On what may we base our political opposition to the oppression and exclusions that are revealed in these new discourses? Are all claims to knowledge and power equally valid?

Many, geographers included (Driver 1992; Gregory 1994), have pointed to a paradox at the heart of the three theoretical positions discussed above: their critique of the exclusion of "others" surely depends for its critical force on acceptance of modern ideals of autonomy, human rights, and dignity. Thus, as Soper (1990) asks, "why concern ourselves with the exclusions from the 'humanist' universal of women, or blacks or gay people, or the insane, or any other oppressed or marginal group, except on the conventional ethical grounds that all human beings are equally entitled to dignity and respect?" (Soper 1990, p. 149).

It seems to me that here we must establish ways of criticizing universalistic claims without completely surrendering to particularism. And surely this is a more general example of the problem that geographers, in their focus on the particularity of place, have been grappling with for many years. Geographers have long accepted a form of relativist argument: that the specificities of local socioeconomic structures, differences in cultural attitudes and in ways of living, are part of the explanation for uneven development (McDowell and Massey 1984; Massey 1984). Drawing on this work, in which universal and particular explanations are neither counterposed nor seen as alternatives, we need to *recast our general notion of relativism* and no longer see it as counterposed to universalistic explanations. Rather, we must see relativism, in the sense of difference, as inescapable. Acceptance that there are different ways of knowing which are historically and geographically specific does not mean that we cannot make judgments about their value. Foucault drew our attention to the ways in which discourses, or speaking positions, are constituted by power relations. Thus oppositional knowledges of the subjugated are inherently and inescapably political, demanding judgment about the validity of different discourses on the basis of political considerations. We might, therefore, accept the multiple claims to power of some groups and reject those of others, in attempts to construct a multiple and inclusive notion of social justice that includes the claims of the non-white, non-masculine subject (Flax 1992; J. E. Young 1992). But as Barrett points out, this second move, one that she terms "transformation," is much more difficult than the first, the transgressive deconstruction of dominant binary oppositions where feminism, poststructuralism and postcolonialism have been so successful. Once the categories of "woman," or "postcolonial subject," have been dismantled, it becomes more difficult to speak on their behalf. There is a tension between challenging the oppositional construction of others and continuing to make political demands on their behalf. A possible way forward is suggested by "black" theorist Du Bois (1989, and see Gilroy 1993) in his theorization of "double consciousness" that simultaneously allows for the deconstruction of the category black and for the need to assert the truth and authenticity of the black identity. Biddy Martin (1988) similarly has suggested a "doubled strategy" for feminist struggles.

Conclusions: Principled Positions

The theoretical and moral implications of a rejection of both universalism and relativism as conventionally defined are complex. It is clear that the exercise of moral responsibility is culturally dependent and temporally specific. But the fact that political choices and everyday decisions are made in relation to a contestable code or set of conventions as to what is "good" or "bad" does not in itself detract from the element of judgment and the assertion of values in our choice. As Haraway (1991), a profoundly thoughtful theorist, has suggested, we need both to accept irreducible differences and yet hold on to what she terms a "successor science project" in which "projects of finite freedom, adequate material abundance, modest meaning in suffering and limited happiness" (p. 187) are the focus of action. As Gregory suggests in his recent book: "If we are to free ourselves from universalising our own parochialisms, we need to learn how to reach beyond particularities, to speak to larger questions without diminishing the significance of the places and the people to which they are accountable" (Gregory 1994, p. 205).

Theorizing diversity does not mean abandoning these larger questions and projects. The challenge now for geographers is to find ways to respond to them.

Note

I avoid the use of the term postmodern in the rest of this chapter, preferring instead to use poststructural to refer to theoretical claims, and reserving postmodern for social and cultural shifts. However, like Rosenau (1992), I recognize that "the terms overlap considerably and are sometimes considered synonymous. Few efforts have been made to distinguish between the two, probably because the difference appears of little consequence" (p. 3).

PART V

Geoenvironmental Change

Introduction to Part V: A Burden too Far?

Very few human societies beyond the simplest have lived in harmony with their natural surroundings, so it has been rare for the environment not to be depleted as a consequence of human occupation. Some societies, such as those which practiced slash-and-burn agriculture, left the environment to recover from their depredations, but even with them the prior state was rarely regained.

More complex societies have taken more from the environment than they have returned to it, and have stimulated very substantial change in their physical milieux (Simmons 1993b) – so much so that in some cases the land has never regained more than a small fraction of its former fertility (as in much of the Mediterranean littoral) whereas in many others its carrying capacity has been severely, and almost certainly permanently, depleted (Blaikie and Brookfield 1987). As a consequence, some societies have collapsed because satisfaction of their material wants was not sustainable. Others have maintained themselves by colonizing new areas, enabling them to replace those products which their home environments had formerly yielded: this colonization frequently involved marginalizing, if not destroying, pre-existing societies.

In many areas, of course, farmers learned how to restore the fertility of at least some lands – by the application of manure, for example – and, through trial and error, they developed methods which reduced the rate of depletion. This rate has been further slowed, though not arrested, in many parts of the earth during the last five centuries or so by improvements in agricultural technology. The increased use of machinery in the past century has assisted this, as with contour ploughing; and the more recent rapid development of the chemical and biotechnology industries has aided the goal of increasing yields from the land, though in many cases with unforeseen consequences whose later realization has created new environmental crises.

Changing Relationships with Nature

Despite this increased dominance of society over nature, and the rapid domestication of great tracts of the earth's habitats, resource depletion has increased apace (Turner et al. 1990). By the late twentieth century many observers believe that, although much remains unknown about the physical, chemical and biological processes involved, an environmental tragedy is looming. At some time in the relatively near future, it is argued, the earth will be unable to meet the demands made upon it and the transformations wrought by human use of its renewable and non-renewable resources will become irreversible. Others claim that this doomsday scenario is overplayed, that we have managed to resolve such problems in the past and there is no reason to believe that human ingenuity will not cope again.

Resolution of this debate may come too late, especially if the pessimists (who may well argue that "it is better to be a pessimist and be proved wrong than to be an optimist and be wrong") are correct. If the doomsday scenario is wrong – at least for the present – then we will lose little, except the time and human resources which might be better expended elsewhere, in preparing to avoid it. Furthermore, our increased appreciation of the environmental constraints to action will assist the development of strategies to ensure sustainable development in the future (Adams 1990). Alternatively, if it is right, then if we don't seek to prevent its realization, our future is bleak indeed.

The pessimists' cause is supported by observation of current trends within society which are increasing the nature and pace with which we are ravaging the earth. First, pressure on the earth's resources is growing because of increasing population. Secondly, demands on those resources are increasing more rapidly than is the population because of greater per capita material expectations: average living standards are increasing (despite their decline in some regions), producing a variant on Malthus's predictions of nearly two centuries ago regarding the exponential growth of demand (Woods 1989). Thirdly, very few parts of the earth's surface remain relatively untouched by these demands – the colonial "escape valves" of previous centuries have been removed – and conversion of increasing tracts to non-productive uses, through urbanization processes, reduces the amount of land available for exploitation. Finally, technological advances have been such in recent decades that not only has our ability to ravage the earth been magnified manyfold but in addition we are ravaging it in new ways, through technologies that enter the core biological and chemical processes that maintain life forms.

Changing those Changing Relationships

Assuming some general validity to the pessimists' arguments, what can be done to end the destruction of the resource base which sustains human life? Several major problems can be isolated.

First, there is a need to win acceptance that a major catastrophe could be on, or just beyond, the horizon, which calls for rapid immediate action if it is to be averted. This involves a major educational task, which is made difficult by the absence of conclusive evidence and exacerbated by the lack of detailed knowledge of how sustainable development could be ensured given the existing pressures on the environment, let alone those likely to come in the next few decades as a consequence of growing demands for increased material standards.

This first group of problems is associated with a second, which is largely political in its structure. Much of the "rape of the earth" to date has been undertaken either by or for the populations of "developed world" countries. With a changing world political order, the governments representing the peoples whose lands and livelihoods have been exploited in these unequal relationships argue that they should not pay the price of resource depletion, for which they have not been responsible and from which their populations have benefited very little. Suggestions that population pressure is a major cause of the oncoming problems, and that as a consequence high priority should be given to introducing and promoting birth control policies, often stimulate the response from "Third World" governments that for them to do so is to accept the basis of the current inequality within the world in an intrinsically unjust economic system (Harvey 1974).

A third group of problems is linked to the common treatment of natural resources as private property. Much resource depletion and most environmental pollution comes about because of individual (including corporate) actions, each of which in itself is a very small, marginal contribution to the growing problems. Those actions take place within a mode of production whose political and other leaders increasingly promote the private ownership of all means of production, including nature: within capitalism, economic survival demands that resource exploiters continually increase their pressure on nature in order to sustain their competitive position in world markets (Johnston 1989b).

How can this deleterious pressure be reduced? Education and persuasion are important, but unlikely to have major permanent impacts, because of the apparently conflicting goals of "saving the earth" (for posterity) and sustaining one's standard of living in the face of increased prices (now). Some individuals and groups may react positively to calls for changes in their behavior (especially in the more affluent countries), as might some relatively altruistic business leaders, but the dynamo of capitalism is against them (Pepper 1993): as a whole, if not as individuals, we are pressed to act against our long-term interests in order to maintain our short-term positions.

The necessary conclusion is that to ensure sustainable development in the future others must require individuals and firms to respond to the imperative to conserve rather than destroy: those others can only be governments, for no alternative institution than the state has the sovereign power with which to insist on actions to protect the environment (Johnston 1989a). Three main government strategies are available: public regulation of activities, to ensure that pro-environment strategies are implemented; taxation of polluters, thereby making it more sensible and efficient for individuals to change their behavior; and state subsidies to assist change to a "greener" set of actions. All are being used. Some policies – such as the reduction of air pollution from coal-burning in British cities since the 1960s (Brimblecombe 1987) – have been very successful, to a considerable extent because people perceive the benefits and are content to comply, including meeting their share of the costs, when they know that everybody else is doing so too.

Despite such successes, however, much environmental use remains either poorly controlled or entirely unregulated, and contributes to continued environmental degradation. In part this is because, in the pursuit of environmental policy goals, governments are constrained by the pressures of the capitalist dynamo: if they make the costs of production more expensive within their territories, compared with other states where there is less environmental control, then they will potentially be making their local industries uncompetitive, so harming the employment chances and living standards of their populations relative to the residents of more "liberal" states. Furthermore, given the global interconnections within environmental systems, free-riding is not possible: if some states fail to implement environmental protection, and cannot be persuaded to do so, then eventually the land, water and air of all states will be despoiled (Johnston 1989b).

Just as individuals within a state may be prepared to comply with environmental legislation when they know that everybody else is, so individual states may be prepared to act if all other states are. But there is no super-state which operates above individual states and to which they are prepared to cede sovereign power – with the partial, strongly contested, exception of the European Union. Thus international action will only come about if there is international agreement, and in many areas of environmental concern there is little evidence – despite a great deal of rhetoric – that such agreement is being reached and enacted. At the Rio Earth Summit in 1992, for example, the US President declined to sign the Biodiversity Convention because it "threatened American jobs": his successor did sign, but Bush's action, like others by a number of states (as with the Law of the Sea and the Ozone Treaty), illustrates governments' unwillingness to sacrifice local economic advantage – especially in the developed countries which have most to lose (Johnston 1992).

In sum, we have an impasse. Many believe that environmental tragedies are rapidly approaching because we have so abused the earth that it cannot continue to support human life, at least at its present magnitude and at the high material standards enjoyed by a small proportion of the total (and coveted by many more). Others hope that this is not so: they await scientific evidence that the stories of the earth's demise are much exaggerated, and press for continued scientific innovation which will reassert human hegemony over nature. Whether the optimists or the pessimists are right may be settled early in the twenty-first century, unless much more concerted action than is currently taking place is taken very soon to boost the optimists' cause.

18

The Earth Transformed: Trends, Trajectories, and Patterns

William B. Meyer and B. L. Turner II

Introduction

The twentieth century, especially its second half, has been one of unprecedentedly rapid change in the earth's environment. Human impacts on the physical form and functioning of the earth have reached levels that are global in character, and they have done so with escalating speed. Human-transformed landscapes on the earth's surface now rival in extent the area of natural or lightly altered biomes; strictly speaking, indeed, there are no "natural" biomes that are not at least lightly altered. Human-induced changes, by and large, have escalated in recent decades and are likely to continue escalating through the twenty-first century.

If the degree of change is novel, so is the degree of concern – scientific and public – that it now raises. Throughout recorded history human societies have recognized their power to alter the earth (Glacken 1967). Never before, though, has environmental change been a matter of more interest than it has become in the past few years, and never before has the interest been so tinged with pessimism. Humankind has become what many have long wished it to be. It is now a transforming force equal in its power and effects to natural forces. Yet that accomplishment is viewed with regret at least as much as with satisfaction. Humankind was expected to become more and more a force rationally and deliberately improving the environment while preserving and cultivating what is valuable in it. It appears more and more to have become, instead, a powerful agent that acts blindly and haphazardly; on a gloomy view it represents a novel force of nature wreaking havoc with the biosphere and hardly more conscious than the weather or plate tectonics of what it is doing. Its ordinary activities have profound and rapid, yet often unintended, unforeseen, and little-understood effects.

Views of Environmental Change

A topical world geography compiled for students at the beginning of the twentieth century would probably have devoted no separate chapter to the human alteration of the earth's surface: not because there existed little notion that the earth was being altered greatly, but because it seemed not to be a major problem nor likely to become one. At the most, such a volume would have dealt in passing with the need to conserve and use wisely the earth's finite mineral and energy stocks. Forecasts at the end of the nineteenth century of what was in store for the twentieth by and large predicted further improvement of the environment by its human users: more deserts and near-deserts to be irrigated, more coal seams opened for use, more useless swamps and forests converted to productive cropland, the oceans' resources harvested ever more efficiently.

Most turn-of-the-century seers also forecast a rosier social future than has thus far arrived. Yet the change in environmental perceptions has been an even graver blow to faith in progress. The bettering of the environment had long been that faith's last resort, the trend to which it could appeal no matter how slowly or how little the behavior of human beings toward one another seemed to improve. The erosion of environmental confidence has been a fairly sudden one. Even at mid-century, a major scholarly assessment of *Man's Role in Changing the Face of the Earth* (Thomas 1956) documented in detail the human alteration of the globe but "displayed a remarkably low level of concern about human impacts" (Lowenthal 1990, p. 124); it did not convey "any feeling of urgency" (O'Riordan 1988, p. 25). Now, the historian Paul Kennedy (1993) devotes a full chapter of his book *Preparing for the Twenty-First Century* to the harmful human impact on the globe. Even within a generally pessimistic volume the tone of the chapter is not hopeful. The problems are grave, Kennedy concludes, and the decisions and reorganizations needed to address them more than superficially are unlikely to be taken. His is not an isolated view among those writing on the subject and may not even be a minority one. The belief that the earth has been seriously damaged, and is today being damaged more severely and rapidly than ever, is at least a far more prevalent and respectable belief than it was only a few years, let alone a few decades, ago. Increasingly the assumption that the earth is being improved requires a defense and an explanation, while the assumption that it is being dangerously degraded requires none. Coping with global environmental change has come to appear one of humankind's most pressing problems.

What is Global Change?

Not all environmental change is global, though how much is depends on how that word is defined. In the strictest sense, no change is fully global, inasmuch as none occurs uniformly across the earth. In the most expansive sense, all change is global, inasmuch as all changes are ultimately connected with one another through physical and social processes alike. Some intermediate definition is required if "global" is to remain a useful distinction.

One such definition is based on the physical character of the change. Human impacts on the environment can acquire a global character in one of two ways. In the first, "global refers to the spatial scale or functioning of a system" (Turner et al. 1990, p. 15). Because of the planet-scale fluidity and connectedness of the system affected, changes in it have the potential to have effects around the world – albeit far from uniform ones. There are two systems that can operate in this way: the atmosphere and the oceans. Global warming induced by greenhouse gas emissions, sea-level rise as a result of such a warming, and the depletion of stratospheric ozone by chlorofluorocarbon releases are the classic examples of such *globally systemic* change. It is likewise through the global systems of the atmosphere and oceans that some trace-pollutant releases spread worldwide from their places of origin:

fallout from atomic tests, DDT and its residues from farmland, lead from automobile exhaust.

A second kind of environmental change can also be considered global: "if it occurs on a worldwide scale, or represents a significant fraction of the total environmental phenomenon or global resource" (Turner et al. 1990, pp. 15–16). Losses of forest, biodiversity, soil fertility, and wetlands widely repeated around the world subtract significant fractions of the net worldwide stocks of the resources affected. Some of these *globally cumulative* changes – deforestation, for example, which releases carbon dioxide – have direct connections to systemic changes as well. So, linked or not, they in themselves may pose threats to the resources on which the habitability of an ever more populous and more affluent globe will depend.

The systemic impacts of human activity – notably, greenhouse climate change and ozone depletion – were central to the emergence of the current scientific and popular interest in global change. It is, however, the cumulative global changes that are better documented and, to date, more significant for human activities.

Documenting Global Change

Documentation of either kind of change, however, is no simple task. Some aspects of the environment fluctuate substantially and unpredictably even in the absence of significant human involvement. Hence the effects of such involvement are difficult to identify. Climate at the global and lower levels is one example. It remains open to debate to what extent global temperature changes in this century are natural fluctuations or the result of greenhouse gas emissions (Karl 1993). Marine fish populations (Hilborn 1990) experience significant variations that are not necessarily human-induced. The parts played by meteorological shifts and human land use in degrading the North African Sahel zone remain matters of scientific dispute (Tucker et al. 1991).

Much is asserted about the degree of past, present, and future change, whether or not much is known. The world has never been so awash in statistics and assertions about the environment as today. Supply has responded to an explosion in demand; government agencies, scholars, journalists, and many others dealing with environmental matters require facts, preferably numbers, to illustrate their arguments and perhaps even to guide them. Yet these users, by and large, are ill placed to ask how accurate, how representative, and how current those facts are. Because environmental change is very much a multidisciplinary and also an extra-academic affair of research, it loses one of the principal advantages possessed by established academic disciplines: that of quality control, imperfect though even that may be (Carpenter 1989). Bad numbers once in print are extremely hard to confine or root out, corrections hard to make stick.

Uncertainties about cumulative kinds of changes in the states or faces of the earth arise from the poor reliability of even precise and officially guaranteed figures. In China, as Smil (1992, pp. 434–6) observes, post-revolutionary accounts of "mass reforestation programs . . . appeared to convey a success story," duly recorded in annual United Nations FAO (Food and Agriculture Organization) land-use statistics as a net increase in China's forest area. "Realities, gradually disclosed since the 1970s, are quite different." Most new plantings did not survive, and existing forest has continued to decline both nationwide and regionally: "by 1988, the forest cover in Sichuan, China's most populous province and . . . one of the country's principal timber bases, was down to 12.6% from 19% in the early 1950s." One of the most widely discussed arenas of contemporary environmental change, that of tropical deforestation, "remains an area in which increasingly sound methodology is applied to very unsound data" (Brookfield 1992, p. 4). As for systemic change, evidence to test the theoretical prognostications of greenhouse warming, for example, will not be available until well into the twenty-first century. Climate systems are complex, involving many feedbacks and experiencing many other changes at the same time as they are being affected by greenhouse gas releases. Thus it cannot be certain that the observed and well-documented increases in greenhouse gases have yet led or will lead in the near future to a significant rise in the earth's average temperature.

Some reasonably reliable generalizations can nonetheless be offered about the extent and trends of human-induced environmental

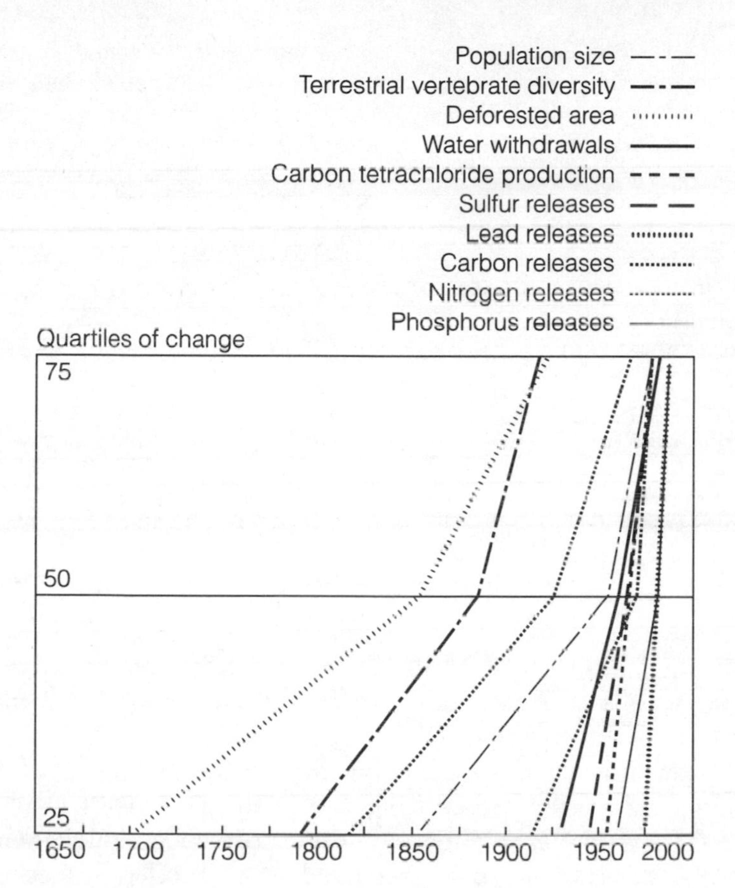

Figure 18.1 Trends in selected forms of human-induced transformation of environmental components.

change, in the global aggregate, drawing on a recent inventory covering the past three centuries (Turner et al. 1990). Indices of change compiled in that effort cover a representative array of major forms of environmental impact (figure 18.1; table 18.1). Taken together they offer a solid empirical base for several assertions. First, human activities are comparable to or greater than natural forces as drivers of many kinds of change. Secondly, most of them have only recently become so. Yet, thirdly, acceleration and intensification of human impact, though frequently and perhaps usually the case, has not always been so. Finally, human impacts have steadily expanded in variety and character, from

Table 18.1 The Great Transformation: selected forms of human-induced transformation of environmental components – chronologies of change.

Quartiles of change from 10,000 BC to mid-1980s			
		Dates of quartiles	
Form of transformation	*25%*	*50%*	*75%*
Deforested area	1700	1850	1915
Terrestrial vertebrate diversity	1790	1880	1910
Water withdrawals	1925	1955	1975
Population size	1850	1950	1970
Carbon releases	1815	1920	1960
Sulfur releases	1940	1960	1970
Phosphorus releases	1955	1975	1980
Nitrogen releases	1970	1975	1980
Lead releases	1920	1950	1965
Carbon tetrachloride production	1950	1960	1970

involving mainly some of the landscape resources of the earth – forests, soils, water, biota – to affecting the material and energy flows of the biosphere.

Human-induced environmental change is, of course, of long standing. Alterations amounting to transformations of local and regional environments go back to prehistory. Occasional environmental disasters may have occurred in premodern times. Sustained and knowledgeable use of the land was probably the norm, however, and clear cases of purely environmentally-induced civilization collapse are rare or non-existent. To say so is not to endorse the "green legend" (Whitmore and Turner 1992), the myth of the environmentally nurturing premodern or precapitalist society that is so prevalent in the literature today. Indeed, one of the authors has documented the kinds of environmental changes inflicted by such societies in the western hemisphere (Turner and Butzer 1992; see also Denevan 1992). It is, rather, to contest another myth: that societies that have profoundly altered their environments in the pursuit of wealth and power have been punished by environmental catastrophe. Many, in fact, have transformed the lands that they occupied, and thrived. Claims of environment-related societal collapses in the distant past are contested and cannot be resolved by the state of the evidence at this time (e.g. Turner 1991).

The modern era of widespread concurrent change in the earth's land cover – species transfer, the conversion of forests, grasslands, and wetlands to cropland along expanding frontiers of change – may be thought of as having been inaugurated by the Columbian Encounter in 1492 (Turner and Butzer 1992). Substantial change in the earth's biogeochemical flows is a product of the Industrial Revolution, beginning in Western Europe in the late eighteenth century.

Set in motion by these processes, the major human-induced changes to date have begun to rival or surpass in their magnitude and rate the operation of natural forces in the biosphere. World forest area has been reduced by some 20 percent and an area of land the size of South America converted from its original vegetation cover to cropping. Global fossil-fuel and industrial mobilization of sulfur, a major source of acid precipitation and other forms of pollution, now approximately equal the natural flow of the element. The important atmospheric trace gases carbon dioxide and methane, both contributors to the greenhouse effect, have been increased by about 25 percent and 100 percent respectively over their preindustrial levels by human actions. Emissions of some trace pollutants released by human activities – a number of metals, for example – now greatly exceed natural flows, while releases of synthetic organic chemicals introduce novel substances that do not exist in nature.

The changes vary both in the lengths of time over which they have occurred and in their current trajectories (figure 18.2). Long-standing impacts on the earth's land cover are represented by a decline in forest area and by terrestrial biodiversity; half of the total impact on each was registered as early as the nineteenth century. Not of such long standing, as a global problem, is human withdrawal of fresh water from the hydrologic cycle, though in some localities it was substantial, as a result of irrigation, millennia ago. It represents a rapidly accelerating process, per capita withdrawals having grown fourfold since 1950. Human releases of key elements of the biosphere – carbon, sulfur, nitrogen, and phosphorus – are all, except for carbon mobilization through land-cover change, relatively recent as global phenomena. Sulfur

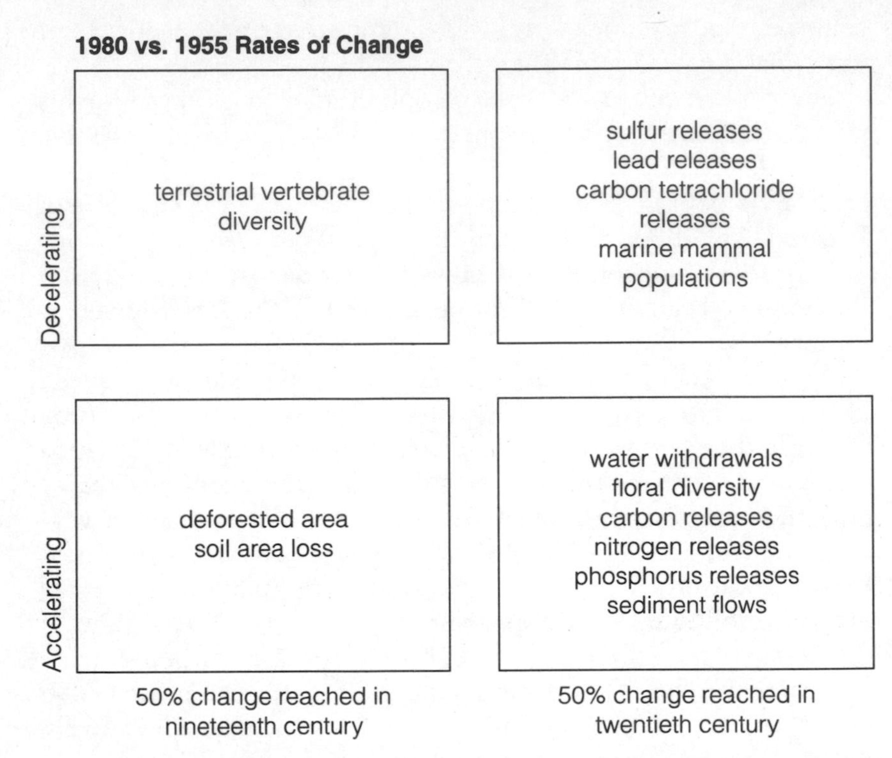

Figure 18.2 Recency and rate of change in human-induced transformation of environmental components.

releases, at least, appear to have been leveling off by the mid-1980s compared with the previous decades, in consequence of fossil-fuel price increases and pollution regulations. Trace pollutant releases – for example, those of metals and synthetic organic chemicals – are quite recent in origin as significant phenomena. Both lead and CCl_4 releases, like those of sulfur, have shown some tendency to decelerate during the second half of the twentieth century, again in part as a result of the regulation to which they have been subjected in some countries; other trace pollutant emissions, though, and perhaps most, have continued to accelerate.

All of these changes are significant not only in themselves but for the further physical changes that they can drive: the alteration of the composition of the atmosphere, for example, may affect the climate, and climate change in turn reacts on forest area and composition. It is not only the magnitude of contemporary change, but its variety, that presents formidable challenges to societies' abilities to adapt to the consequences.

The Meaning of Global Change

Environmental change as defined in natural-science terms is, from the human point of view, neither good nor bad in itself. It is, as Zimmermann (1951) described the nature of "natural" resources, merely "neutral physical stuff." Environmental change is neither environmental improvement nor environmental degradation per se. The characteristics of the society interacting with physical phenomena are what make those phenomena either resources or hazards. This does not mean that the physical characteristics of environmental change can safely be ignored or given only superficial attention, for these characteristics matter a great deal. A close acquaintance with them is necessary for an assessment of the meaning of change, but it is not sufficient.

For most policy-makers, and probably for most of the public, the "bottom-line" concerns over global change are less those about its alteration of nature per se than about its economic and social consequences. How will the settings for economic activities be affected by substantial human alterations of the biosphere? How much would it cost to prevent those alterations? Deciding how desirable or undesirable an environmental change is is no simple task. Interpreting the consequences of greenhouse-gas-induced global warming for yields of current cropping systems requires combining a large variety of impacts: atmospheric CO_2 enrichment, changes in temperature, changes in water availability, changes in extreme weather events, changes in soil, and changes in pests and diseases (Rosenzweig and Hillel 1993). All are themselves uncertain, and results based on them must also be combined with speculations on future land uses, technologies, markets, and resource availability to produce meaningful projections or forecasts of the important human consequences.

In most environmental changes, even ones that overall are more harmful than not, there are individual winners as well as losers. Drought is a frequent form of natural climate fluctuation that would probably be increased in some areas by global warming. It often benefits farmers living in other areas, or farmers in the areas affected who have access to irrigation water, because reduced overall production raises agricultural prices. Intervening to prevent change would also produce winners and losers. Environmentalists have proposed some courses of stringent carbon emissions regulations that economic analyses suggest would cost more for society overall, and much more for certain sectors of society, than would adapting to the changes as they occur or pursuing more moderate courses of regulation (e.g. Manne and Richels 1992; Nordhaus 1993). Environmental-hazard studies suggest that the poor and unempowered are the most vulnerable to these impacts because they have fewer options for adapting. Yet it is not clear that the poor and unempowered would not also disproportionately bear the costs of action taken to prevent these impacts.

But selecting the "best" (by whatever criteria) responses to global environmental change is complicated not only by these distributive issues but also by the unprecedented scale, rates, and uncertainties of the physical changes in question, be they systemic or cumulative in kind. The cost of guessing wrong may be catastrophic. The size of human-induced change may be so great as to outstrip the ability of individuals and societies to adapt adequately to it. The whole course of environmental history, and particularly its modern course, suggests that "surprises" – unknown and unpredicted environmental impacts – will continue to occur. As human-induced change increases in scale and rate it seems probable that they will occur more frequently. The CFC–ozone relationship is a good example, one in which a new chemical compound thought to be environmentally benign rapidly accumulated in the atmosphere and began to erode the shield of ozone that protects the earth's surface from ultraviolet radiation. We are ignorant of such potential threshold levels and non-linear relationships because the conditions that produce them lie outside our experience. Even economists who doubt that stringent and immediate measures to curb carbon emissions would, on the basis of current knowledge, be cost-effective, acknowledge that "the potential for catastrophic surprises" raises questions that cannot be conclusively answered by current models or by purely

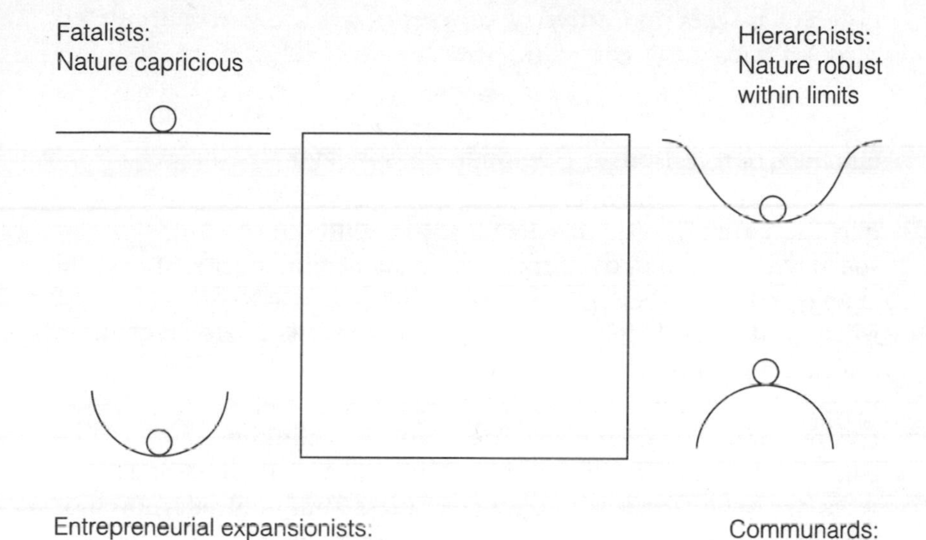

Figure 18.3 Risk and blame.

economic reasoning. Such efforts, as Nordhaus (1993, pp. 23, 24) writes, can inform but cannot dictate the political decision of "how to balance future perils against present costs."

Weighing heavily in that decision will be different assumptions about the nature of nature and about how it will react to human impacts: most broadly, the four "nature myths" discussed by Douglas (1992) (figure 18.3). Some see natural systems as resilient, little affected even by powerful shocks; some see them as resilient within limits but vulnerable when those limits are exceeded; some see them as inherently fragile and likely to react sharply even to mild pressures; and some see them as capricious and largely unpredictable. The first will see little need for restraint in exploiting and altering the environment, but the second and third will see a need for caution to avoid disaster; the fourth may display a fatalism that sees concern and regulation as futile. The four views, Douglas suggests, reflect different social bases, the view of nature as resilient, for example, reflecting a relatively unconstrained and entrepreneurial culture; that of nature as extremely fragile, an egalitarian and sectarian one. In any case, the same scientific data about what is occurring now will suggest different futures and different courses of action when interpreted through such different world views.

Global Change from a Regional Perspective

It is increasingly recognized that understanding the physical and social patterns and dynamics involved in global environmental change requires regional as well as global assessments (Turner, Moss, and Skole 1993). The trends and patterns of global changes are aggregations of regional and local ones that display great variation. Releases of carbon dioxide, for example, are strongly concentrated in the industrialized world, including China, while rural areas with intensive wet-rice production, such as Southeast Asia, are major sources of methane, a less abundant but more potent greenhouse gas. The global pattern of increased deforestation and cropland expansion is not evident in Western Europe and the United States, areas in which much former agricultural and other cleared land has reverted to tree cover and where the land may be a net absorber of carbon from the atmosphere.

The human causes of changes in the biosphere also vary widely between regions and localities. At the global aggregate level, most changes correlate well with the variables in the "PAT" formula (Ehrlich and Holdren 1971), which ascribes human impact on the environment to the product of population, per capita resource use, and the technologies by which the resources are used. At the regional level, studies demonstrate the importance of other factors – institutions, policy and political structure, trade relations, beliefs and attitudes (Meyer and Turner 1992). At sub-global scales these forces often overshadow the PAT drivers of change.

The human consequences of global environmental change also tend to differ throughout the world. A doubling of atmospheric carbon dioxide may have a major fertilization effect on some crops but little or none on some others in which the chemical process of photosynthesis follows a different pathway (Rosenzweig and Hillel 1993). Likewise, global warming, should it occur as forecast by current general circulation models (GCMs), would extend commercial crop production further northward in parts of Russia, Japan, and Canada, while increasing heat and water stress and substantially lowering yields on croplands nearer the equator. Putting these and other factors together, the food supply and economic impacts appear to be modest globally but severe for the tropical world (Rosenzweig and Parry 1994). That is the sector of the globe, moreover, where countries, organizations, and individuals are most constrained in their abilities to respond. Successful adaptations may require investment – in farming systems, in infrastructure, and in policy innovation – that can least be afforded by those lands that may most require it to deal with the challenges of global climatic change. It is also the developing (mainly low-latitude) world where agriculture and other climate-sensitive activities remain major elements of overall national economies, and where cumulative changes degrading land resources are most severe in any case.

Comparative case studies at finer scales can further help disentangle what is and is not shared between the global and subglobal scales and between different regions. These studies may range from comparative assessments of single facets of change (e.g. trajectories of deforestation) to more holistic assessments. An example of the latter is the Project on Critical Environmental Zones (Kasperson, Kasperson and Turner 1993). It examined nine regions of the world where pressures of change and physical and socioeconomic vulnerabilities appear to have come together to produce environmental crises of unusual severity, jeopardizing continued use and occupance at existing or projected levels of population and standards of living.

Only in one of these regions was the wealth or well-being of the region's population clearly declining as the result of adverse environmental changes. That region is the Aral Sea bordering Kazakhstan and Uzbekistan in former Soviet Central Asia. Water withdrawal from the tributary rivers for irrigated cotton production steadily reduced and continues to reduce the Sea's water level and surface area (Micklin 1988), creating saline conditions that have destroyed the once lucrative fishing industry and rendering shipping facilities useless as the shoreline recedes (Kotlyakov 1991). The climate has become drier and harsher. Large dust storms, carrying salts from the exposed sea bed, have become more frequent and more damaging to agriculture. Water contamination by salts, fertilizers, and pesticides has seriously affected human health.

In each of the other regions, environmental change, though rapid and substantial, has not yet reduced the overall wealth and well-being of the population, severely though it may have affected some subgroups. In many of them, it appears likely to do so in the near future, given the range of feasible responses. If only the Aral basin fitted the project's definition of environmental criticality; the Basin of Mexico, the Sundaland rain forest of Indonesia and Malaysia, the Ukambani region of southeastern Kenya, and the Ordos Plateau of northern China were classed in whole or part as environmentally endangered regions: ones in which trajectories of environmental change put the sustainability of current human uses and standards of living in question in the short- to medium-term future. Amazonia, the Llano Estacado, and the North Sea basin also incurred major environmental changes. Yet, overall, either the rates of environmental change were declining, the changes themselves, though substantial, did not seriously undermine the livelihoods or the health of the regional populations, or the wealth of the region provided a series of feasible options for its sustained use.

These nine regions are only a sample of an increasing list of areas where human use has so drastically altered the environment that alarm has been raised about their long-term occupance and about the cumulative impacts at the global scale of a large number of regional environmental catastrophes. Such alarm cannot and should not be taken lightly; many people in many regions will suffer because of the environmental changes wrought now and their inadequate access to the means to alleviate the impacts of, or adapt to, these changes. Nevertheless, if confident predictions of the conquest of nature have not been borne out in the twentieth century, the literature of environmental history abounds in prophecies of doom that have also failed to come true.

The lessons from such comparative case studies are several. There exist large regions where environmental degradation is seriously jeopardizing the sustained livelihood and/or health of their inhabitants, where its rate is outstripping the ability to adapt, and where the medium-term costs either of permitting it to go on or of responding are likely to be enormous. Management response has typically not been impressive, even where the problems were clearly foreseen (as in the case of the Aral Sea, for example); a do-nothing response was often preferred, though it greatly increased the eventual costs. The fact that global environmental problems require negotiations and agreements among many nations would only complicate matters at the higher level and make response more difficult. Yet there are also areas where adaptation, often hard to foresee or predict, has deflected what would once have seemed an inexorable trajectory toward criticality.

Conclusions

Though the patterns, sources, and impacts of environmental change are not uniform across the globe, humankind has so altered the physical conditions of the earth that we must recognize overall an "earth transformed" and an earth certain to be transformed further. These changes will find different expression and have different consequences in different regions, but they will be significant almost everywhere. The twenty-first century is likely to witness an earth in which all lands are formally managed; few, if any, frontiers will remain. The demands for resources will increase, even should they change in character owing to technological changes, and they will mean increasing pressures on the earth to fulfill them. The question for humankind is whether it will be able to satisfy these demands and adapt to the foreseeable and unforeseeable consequences without destroying the only home it has.

19

The Earth as Input: Resources

Jody Emel and Gavin Bridge

> The plot of *Black Indies* (by Jules Verne) arises from one miner's refusal to believe that the Aberfoyle pit is exhausted. Early in the book, young Harry Ford exclaims to James Starr, "It's a pity that all the globe was not made of coal; then there would have been enough to last millions of years!" The older, wiser engineer responds that nature showed more forethought than human beings by forming the earth mainly of stones that cannot be burned. Otherwise, Starr comments, "the earth would have passed to the last bit into the furnaces of engines, machines, steamers, gas factories; certainly, that would have been the end of our world one fine day!" (Williams 1990, p. 197)

Introduction

One distinguishing characteristic of the Fordist industrial order was the shift from coal and steam to newer technologies based on electricity, oil, and petroleum. Intriguingly, the material basis for Fordism has continued to be the material basis for post-Fordism. Despite the cultural, political, and economic upheaval of these "New Times," mining, damming, timber cutting, and fishing go on much as before. Changes in ideas about this massive appropriation of nature have occurred and new ethical dilemmas have emerged in the globalization of the economy and the fragmentation of production/consumption spaces. Nevertheless, the character of materials-development "crises" appears to have changed only superficially in the transition from previous modes of regulation.

The failure to alter substantially the dependence on non-

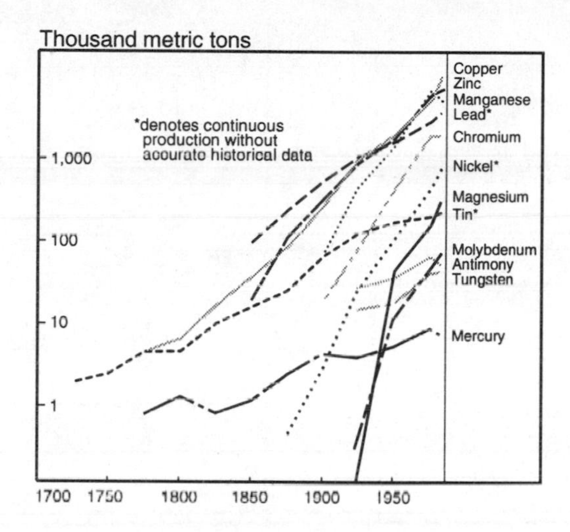

Figure 19.1 Annual worldwide production of selected metals from 1700 to 1983.

renewable resources of current modes of production constitutes a crisis that has been continuously postponed, primarily through the spatial expansion of production capital into hitherto undeveloped areas. This expansion, coupled with the minerals production cycle, creates enduring environmental and social crises in extractive localities and regions. Although the relative importance of the extractive (or primary) sector has decreased over time for many countries, the volumes extracted and the number of locations involved in extraction have increased. The long-term trend in materials extraction is one of increasing output (figure 19.1 – from Ayers 1992). Figure 19.2 illustrates the more recent history of selected major commodities. Global water withdrawals and annual world catch of fish are also on the rise (United Nations Environment Program 1991; World Resources Institute 1991).

Gains have been made in the efficiency of use and recycling of many exhaustible resources. In 1990, for example, recycled material provided nearly 75 percent of the lead and over 50 percent of the iron and steel consumed in the US; 30 percent of global aluminum demand was met with recycled material. Substitutes such as plastics, ceramics, and other composites are competing with metals in cars, airplanes, building construction,

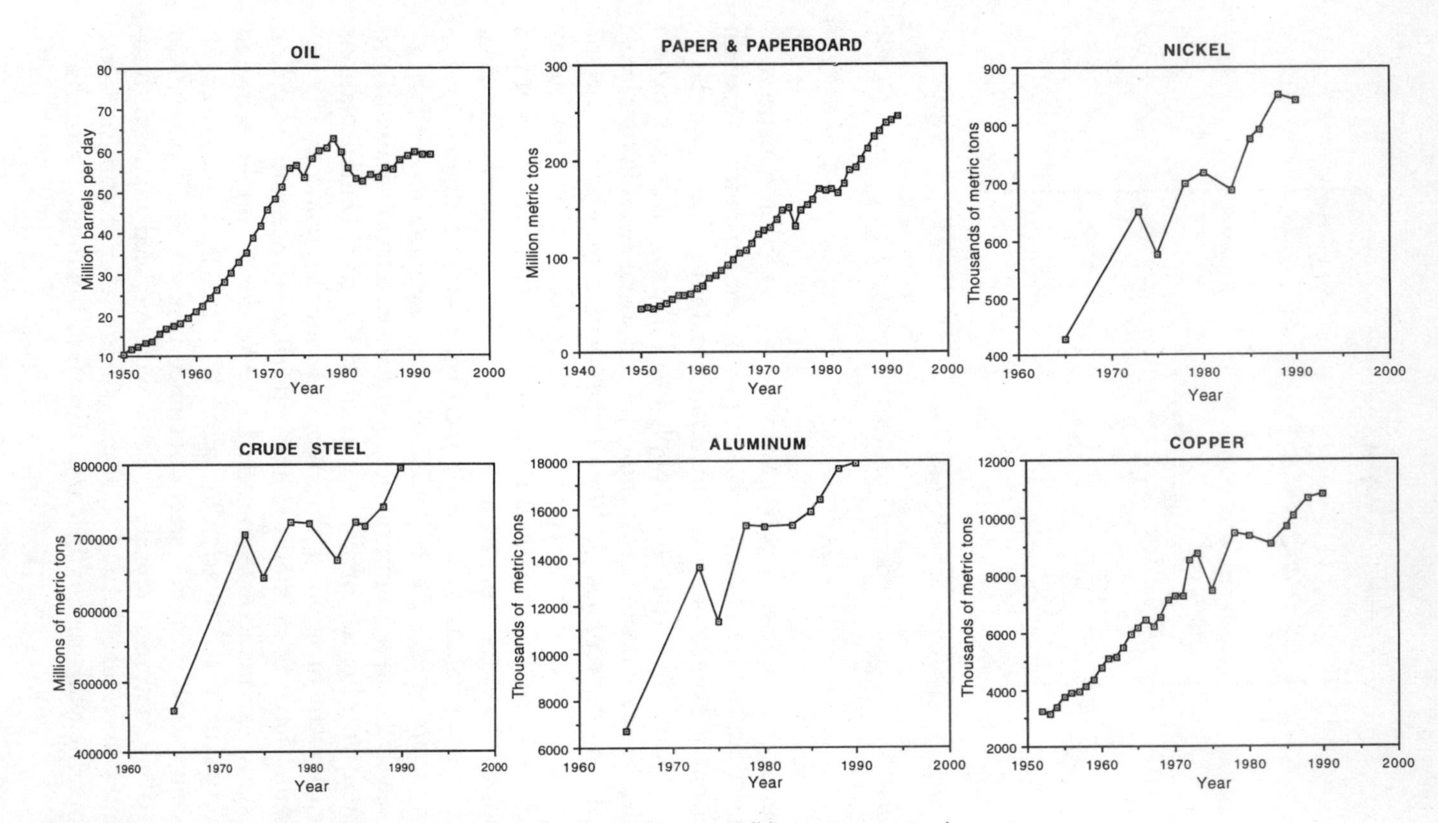

Figure 19.2 Annual worldwide production of selected commodities: recent trends.

communications, and pipelines. Economic restructuring, from heavy industry to service and high-technology sectors, has also reduced industrial requirements for metals and other materials. Since the energy crises of the 1970s, OECD countries have halved their energy consumption per unit of output within the chemicals industry (OECD 1993): raw-material requirements for a given unit of output have fallen an average of 1.25 percent per annum since the early part of the century; yet, despite all of these innovations and changes, overall energy consumption increased by 20 to 30 percent over the following two decades.

In the economies of the "South," minerals use is growing. A principal feature of global economic restructuring has been the relocation to the South of many of the materials-intensive heavy industries such as iron and steel, textiles, industrial chemicals, and petrochemicals. From 1977 to 1987, the Third World's share of aluminum and copper use grew from 10 to 18 percent, and that of zinc from 16 to 24 percent (Young 1992). The rise in total energy and oil consumption in the newly industrializing countries and elsewhere in the Third World could be twice their 1990 levels by 2010 owing to industrialization, transportation needs, urbanization, and population growth (OECD 1993).

Intractability and Biophysical Limits

We have a global resource supply crisis for two reasons: the intractability of our exhaustible mineral economy, and the existence of biophysical limits to minerals development. Intractability derives from a number of factors which reflect not only the existence of fixed capital in infrastructure and social organization, but also prevailing norms and values formed over hundreds of years.

Intractability can be understood from the standpoint of three interest groups involved in the production process: consumers, producers, and the state. Producers have vested interests in the status quo. Those enterprises and regions presently involved in resource production receive income from current patterns of resource use. Regardless of the ecological degradation and the social costs to communities, economic reliance on single-commodity production promotes intractability. Around one-quarter of all the developing countries generate at least 40 percent of their export earnings from the minerals sector (including both hydrocarbon producers and the hard-mineral exporters). Even in some diversified industrialized countries (like the US, Canada, and Australia), natural resources continue to be important sources of manufacturing inputs and revenues. The US share of industrial output has risen among OECD countries, largely within sectors associated with natural resources. Textiles, for which US irrigated cotton production provides an advantage, and wood products, dependent upon logging US forests, are significant in this regard (Dollar and Wolff 1993). The problems of reducing these resource-based revenues in dependent economies are obvious.

For many consumers, current standards of living are dependent on continued high levels of resource use, and consumption is on the rise globally. Since mid-century, world per capita consumption of cars and cement has quadrupled, plastic use per person has quintupled and air travel per person has multiplied 33 times (Durning 1992, p. 29). In Chinese cities, for example, two-thirds of the households have washing machines, one-fifth have refrigerators, and half have televisions (*China Daily*, 15 February 1988). The most wealthy of global consumers (about 20 percent of the total population) are accustomed to cars and airplanes, processed food and meats, appliances, climate control, and enormous quantities of non-durable goods. These consumption trends are not expected to decrease in the near future.

Economic development requires access to cheap, plentiful energy in a highly intensive (rather than dissipative) form (namely the fossil fuels). High rates of materials usage are necessary for infrastructure development – roads, bridges, water supplies, houses, and factories. The Green Revolution, for example, depended upon high input rates of fertilizers, pesticides, and water. Alternatives are not without impact either. Hydropower is clean at the site of electricity usage but ecologically and socially devastating at the production site – witness the James Bay Project in Quebec and its effect on the Cree and Inuit peoples. James Bay, at the southern limit of the fragile Arctic vegetation zone, is the home of some 5,000 Cree, Inuit and Innu people. Phase I of the James Bay Project, built between 1971 and 1985, directly affected an area of 175,295 square kilometers. The second phase will cut one river's flow by 75 percent, while increasing another's 10 times. This project will destroy some of the most productive Cree and Inuit hunting areas as well as portions of the central flyway for many species of North American migratory birds. Phase I caused mercury contamination of fish and people, the drowning of 10,000 caribou during a flash flood, and the loss of homes and hunting lands. Similar problems of equity and ecological change attend other energy alternatives.

Any government which curbs resource availability in such a way as to threaten current patterns of consumption risks removal from office. In fact, the "state" has generally tried to stimulate production and consumption. Disposal of public lands, price ceilings on gas and oil, massive water development projects, and forestry and minerals development on public lands, are examples. Newer efforts at regulating environmental degradation, worker health and safety, and gasoline taxes, are examples of contradictory policies that may tend to reduce selected resource development and use. Whether they are successful in mediating the resource crisis depends upon such factors as spatial displacement of the problem, substitution impacts, technological and institutional innovations, and resistance to regulation (one example is the resistance of both consumers and producers of coal to carbon taxes).

The second component of the resource crisis is the biophysical limit. Cleveland et al. (1984) argue that technical improvements in the extractive sectors have made available previously uneconomic deposits only at the expense of more energy-intensive forms of capital and labor inputs. Resource development has historically taken place on the most highly concentrated, and most easily accessible mineral reserves: the more highly concentrated a mineral, the less energy is required to extract, refine, and process it into a unit of usable product. Over time the average grade of mined minerals declines; the result is increased energy costs and increased disruption to progressively larger areas. Thus the increase in volume indicated by figures 19.1 and 19.2 has taken place against the backdrop of progressively declining grade. Copper, for example, has long been one of the most important industrial minerals on account of its excellent properties as a conductor of heat and electricity. The global output in 1991 was 9,100,000 tons, 570 times that of 1800. Four hundred years ago the average grade of copper mined was 8 percent; now the world average is 1 percent (Young 1992). This drop in grade has directly increased the amount of environmental disturbance, since more ore must be processed to extract the same amount of copper.

Relocating the Crisis – Out of Sight, Out of Mind

The primary means of resolving or postponing the global resource crisis is the spatial expansion of extractive industry into increasingly marginal areas. Raw-material extraction has historically been a feature of marginal or frontier regions, and has in many cases defined the nature of the frontier economy (Innis 1967; Bunker 1985). In the late twentieth century, the margin for extractive production takes two forms: the last remaining continental frontiers such as Alaska, Amazonia, or Siberia; and the internal frontiers within core regions, where poor political representation and lack of economic opportunity lend weight to arguments of internal colonialism.

A distinguishing feature of recent resource extraction has been the development of materials at progressively higher latitudes. There has been a marked shift northward in petroleum and gas production sites – especially in the US, Canada, and the former Soviet Union. Development in these cold regions means substantially higher initial capital and thermodynamic costs. In the 1980s, exploration costs off Labrador and Greenland, impeded by ice floes, were ten times higher than costs in the North Sea and twenty times more than those in Saudi Arabia (Chisholm 1982). Getting resources out of these remote areas also requires substantial capital and energy expenditures (e.g. the trans-Alaska pipeline).

Alaska, located astride the Arctic Circle and on the northwestern edge of the North American continent, is an exemplar of this trend toward development of the North. The boom that transformed Alaska's economy came from the Prudhoe Bay oil discovery in 1968. Fully developed during the 1970s, Prudhoe Bay now supplies as much as 85 percent of the state's budget, and 25 percent of domestic demand in the United States (United States Bureau of Land Management 1991). To the east of the Prudhoe Bay field is the protected Arctic National Wildlife Refuge (ANWR). Established in 1960 to maintain viable populations of large mammals, notably the internationally significant porcupine caribou herd (Trustees for Alaska 1993), its 19.5 million acres of tundra have been off-limits to extraction despite the fact that the refuge may contain similar oil wealth to that at Prudhoe Bay. The ANWR represents one of the last remaining large tracts of land in the world that has experienced very little use by societies.

These northlands may be less ecologically resilient than temperate or tropical climes. The environment of the arctic tundra shows low rates of growth, there is limited species diversity in comparison with other biomes, and a high degree of annual fluctuation in species number has been noted (Dunbar 1974). Development threatens such environments through its effects on the reflectivity of the surface, and through pollution of water and air. The high reflectivity of ice, snow, and bare rock limit the absorption of heat energy at the surface, preserving deep layers of permafrosted soil and matrix. Removal of vegetation, wholesale stripping of surface material, spillage of oil, and particle deposition all have the potential to increase surface absorption and thus locally melt the permafrost. Practical solutions to this problem have involved the construction of all facilities on gravel insulating pads. This in turn has placed large demands on local gravel supplies, often located in river beds. Thus, mineral extraction has greatly modified northern ecosystems and their ability to maintain wild populations of waterfowl and large mammals.

The spatial drive to the ecological and political peripheries of the world is bringing some of the last remaining wilderness areas into the global economy. Despite the mineral sector's poor performance as the initiator of sustainable growth and community formation, its provision of infrastructure has opened up frontier areas for development, planned or otherwise. Within these areas are many or most of the world's 250 million or more "indigenous" or "first" peoples. Many of the so-called indigenous cultures residing in these "ends of the earth," do not identify with the nation state or the mainstream culture of the political boundaries within which they reside. Many – like the Cree in Quebec, the

Penan and the Martu in Malaysia – have protested against resource development on their lands (see, for example, Burger 1990). Their protests are supported by various international alliances (i.e. Cultural Survival, World Council of Indigenous Peoples, National Indian Brotherhood, Assembly of First Nations) and, to some extent, by the United Nations. In similar fashion, people with aesthetic and spiritual values regarding "nature" have developed social movements to protect remaining wilderness areas and remedy environmental degradation. But at the same time as these new voices are heard and supported by the media, the state, and other institutions within many nations of the globe, the process of production and consumption separation continues to expand and deepen across the globe. This separation makes it more difficult politically, economically, and ethically for any one group to make substantial changes in the existing regime of global production. The social and ecological disruption from massive resource development may be unknown to consumers because the chains of commodification are too long and complex. The rural–urban split has long afforded some separation of production and consumption; but now there exist vast spatial separations that involve many nations at once (e.g. the "global factory"). Ironically, at a time when many people are socially attuned to supporting people and environments threatened with exploitation from resource consumption, they are (or feel) more powerless to alter the system. At a time when we are most supportive of the new rhetoric of "sustainable development," we are complicit in the destruction of the only remaining sustainable production systems on the planet.

The Local and Regional Crisis

The trading and utilization of raw materials takes place at the global scale. Regardless of the extent to which commodities have become global, they can, however, only emerge out of locally-based extraction systems (Bunker 1985). Often remote and sparsely settled, these localities are adjuncts to the urban-based global economy. While the benefits of resource extraction (in the form of the utility of the derived material) are diffused across the globe, many of the negative social and ecological impacts of extraction fall heavily on the local community.

Mineral extraction has its primary effects at the local scale, disrupting soils, vegetation, and drainage patterns, producing significant levels of water and air pollution, reducing landscape values, and potentially visiting economic and social losses on the workforce and host community (Rees 1990, p. 51). The Global 2000 Report estimated that in the final quarter of the twentieth century over 60 million acres would be disturbed by the mining industry (US Council on Environmental Quality 1982). Disturbance of the ground cover mobilizes minerals, which enter water courses through run-off, mine drainage, or seepage from tailings ponds. Mechanical and hydraulic techniques are particularly destructive, involving the large-scale removal of overburden. Hydraulic gold mining in Brazil, practiced not by highly capitalized corporations but by thousands of small-scale producers, has resulted in serious sedimentation and health problems in downstream communities. Young (1992) estimates that pollution from smelters worldwide has created several biological wastelands: one of the most infamous is the 10,400 hectare area surrounding the Sudbury nickel smelter in Ontario.

In addition to the ecological problems, local communities or regions are subject to two broad sets of social and economic problems. First, hinterland or peripheral resource-rich communities have little political or economic autonomy. This tendency to become an enclave is due to the intersection of a number of factors (Auty 1993). Backward, forward, and final demand linkages, important to regional development, are typically weak. Minerals production is characterized by a high capital to labor ratio, and so employs relatively few people but requires large amounts of capital. Capital is typically provided by external sources, through state investment in state-owned enterprises, state investment in infrastructure provision to encourage private investment, or wholly private investment. Only modest production linkages exist between the extractive enterprise and the locality, as few upstream supply or downstream processing facilities are established. As a result, a large fraction of the earnings from the exaction flows from the region to service external capital investment.

In many circumstances the fiscal benefits of hosting an extractive enterprise are curtailed by a lack of ownership rights to the mineral resource. In some cases these rights are undefined or ignored, or the host community may not own the mineral rights, which historically have been retained by the state. The lack of rights to minerals can be a serious obstacle to receiving benefits from extraction; in such cases royalties or rentals are denied, and the principal mechanism for retaining some of the value of mineral extraction is taxation. This problem can be exacerbated by an area's political marginality. Describing Japan's investment in copper extraction in Papua New Guinea, Iran, Peru, Indonesia, and Zaire after the depletion of domestic reserves, Bunker (1991) explains that big mines located at a great distance from markets tend to be in areas with sparse populations that have minimal integration into capitalist political, economic, and legal systems, and "limited access to the technical information required for effective rent bargains or for environmental or social regulations."

The Western Siberia region of Russia is a good example of a dependent economy. The contemporary position of Siberia in relation to the Russian and world economies continues its historical trend as a resource periphery, in which its plentiful resources are extracted and exported from the region. High-quality coal resources have been developed in the Kuzbass basin since the 1930s, and the region's resources played an important role in Stalin's rapid industrialization of the country. Since the 1970s, the massive oil and gas resources of the northern Tyumen region have been developed. The Kuzbass coal field illustrates the extent to which production is embedded in the existing social formations of the region. The estimated 230 billion tons of coal reserves in the Kuzbass have been the foundation for its specialization in the mining, metallurgy, and chemical industries. By the end of the 1980s the region was supplying over one-third of Russia's total coal output, and over two-thirds of its output of high quality coking coal (Institute for Economics and Industrial Engineering 1993). The region's economic output and employment are dominated by mining, processing, and mineral-based production industries; not only does coal mining provide jobs for the community, but, in many cases, social services such as housing, medical care, and public amenities have been provided through the mines. Many of the region's 3 million people are dependent on the continued production of coal for the preservation of their daily lives. Economic restructuring, however, has exposed the non-profitability of up to 80 percent of the mines (Institute for Economics and Industrial Engineering 1993). Specializing in mining from the beginning, the region has neither diversified to provide alternative employment nor trained its workforce in widely usable skills. Mindful of the social security needs of the Kuzbass population and the energy requirements of Russia as a whole, a de facto decision to support continued production has been made.

Boom and Bust

The mineral cycle is of acute significance to the host community because "boom and bust" translate into social dislocation and depression for many extractive areas. The initial extraction phase is associated with an influx of population, which can be both large and sudden, placing a strain on existing social facilities. The demographic characteristics of the immigrant population (typically young adult males, or young families) can make for significant personal and social problems, as both immigrants and host community struggle to adjust. Once the resources are gone, or are no longer economically tenable, the population experiences unemployment, a decline in living standards, and the need for out-migration. These problems have been well studied (Gilmore 1975; Stangeland 1984; Scydlitz et al. 1993).

For a locality which has not extensively been involved in economic activity and that has had a shallow tax base, the influx of wealth from an extractive development (whether it be a significant proportion of the total revenues or not) can be destabilizing (Auty 1993). The wholesale dependence of many resource-producers on a single commodity renders them vulnerable to world market vagaries. In producer-states (such as the oil economies of OPEC), resource revenues accrue directly to state treasuries, giving the state responsibility for the realization of resource rents. Downward adjustments to the flow of resource revenues stress the state's ability to meet its domestic and international commitments, and can force the imposition of austerity measures on the population (Watts 1984). The macroeconomic impacts of large mineral wealth have led some analysts to see a rich resource endowment as "curse" rather than benefit (Gelb 1985, 1988; Auty 1993). The impacts at the sub-national level caused by the sudden but unsustainable influx of wealth into an extractive community can be equally ambiguous. For example, prior to the production of oil from the Prudhoe Bay field, the native Inupiat who inhabit the region had no formal local government and hence no local budget. The establishment in 1972 of a municipal government enabled the taxation of all activities on the North Slope. By 1987 the Inupiat had access to funds from the taxation of an estimated $13.6 billion of oil-related property (Strohmeyer 1993).

The problems of booms are not unique to mineral regions, but are particularly acute there for three key reasons (Auty 1993). First, mineral prices are highly volatile owing to the short- and medium-term rigidity of mineral markets, caused by the difficulty of adjusting investments with large fixed costs to changes in demand. Secondly, the financial windfall from development accrues to the government (typically the national rather than the local state). And thirdly, the small workforce which receives wages can develop into a labor aristocracy, with spending power beyond the means of the general community.

The Timber Boom in Indonesia

Indonesia has played host to a number of resource booms in the years following the political turbulence of the mid-1960s. The macro-economic impacts of a timber boom from 1968 to 1972 were replicated by two oil booms, one in 1973–4 and the other in 1979 (Gillis 1987). Each of these resource-extraction episodes was characterized by enclave formation and budgetary problems at the provincial and local level, caused by a rapid increase in resource receipts and limited linkages with the pre-existing economy (Daroesman 1979).

While a potentially renewable resource, the forests of East Kalimantan have, like many other tropical forests, been "mined" by intensive exploitation of timber products. The overlaying of a highly capitalized, vertically integrated forest industry on the pre-existing smaller-scale system, in which native producers sold to merchants, has been disruptive to the forest community and its environment. The ecological effects of such extensive deforestation at the local, regional, and global scale have been discussed at length elsewhere (Williams, M. 1990). The social impacts, both as a result of ecological degradation and directly as a function of capital and labor flows associated with large-scale, rapid commercial forest-resource extraction in East Kalimantan are similar to those of other non-renewable resources. Low levels of job creation; mechanization and the requirements for higher-skilled labor; light taxation of profits and exports; the deliberate undervaluation of export receipts; and limited domestic processing, have reduced the stream of potential benefits from forest-resource development (Gillis 1987, 1988).

The timber boom in Indonesia illustrates the two principal challenges facing extractive-resource localities: the wresting of a "fair share" of economic rent by interest groups at central, provincial, and local levels; and the application at each level of windfall wealth to satisfy both short- and long-term development needs. Yet the Indonesian case also illustrates that more local levels of control do not automatically improve stewardship of the resource. Up until 1971, provincial governments retained full autonomy over the granting of concession areas up to 10,000 acres, and 70 percent of the license fee and royalty receipts flowed to the provincial government. Such generous revenue provision for the provinces spurred the rapid and indiscriminate granting of concessions for deforestation.

Conclusions

We have outlined a rather dim view of prospects for the "earth as input." This perspective is based on forecasts of increasing natural-resource production and consumption, despite large improvements in efficiency of use and materials recycling. Technological innovation, substitution, and attempts at "closing the materials cycle" through secondary sourcing have been offset by increasing demand. Trends in the industrialized nations toward "dematerialization," have not been sufficiently large to counteract increases in material consumption in both the developed and developing countries. Increased extraction of lower-grade ores and the opening up of new areas for extraction have delayed global resource depletion, but in so doing have exacerbated the impacts at the sites of extraction, where local and regional effects on the ecology and social structure can be acute. The global spread of extractive industry, and its location in regions that are now considered to be of ecological significance at the global scale, constitute this as a crisis which transcends the local scale. Contrary to the suggestion that we are now undergoing a major transition in forms of economic, political, and social life, patterns of resource extraction continue to exhibit much continuity with the past.

Yet political and ethical changes have taken place which allow the experience of alternative voices, valuation of the environment, and patterns of development. The ecological and public-health impacts of resource extraction have gained in visibility. Legislation requiring environmental impact statements and emissions regulation, plus whistle-blowing by labor unions, environmental groups, and the popular press bear testimony to heightened concern for the environmental ramifications of resource extraction and use. Such a trend is welcome: not only does it give voice to the unspoken, but it reframes resource extraction as a societal problem. Illustrative of this trend is the emergent field of industrial ecology, which has begun to discuss resource use within the context of global environmental change. This approach begins with an appreciation of current environmental impacts derivative of resource extraction, and moves to examine the fixity of resource use within industrial societies, seeking ways to reduce material flows. In sum, the intractability and biophysical limits that constitute the resource crisis must be addressed, but the magnitude of the problem should not paralyze or polarize discussions of alternative futures; rather, the crisis can be the foundation for a creative discussion of future options for those producers, consumers, and localities engaged in resource development and use.

The Earth as Output: Pollution

Malcolm Newson

Introduction: The Geography of Pollution – Revealing Problems and Solutions?

Pollution is an ideal topic for geographers, and not only because it is now realized how wide-ranging are its causes, consequences, and "cures" (involving thought and action by academic communities in science, social science, and the humanities). The plentiful map images to which we are daily exposed in news media reveal inherently geographical patterns and processes of pollution, such as the "Chernobyl Cloud," the "Exxon Slick," and "Death on the Rhine." Indeed, these images are so strong, but fade so quickly, that the "issue-attention cycle" (Downs 1972) is often accused by environmentalists of diverting attention toward the pollution *incident* and away from the *continuous* shedding of the wastes from our (otherwise) sophisticated society into environmental media whose capacity we know little about. Deeper analysis of media coverage of environmental issues (e.g. Hansen 1993) reveals much about the cultural and political aspects of the current crisis; geographers are able to balance these aspects with the physical-science dimensions of pollution.

It is the first purpose of this chapter to transcend the issue-attention cycle by briefly reviewing the contemporary space–time signature of global pollution, going beyond the simple conclusion that "It's all geographical," to match the patterns and processes of pollution to some inchoate theoretical structures by which to judge its seriousness. Secondly, it is essential to ask whether the

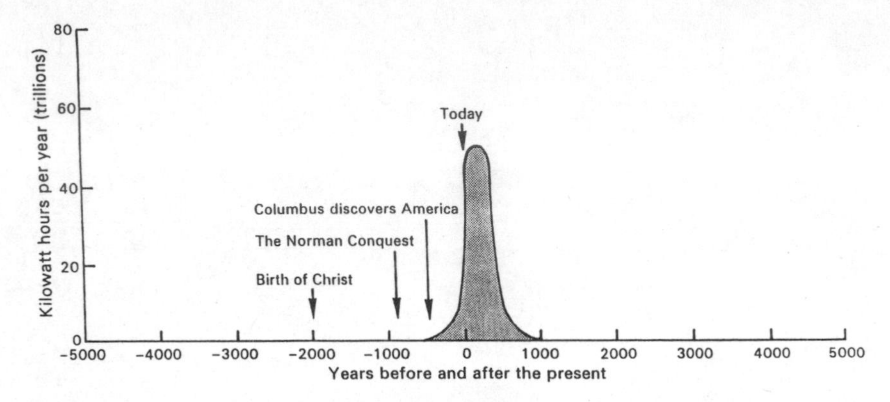

Figure 20.1 Historical and expected fossil-fuel use.

sociopolitical forces highlighted by a space–time framework point to future styles and outcomes of environmental decision-making on the control of pollution problems (at a cascade of scales from global to individual). Scale is a persuasive (but not a pervasive) concept in environmental policy-making. Haber (1993) has stated that "The scale problem is also one of the main reasons for the failures or slow progress of international environmental politics. . . . Successful and trustworthy environmental politicians distinguish themselves by an elevated scale-mindedness" (p. 42). Can we geographers be judged in the same way?

"Think Globally, Act Locally" – Which for Geography?

A typical environmental scientist, especially one taking a longer time perspective than that characteristic of politicians, might well claim that the Earth's air, water, and soil/rock media can cope with the outputs of *Homo sapiens*, a species which has failed to occupy the whole planet (leaving huge wildernesses to harbor genes and provide dilution) and whose damaging carboniferous energy era will soon be over (see figures 20.1 and 20.2 and table 20.1).

However, those who live in a smog-polluted city basin, and have their local aquifer polluted or their river basin degraded, are

Table 20.1 Wilderness distribution by continent

Continent	Wilderness km²	Wilderness %	No. of Areas
Antarctica	13,208,983	100.0	2*
North America	9,077,418	37.5	85
Africa	8,232,382	27.5	434
Soviet Union	7,520,219	33.6	182
Asia	3,775,858	13.6	144
South America	3,745,971	20.8	90
Australasia	2,370,567	27.9	91
Europe	138,553	2.8	11
World	48,069,951	32.3	1,039

* This is really one contiguous block divided in two only for purposes of biogeographical classification.

demanding action based upon these environmentally-defined units. In the throw-away society characterized by the rapid innovation and designer obsolescence of the late twentieth century there is, however, said to be no such place as "away." Decisions about where and when pollution can occur are therefore concentrated at scales from national to local, where gradients of inequity (in terms of outcome) are easiest to judge. The impact on communities or regions of waste disposal, in the face of universal NIMBY ("Not In My Back Yard") attitudes, appears to be to create a rapid learning curve in the principles of sustainable development!

The question must be asked, therefore, whether today the "big issue" of pollution and its control is scientific and *global* or political and *local* in scope – operating at the scale of the problems perceived by non-technical (often commercial) actors. Furthermore, the perception of the problem will depend heavily on those factors which also control our perception of risk; we have no international culture of risk to rival that possessed by individuals and communities. This is well known to geographers studying natural hazards (see Cutter 1993); central to such perceptions in the

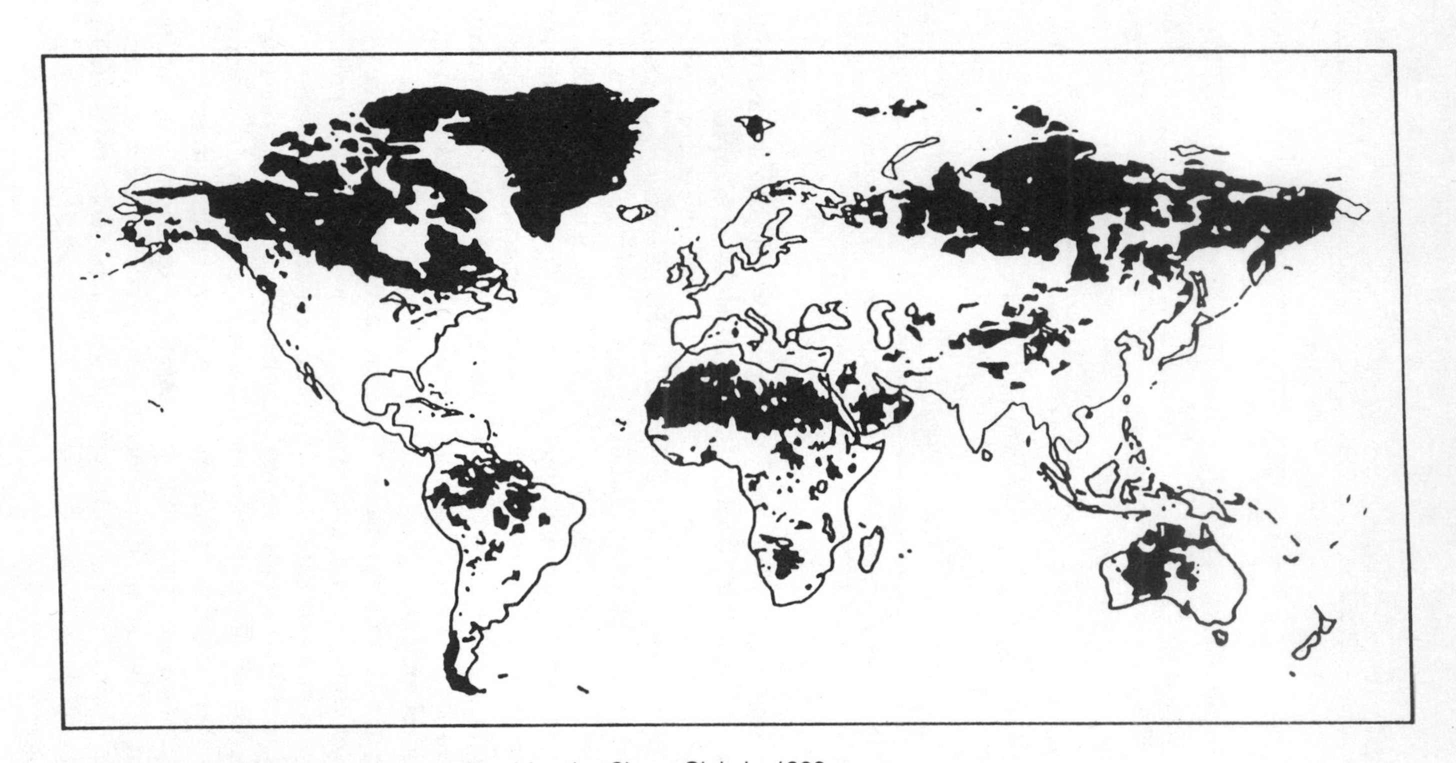

Figure 20.2 Wilderness areas as defined by the Sierra Club in 1988.

pollution field is the notion of "harm" or "damage" done by pollution to an increasingly wide range of environmental receptors. It will be argued below that what links the local action with the global thought is uncertainty and therefore risk-taking.

There is also a link on another scale – the pervasiveness of financial pressure to innovate and to develop resources under the capitalist system; failure to recognize this partly explains the negative aspects of the United Nations Conference on Environment and Development ("The Earth Summit") in 1992. It is to the credit of those who drafted "Agenda 21" for the Conference follow-up that they homed in on local actors and communities and showed sensitivity to inequity, risk, and economic systems in this plan for sustainable futures.

Where's the Harm? The Global and Local Geographies of Pollution

"Remapping the world" bespeaks, in terms of pollution, images at a variety of scales, to be used as a guide to the causes and seriousness of pollution in the late twentieth century. Chapter 18 has already illustrated the importance of the historical background to these spatial presentations.

At the *global scale*, figure 20.3 shows how industrial development patterns based upon fossil fuel have established a North–South contrast in the acidification of the atmosphere. Cross–Equatorial dilution and dispersion is not available and as a result the capacity of sensitive parts of the northern hemisphere has been exceeded, producing harm to ecosystems and cultural heritage to such an extent that geographically-based controls on acid emissions have now been introduced (see below).

The contrast in terms of spatial and temporal impacts between the output of acid pollutants and that of carbon dioxide, also from fossil-fuel combustion, illustrates a key component of the geography of pollution: acidification is an international (transboundary) problem but not yet a global one; it can be technically mitigated at the end of the pipe and by land management. Harm is reversible (over time) if the emissions of the pollutant are reduced. Outputs of carbon dioxide, together with other "greenhouse"

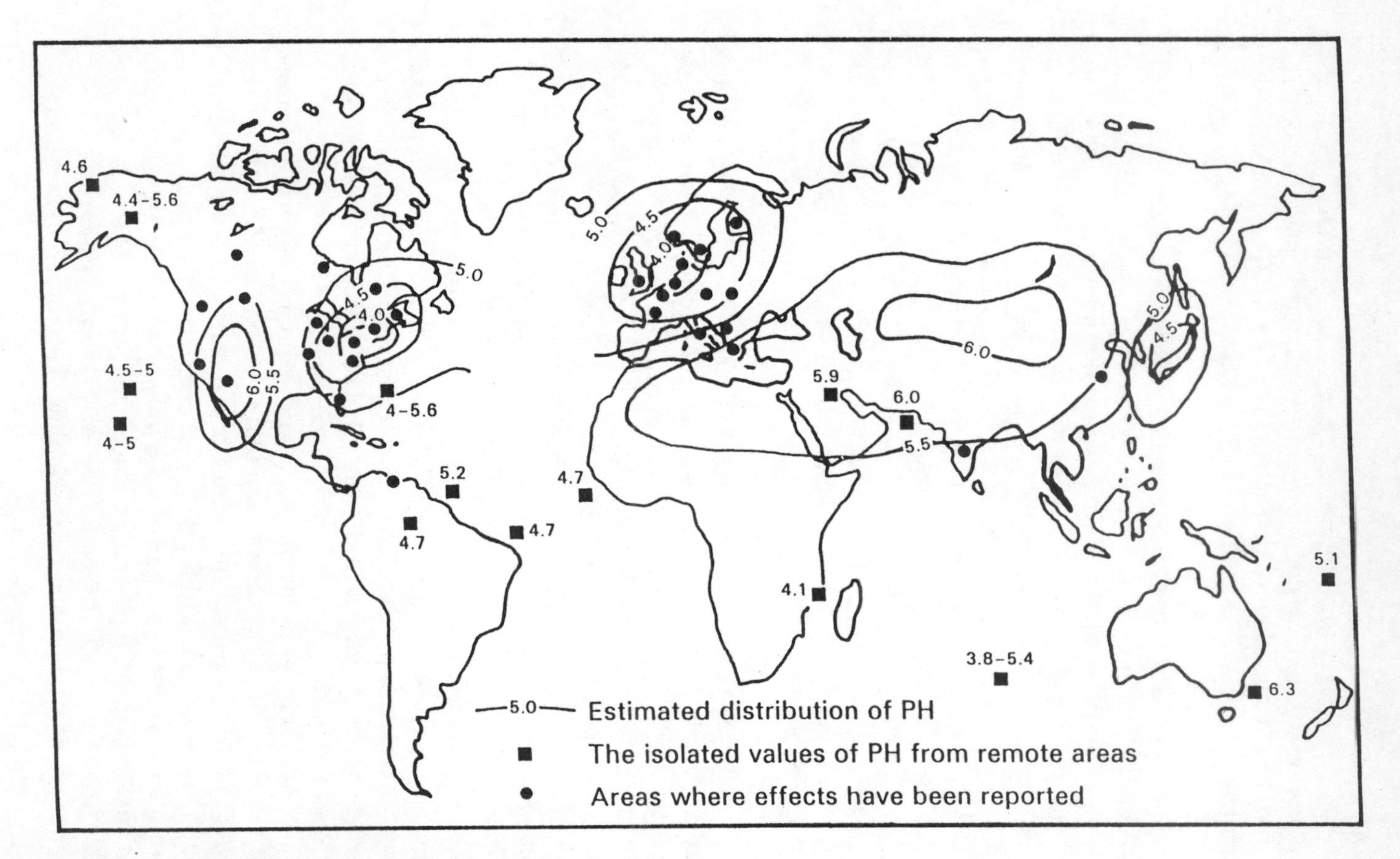

Figure 20.3 Global patterns in pH (acidity) of precipitation.

Table 20.2 Emissions of carbon dioxide to the atmosphere from human activities in gigatonnes per annum (the negative value for Europe indicates a net sink for CO_2).

	from fossil fuel	*land use change*
North America	1,481	112
Europe	1,268	-28
Asia (plus former USSR)	1,950	722
South America	135	579
Africa	141	367
Oceania	63	32

gases, can, however, have an impact on the whole globe by pervading the atmosphere, thus influencing both climatic patterns and sea level (Houghton et al. 1990). We cannot "back-track" the CO_2, as we can acidity, and we are extremely uncertain about the pattern of its effects, particularly on climate. Sea-level rise is more easily predictable – it has already begun! Furthermore, any assessment of current or anticipated harm from carbon dioxide pollution must admit that there is drastic global inequality between those who have utilized the carrying capacity for this gas (table 20.2) and those who will bear the brunt of the impacts (and whose future development, including that of human rights, would be compromised by forgoing further use of that capacity – Grubb 1990).

One perceptual problem of the rising atmospheric content of carbon dioxide is that we are not dealing with a problem of toxic harm. Where the latter is at issue "life and death" becomes the perceptual avenue for risk assessment, even when the assessment is made for non-human biota. Whilst acidification of the developed northern hemisphere has, as a major corollary, the potential release of hitherto insoluble human toxins such as heavy metals into rivers, the first steps toward control involve understanding the *critical loads* beyond which ecosystems are harmed (Hornung and Skeffington 1993). Identification of source–pathway–target linkages has been well researched and the geographical patterns of each link have now contributed to a regulatory framework which could soon become an economic framework (London

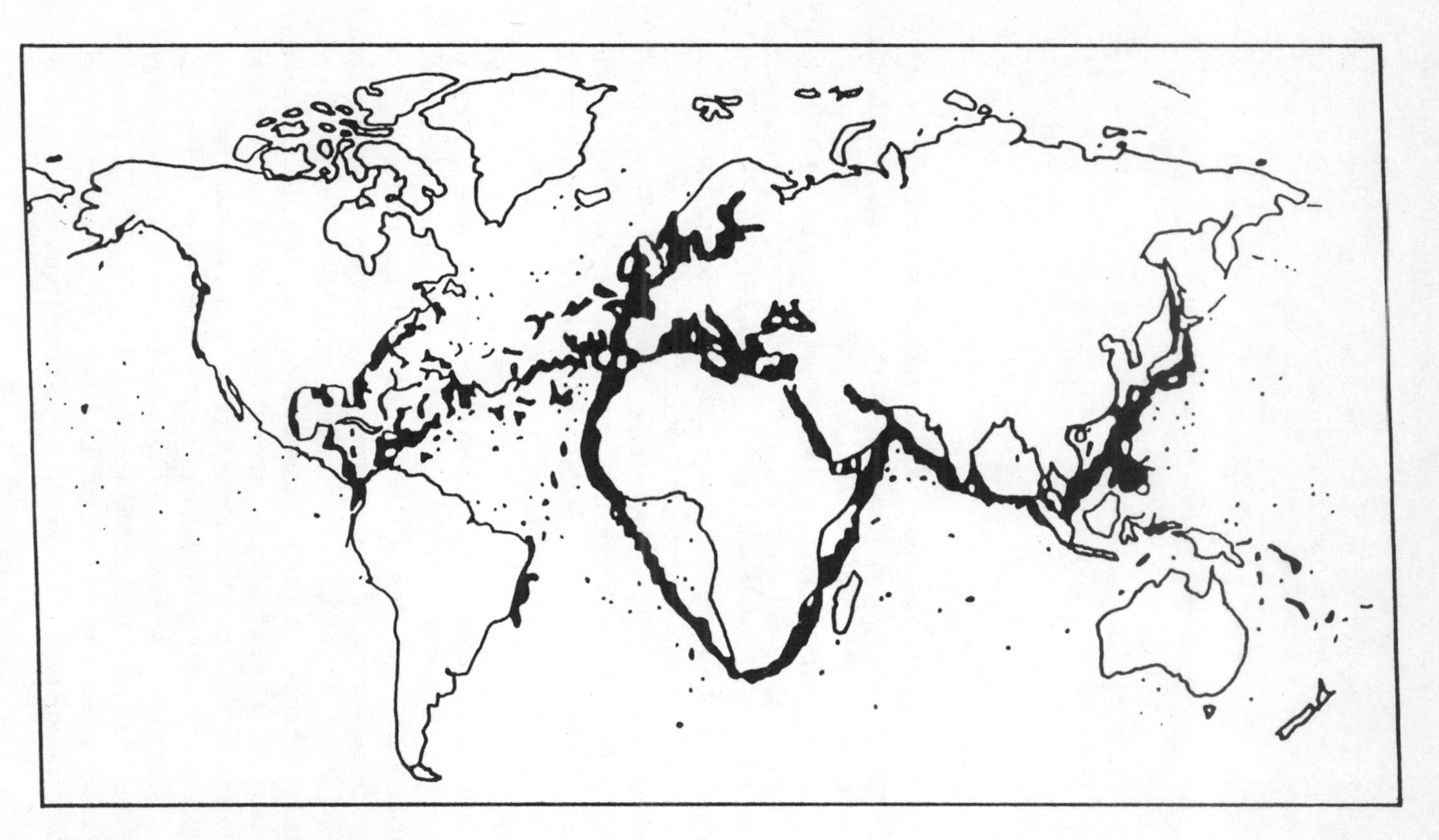

Figure 20.4 Oil contamination of the oceans.

Economics 1992). Acidification can, with time, be reversed. There are now well-publicized (though much criticized) cases of the restoration process and rehabilitation of habitats in acidified areas. Acidification thus shows signs of moving from global cause (development using fossil fuels) to regional or even local solutions, whereas meaningful carbon dioxide controls are still awaited.

Figures 20.4 and 20.5 show how another feature of global development yields a pattern of trade and transport during which pollution inevitably occurs; within this pattern of oil slicks there are both steady, long-term abuses of the diluting capacity of the oceans and local accidental events: pollution *emergencies*. Whilst good practices can reduce the background rate of pollution, no human management system in a hazardous field is ever risk-free – pollution emergencies will continue because of human error alone (one of the reasons for growing international suspicion of nuclear power and its waste stream). Whilst media debate is rife about the true long-term impacts of maritime and coastal oil spills, it must be said that we have not yet experienced "the long term"; optimists point to "recovery" of oiled birds but pessimists suggest that oiled mangroves and corals may never recover.

Continuing with pollution emergencies, it is important to illustrate within our map iconography those important ambient processes and capacities upon which we rely to contain, dilute, or disperse pollution. Figure 20.6 illustrates the importance of atmospheric circulation in diluting and dispersing (but also distributing) the accidental release of radioactivity from the Chernobyl nuclear reactor in 1986 (Gould 1990). The resulting patterns of deposition in Britain are the result not of a government-sponsored monitoring exercise but of the rapid-response activity of ecologists armed with their land-use system maps. Here we see the failures of toxicity testing in environmental systems at their risk-generating best (see below)! The UK system for tracking radioactivity through farming systems was based upon *lowland* soils and farming systems; as a result, the *upland* sheep farmers of North Wales and the English Lake District were surprised to learn that in *their* soils and *their* grazing patterns harm was done by way of accumulation of radionucleides (Wynne 1990). Dilution and dispersion in this case will take many years, and farmers in some parts of Britain still face restrictions on the sale of livestock.

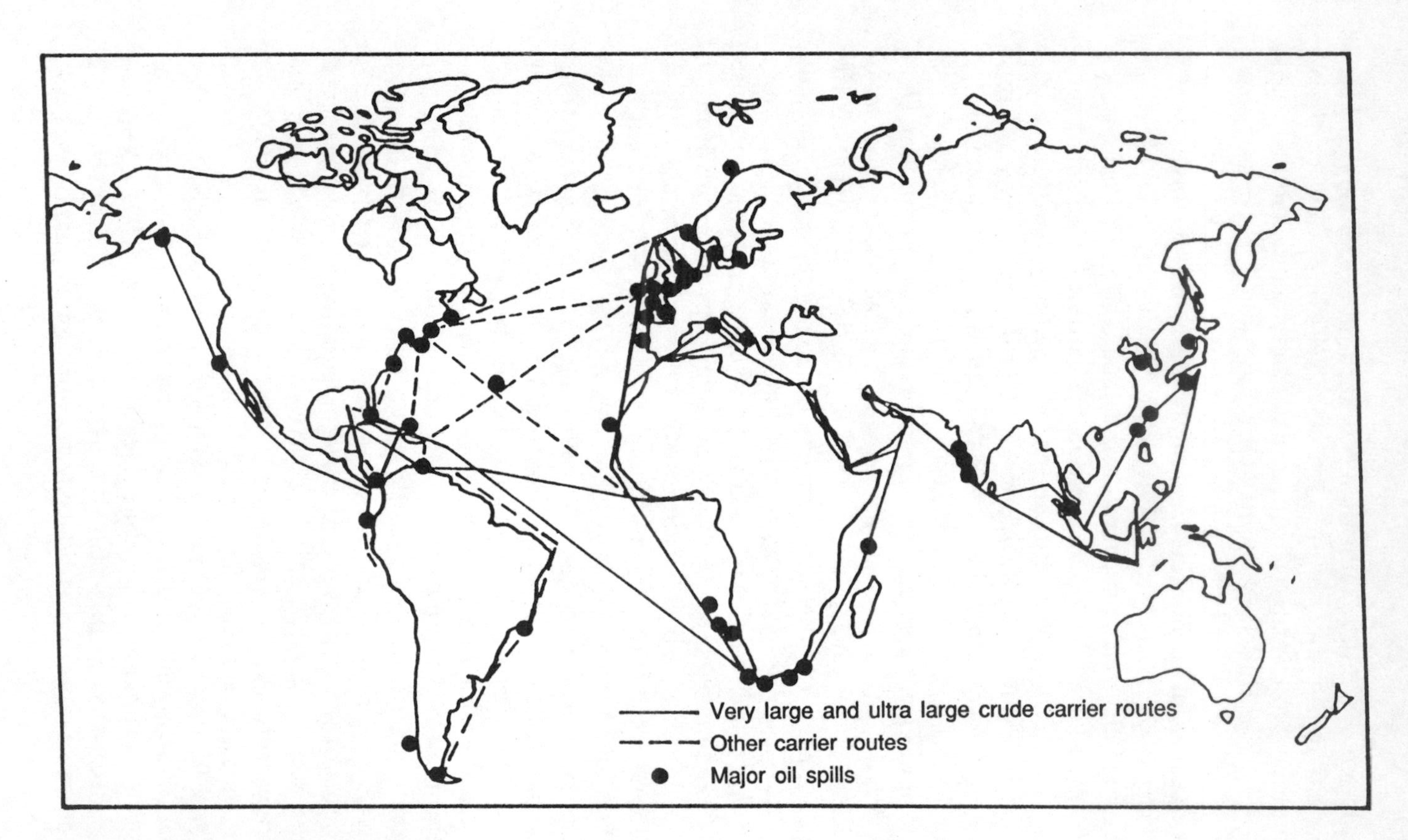

Figure 20.5 Major oil spill incidents.

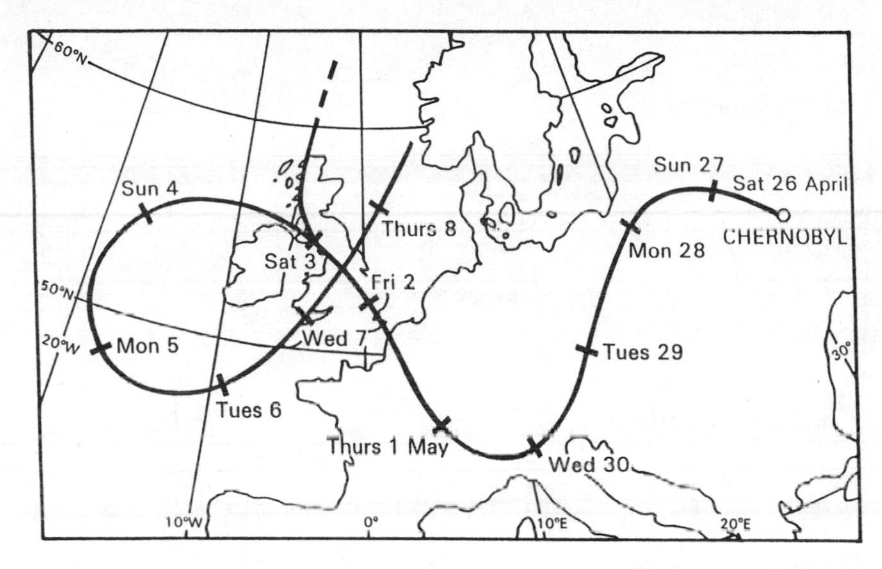

Figure 20.6 The spread of the Chernobyl-derived radiation cloud across western Europe, 1986.

Finally, figure 20.7 shows the geographical imprint of a more local but more immediately catastrophic "accident" – release in December 1984 of methyl isocyanate gas from the Union Carbide plant at Bhopal. Shrivastava (1992) reveals how this form of "accident" is a kind of intended pollution, by such revealing phrases as "The technological preconditions for a major accident were embedded in the design of the Bhopal plant" (p. 45), and "several thousand, for the most part illiterate, people were living in shantytowns literally across the street from a pesticide plant. They had no idea how hazardous the materials inside the plant were, or how much pressure the plant was under to cut losses" (p. 3). Three thousand people and 2,000 animals died; 300,000 people were affected. This was not an accident unless you add the clause "waiting to happen."

To summarize, we have the following principal causal processes of global pollution revealed by the patterns: production, consumption, trade, development, and "accident". We also realize the complexity of the carrying-capacity argument, with some pollutants, like maritime oil, being obviously persistent in space (but we are told harmless) and others, like the Chernobyl cloud,

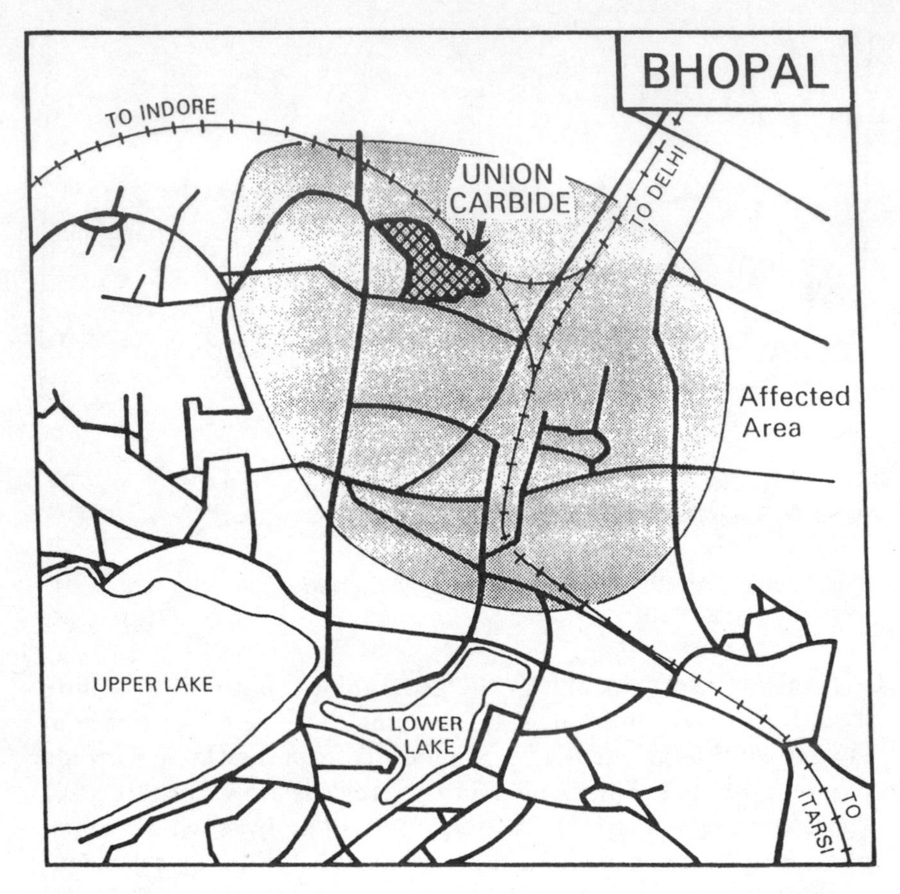

Figure 20.7 Location map of the Union Carbide plant and the pollution plume, Bhopal, India, 1984.

temporary, invisible, yet surprising and overwhelming to the carrying capacity of far-flung ecosystems.

Pollution and Risk – the Crisis of "Harm"

Pollution is, as anthropologists tell us (e.g. Douglas 1966), culturally defined. Nevertheless, we need to look at two modern, apparently technical, definitions in order to set the parameters (scientific, ethical, social, and political) of any judgment about the seriousness of pollution and, thereby, the need for its control.

Pollution has been defined by Murley and Stevens (1991) as:

> The introduction into the environment of substances that are potentially harmful to the health or well-being of human, animal or plant life, or to ecological systems.

Holdgate (1979) extends the definition to introduce the environmental processes upon which we depend when we discard waste into any medium as receptor: dispersion (influenced by both the medium and the properties of the waste, such as density, solubility, and diffusion coefficient) and removal by physical or biological agencies. His geography of pollution reduces to sources, pathways, and targets: "where emissions of a substance to the environment are tolerated, controls need to be adjusted so that targets are not unduly hazarded (just what constitutes undue hazard depends on the nature of the target and the value set upon it)."

The emphasis on harm and hazard is critical to differentiate *pollution* from *contamination* – the lesser evil (a fact unknown to newspaper editors) of introducing substances to the natural environment in greater concentrations than their natural background. From this basic separation we can fan out into space–time, aided by the separation of source–pathway–target.

The spectrum of environmental conditions (stressed by Holdgate) between contamination and pollution introduces the concept of *environmental capacities*. Contamination and *Homo sapiens* are inseparable; the activity rate and creativity of our species increases our dependence upon the capacity of the main environmental media – water, air, land – to accept our wastes. A threshold is crossed, both physically and socially – at any geographical scale – when this inevitable contamination becomes avoidable pollution. If we modify the term "environmental media" to "environmental *receptors*" we can include alongside water, air, and land the equally vulnerable categories of health, biodiversity, heritage, landscape – each of which can suffer harm. Our problem is that, whilst we have a good level of understanding on the processes of dilution, dispersion, containment, or conversion operating in water, air, and land, we have yet to understand the processes operating in these more catholicly-defined receptors, except to simplistically define an economic cost of "damage" by attempts to quantify human values financially (see Pearce et al. 1989).

Reductionist science has faced the challenge of assessing the capacity of environmental *media* in two main ways. We have examined the toxicity of wastes from a highly anthropocentric position (millions of laboratory animals have died helping us do this!). *Toxicity testing* is an extremely crude area of experimental science, by the evidence of its own practitioners (see authors in Nriagu and Lakshminarayana 1989); even when their techniques are yielding usable results for one group of substances the rate of innovation in the chemical field is so great that testing-bottlenecks abound.

Secondly, science has explored the patterns of pollution in *small* units of land, water, and air, mainly in response to *point* sources. For this reason we have been shocked by incidents such as Chernobyl with their very extensive repercussions (even though long ago Rachel Carson warned us about pesticides in the Arctic and Scandinavians successfully complained about "imports" of sulphur dioxide from the rest of Europe). We have further been surprised by the widespread impacts of pollution from *diffuse* sources such as agriculture and by longer-lasting and trans-generational impacts such as those claimed for some nuclear power plants.

Clearly laboratory tests *in vitro* have a (rightly) restricted impact on human perceptions of harm and therefore of risk. "Catch 22" is that, once *in vivo* evidence is available, the harm has been done, at least to a sample of the population, or to non-human organisms which have taken the traditional role of the miner's canary.

Ethical debate about environmental management rightly centers here, rather than with the headline-grabbing pollution incident. What rights do we have to avoid harm? Philosophies of the human/nature interface concentrate upon the moral choice of how far "down" (I prefer "along") the evolutionary spectrum to extend the *concept of rights* (see Taylor 1986). Our definitions of toxicity, harm, and hence pollution, will all depend on this fundamental decision. Science can be helpful here by adding the dimension of time via its division between acute and chronic effects, i.e. between impacts caused temporarily by high *concentrations* of a pollutant and those over much longer time scales resulting from *loadings* and *accumulation*. Furthermore, risks cannot yet be reliably assessed, especially those of long-dated impacts: there is little hope of sustainable development until this can be achieved – unless we take a precautionary approach based upon communities and their corporate risk strategies (see below); the social-science community is already addressing the impact of uncertainty upon decisions made in advance of the environmental changes anticipated from the enhanced "greenhouse effect" (O'Riordan and Rayner 1991; Wynne 1992). We are dealing with the dichotomy identified by Habermas between instrumental rationality (linked to toxicity tests) and value rationality (rights to avoid harm and have risks distributed fairly).

Global Harm – Relax: Gaia is Pro-life

Science is both part of the problem – when it is reductionist – and part of the answer – when it is holistic – in the environmental crisis. Reductionist science, as the hand-maiden (sic) of technology in the capitalist world economy (in both its liberal and its Marxist forms), has led to linear concepts of "progress" based upon production; the generation of waste to polluting levels has been inevitable. Of the environmental sciences, ecology has experienced more holistic phases than most (though even it has reductionist tendencies, as revealed by Worster 1985, and a contemporary dilemma of methods – Peters 1991). The theory of *ecosystems* has stressed that the impacts of pollution may well be distributed in space and time, not only by environmental media, but by the connectivity of systems, e.g. through food chains. Ecology has also offered a defense of the ethical extension approach to harm because it stresses the supportive role of biodiversity in repairing and restoring mechanisms of global equilibrium (Reid and Miller 1989). Yet the political extension of ecological principles (including hierarchies and predation) through *ecologism* has been a feature of rightist rather than leftist tendencies in the twentieth century (Bramwell 1989).

The most serious theoretical contribution by holistic science to the global assessment of pollution impacts in recent years has been the Gaia Hypothesis of James Lovelock (Lovelock 1979). Lovelock, who is fêted by some Greens for the transcendental qualities of his holistic vision, is in fact very optimistic about technology in the future. The global homeostasis achieved by the multiple feedbacks of Gaian cybernetics will, in Lovelock's view, always maintain the Earth's environment as a fit one for life. He does not take a local or regional ecosystems approach, but rather, one which stresses the huge capacities of the global *spheres* – biosphere, atmosphere, hydrosphere – their storages and cycles and their teeming microbiological life, which he deems far more biologically important than macroscopic organisms. The forms of life which co-evolve with the chemistry of the global spheres may not always include humans in a prominent role, but he has optimism even for our species. His grounds are that even the priority pollution-control problems – "Chainsaws, cows and cars" – currently show no signs of swamping the absolute capacity of atmospheric systems to repair and restore. Rather, it is the danger of unstable behavior (crossing the fine thresholds of system state) as capacities are *approached* which Gaia warns us about.

Much more work remains to be done on the Gaia Hypothesis but it has proved its *scientific* worth by allowing a debate about global-scale pollution-control priorities and the importance of atmospheric and ocean capacities. Global scientific research programs are now widespread. Gaia has also focused some attention on the behavior of natural systems close to their homeostatic limits: the potential for abrupt, threshold changes is now a central area of research on the enhanced "greenhouse effect." The rapidity with which, for example, protocols for ozone control were adopted internationally is not just a lesson on the processes of diplomacy (Benedick 1991) but a manifestation of a new global concern with the elements and their stability. However, the case of ozone controls is very unusual in that science spoke directly to power, with great certainty. Developed-world capitalists seemed keen, at the moment of ozone "crisis," to make use of their hitherto hidden innovative skills in this field (witness the great spray-can switch!), and their consumers were convinced enough to refuse the old damaging products.

Thus, the main signal coming to us at the global scale is that of the uncertain risks of approaching threshold conditions in relation to capacities to absorb pollutants. Because there is not yet a global polity able to assess these risks, with a common perception and purpose, we must look to more local perceptions and polities and their relationship to the one human global process – that of international capitalism and its production processes.

What are the Limits? Rates of Growth, Limits to Capitalism

In the 1970s the fast-developing power of computer models was harnessed to create predictions about future resource use and pollution, given the headlong economic growth of the era. Doom-laden predictions emerged, soberly summarized as "limits to growth" (Meadows et al. 1972); these limits were quickly challenged by technocentric optimists. The limits were cast more in terms of resource depletion than pollution and a recent re-run of the models (Meadows et al. 1992) has maintained a relatively optimistic perception by stressing the part being played by the slowly and variably reducing pollutant emissions and hence ambient concentrations/loads in the developed world. Nevertheless, if growth means growth in the number of pollution sources it remains a danger to carrying capacities. Meadows et al. conclude that:

> The limits, let us be clear, are to throughput. They are speed limits, not space limits, limits to flow rates not limits to the number of people or the amount of capital (at least not directly) . . . down is the direction that throughputs will have to go, by human choice or by strong and unpleasant natural feedbacks. (p. 99)

Even the optimistic Gaian scientists question whether Gaia can cope with the present *rates* of environmental change. As put by the editors of a recent compilation volume about the Gaia hypothesis (Schneider and Boston 1991):

> not only is it premature to judge the extent to which biological homeostasis has been validated, it also appears that many of the

> processes by which such feedback may exist take place in *geological timescales*. (p. xx [my emphasis])

Schneider and Boston again:

> Therefore, whether to risk radically modifying the Earth in pursuit of human goals is not a scientific question per se; rather, it is a fundamental *political value choice* that weighs the immediate benefits of population or economic growth versus the potential environmental or societal risk of a rapidly altered Earth. [my emphasis]

Here at last we have a dimension which can reconcile the global with the local, but what values are involved and whose politics? I would add that it is also a fundamental *ethical* value choice – ethics and politics do not necessarily correlate in an era of sound-byte, spin-doctored opportunism. The biodiversity of world politics has itself declined markedly; the domination of the capitalist system makes it extremely difficult to examine the potential of a genuinely green political dimension: Naess (1989) and Paehlke (1989) identify the need for a political eigenvector at 90 degrees to that of labour–capital. However, as Weale (1992) stresses, neither the ethical stance struck by an extension of rights to non-human biota nor a rival political dimension in favor of precautionary "speed limits" is likely to break through when both approaches to pollution control breach the conventional structure of politics in terms of *interests*. Curtailing rates of growth to be compatible with long-term carrying capacities strikes vividly against the growth imperative of national governments (see, for example, "Impacts of 'Greenhouse' gas control strategies on UK competitiveness" – Pezzey 1991). Another infringement noticed by our political referees is that the green dimension will increasingly make it extremely difficult to separate domestic and foreign policy.

However, Weale introduces the term "Ecological Modernization" – a reconceptualization of the relationship between the economy and the environment in which environmental protection becomes a growth promoter, not a burden. This view tends to play into the hands of the science–technology–capitalism–innovation system of the developed world. Many industrial leaders in the developed world are alerted to pollution only because it represents industrial bad practice; they can now qualify for certificates of good environmental practice, carry out audits, and have special logos (doves and leaves are common!) on their goods, but they continue to rely at a largely undiminished rate upon the Earth's carrying capacities for their waste.

International capitalism has considerable impact on modern geographical patterns (Storper and Walker 1989); can this impact be reconciled with the spatial patterns necessary for sustainable environmental management, one component of which is pollution control? Slowly "Ecological Modernization" is coming to mean the privatization of environmental management within a suitably rigged market whose ethical pillars are "stewardship" and "sustainability."

Certain characteristics of ecological modernization can form the basis of a pollution-control policy within world capitalism if the scale units are chosen with respect to the problems and the problems prioritized. River basins have shown this form of environmental politics in many parts of the world (Newson 1992b). We therefore end this excursion locally.

Pollution: Just Act Locally

As Johnston (1989b) reminds us, we cannot undertake any practical policy-making on environmental issues without concerning ourselves with the nation state as a unit and the world capitalist system as a process. This neglects the fact that cities and city-regions are often the major causes of the pollution problems which our policy may wish to cure or correct. Environmentalism may well need to choose its own spatial organization, and in this it will battle further with capitalism, especially in the city.

Can we have environmental ethics which base the choice of right or wrong actions upon their community impacts? Those seeking a rights extension, such as Taylor, are very skeptical about such ecological attitudes to harm. The argument of ecosystem health can be just as anthropocentric and hierarchical, through notions of "stewardship," as that based directly on human self-interest. If, by contrast, we extend from a concern for human rights to one of biotic rights we have a much more concrete, though more difficult, basis for judgment on the necessary policies for management. We also have an opportunity to reintroduce the dimensions of culture and spatial units which geographers find so fascinating.

In the early days of the New Environmental Age, philosophers and pragmatists of green approaches settled on the need to set new scales and networks for living (e.g. Schumacher 1973); this obsession failed (Pepper 1991) largely because it was a rural obsession. Of what value is the commune vision to a "risk-society" (Nohrstedt 1993) when any map of pollution risk is bound to show urban peaks? Murray Bookchin has always stressed (e.g. 1986) the human-rights aspects of pollution and other forms of environmental degradation, seeking a "red/green" communalistic solution around reconceptualized city polities.

The uncertainty of environmental science also has a spatial relevance. As many analysts now tell us, the postmodern element of the environmental crisis is the inability of science, and its hand-maiden technology, to provide traditional mechanistic explanations and therefore remedies (despite news coverage of plans to girdle the globe with mirrors to protect us against the enhanced greenhouse effect and global warming). In the words of O'Riordan and Rayner (1991): "Under . . . conditions of high uncertainty and high societal stakes, risk management decisions cannot be justified according to purely technical criteria or even clinical judgement" (p. 98).

Decisions therefore need to seek acceptance and be interactive with a concerned population in a way that conventional budgetary and foreign-affairs political decisions cannot be. Since pollution is culturally defined, remedies need to be culturally adapted. Such adaptation is said to be the secret of the supposedly harmonious environmental relations enjoyed by indigenous peoples. In seeking sustainable pollution-control policies the polity for our decision-making must be structured to make most use of citizen roles (Portney 1991). Principle 10 in the Rio Declaration on Environment and Development states that "Environmental issues are best handled with the participation of all concerned citizens at the relevant level."

Politicians are rushing to sign up for environmental management units such as bioregions (Gore 1992) or regional community-based power structures for environmental management (Taylor A. 1992). If capitalism is forced into similar units, as a basis for pollution control via the use of economic instruments (in which industry is allowed to trade permits to pollute into regional "pollution bubbles"), we may well see a heady mixture of sustainability and subsidiarity gaining credence.

We are not yet a global "EMU" (environmental management unit); within the Gaian milieu of the spheres there is a myriad local and regional cultural traits that are of relevance to the risk-society and its long-term success or failure in controlling pollution. Here is the proper focus for the phrase, "It's all geographical."

21

Sustainable Development?

W. M. Adams

Introduction

Tight and complex links exist between development, environment, and poverty. The 1992 *World Development Report* opens with the assertion that "the achievement of sustained and equitable development remains the greatest challenge facing the human race" (World Bank 1992, p. 1). The poor are more numerous than ever before (1.1 billion in 1990), and measures of poverty worsened in much of the world, particularly sub-Saharan and North Africa, Latin America, and the Caribbean. This income gap between rich and poor countries is widening. Between 1960 and 1989 the share of global GNP of the countries with the richest 20 percent of world population gew from 70.2 percent to 72.7 percent; the share of the countries with the poorest 20 percent of world population fell from 2.3 percent to 1.2 percent (UNDP 1992). The poor often endure degraded environments, and in some instances contribute to their further degradation. Urban air and water pollution are both rising rapidly, even in those countries in which economic growth is taking place, and the degradation of agricultural, forest, and wetland resources is extending the depth and breadth of deprivation in many rural areas. Enduring problems, for example the lack of clean drinking water, are getting more and not less serious: 2 million children die of intestinal diseases due to unclean water each year (World Bank 1992).

The reciprocal and synergistic links between poverty and environmental degradation forces what Blaikie describes as the "desperate ecocide" of the poor (Blaikie 1985, p. 138). Access to and control over cultivable land, fuelwood, or other usable attributes of nature is uneven. Blaikie emphasizes the political dimensions of rights over resources, stressing the need for those seeking to understand environment-development problems to explore the links between environment, economy, and society that he calls "political ecology." Blaikie and Brookfield argue that "land degradation can undermine and frustrate economic development, while low levels of economic development can in turn have a strong causal impact on the incidence of land degradation" (Blaikie and Brookfield 1987, p. 13). Poverty and environmental degradation, driven by the development process, interact to form a grim world of risk and hazard within which urban and rural people are trapped. Understanding the reality of this world, and the environmental, economic, and political factors that create it, lies at the heart of the widespread contemporary concern for sustainable development.

The phrase sustainable development is now widely employed, in the fields of policy and political debate as well as of research. It seems to contain the potential to unlock the doors separating disciplines, and to break down the barriers between academic knowledge and action. It does this partly because the phrase is at the same time superficially simple and yet capable of carrying a wide range of meanings, and supporting sometimes divergent interpretations. Both radical environmentalists and conventional development-policy pragmatists have seized the phrase and used it to express and explain their ideas about development and environment. In the process they have created a powerful new phrase in the lexicon of development studies, and a theoretical maze of remarkable complexity (Adams 1990; Lélé 1991).

Sustainable development has many definitions (e.g. Pearce et al. 1988). The most commonly quoted and influential is that of the Brundtland Commission in *Our Common Future*, "development which meets the needs of the present without compromising the ability of future generations to meet their own needs" (Brundtland 1987, p. 43). Notwithstanding the rhetorical and slightly vague character of this definition (attributes it shares with many others; see for example the discussion in Lélé 1991), it has proved to be popular and compelling for those concerned about poverty, and inter- and intra-generational equity in human access to nature and natural resources, and for those concerned to address the use of nature itself, and the conservation of habitats and species.

Sustainable development was also the driving force behind the United Nations Conference on Environment and Development (UNCED) in Rio in 1992 (the "Earth Summit"). Attended by 128 Heads of State and some 178 governments, it was also a forum for a vast range of non-governmental organizations that strove to capture the headlines and influence the debate through the parallel Global Forum (Holmberg et al. 1993). The media storm was intense, and in many cases targeted the glaring contradictions of contrast between privileged delegates and those living in Rio de Janeiro's *favelas*. Analysis in the immediate aftermath argued that the conference had failed to live up to the hype that had preceded it, and hopes that it would bring about a new environmental world order. Many observers commented bitterly about the deep conflict of interest that was revealed at Rio between industrialized and non-industrialized countries, epitomized by the obvious lack of enthusiasm of the USA for the whole UNCED process.

The conference certainly failed to realize the hopes of environmentalists in many countries and of the governments of Third World countries, and it also fell short of the stated aims thrashed out in the various "Prepcoms" (Preparatory Committees) that preceded it. However, the cumbersome and costly UNCED process did yield some fruit. The Rio Declaration was signed (admittedly after getting deeply mired in trivial debates about the exact wording of the text); a set of Forest Principles was agreed upon, as was a Biodiversity Treaty (which the USA refused to sign at the conference), a Climate Treaty, and the action program *Agenda 21*. This is a vast document, containing 40 separate chapters and amounting to a volume of more than 500 pages. It was drafted and argued over minutely by government bureaucrats, and is both a fantastic compendium of ideas, issues, and principles, and a hard-won consensus that, it claims, "reflects a global consensus and political commitment at the highest level on development and environmental cooperation" (Holmberg et al. 1993). For better or worse, *Agenda 21*, and the mountain of other printed words generated by the UNCED hothouse, defines the practical contours of sustainable development for the rest of the twentieth at least.

Environment and Development in International Debate

Even before UNCED, there was no doubt that the environment was widely seen as a central issue in development, by governments, researchers, aid donors, and many individuals. This concern began to grow in the 1960s and 1970s, and is well demonstrated by the way the ideas were developed and incorporated into developmental and environmental debates by international organizations. Throughout this process, the idea of sustainability was pushed most strongly and effectively by scientists (like Sir Julian Huxley, first Director of UNESCO), particularly by ecologists engaged in international scientific collaboration (Worthington 1983).

Links between ecology and development were fostered in 1964 by the establishment of the International Biological Program (IBP) to consider "the biological basis of man's welfare" (Waddington 1975, p. 5). The Scientific Committee for Problems of the Environment (SCOPE) was established in 1969 under the International Council of Scientific Unions, and a few years later, in 1971, the Man and the Biosphere Program (MAB) was launched by UNESCO. This latter initiative grew out of the "Biosphere Conference," the *Intergovernmental Conference of Experts on a Scientific Basis for Rational Use and Conservation of the Biosphere*, held in Paris in 1968, and focused for the first time on the growing engagement by conservationists and environmentalists in the First World with the development process in the Third World (Caldwell 1984).

Through the 1960s, environmentally orientated international organizations realized with increasing clarity that they could not influence decisions about the use of natural resources in the Third World unless they were prepared at least to adopt the (for them) novel language of development (Boardman 1981). UNESCO became involved in a review of natural resources in Africa at the request of the Economic Commission for Africa in 1959, and in the early 1960s IUCN (the International Union for Conservation of Nature and Natural Resources) launched the African Special Project to promote wildlife conservation in the continent. In 1968, IUCN joined UNESCO, other UN Agencies (FAO (Food and Agriculture Organization) and UNDP [UN Development Program]), the Conservation Foundation, and the World Bank in organizing a conference at George Washington University in Virginia on ecology and international development, published as the landmark text, *The Careless Technology* (Farvar and Milton 1973). This was followed by a handbook for development planners, *Ecological Principles for Economic Development* (Dasmann et al. 1973).

The United Nations Conference on the Human Environment, held in Stockholm in 1972, is usually identified as a watershed in the emergence of sustainable development, although it was only partly, and belatedly, concerned with the environmental and developmental problems of the emerging Third World (Adams 1990). The growing influence of concern for the environmental aspects of economic development was by no means accepted unopposed. "Developing" countries saw discussions of global resource management as an attempt by industrialized countries to take away control of their resources (Biswas and Biswas 1984), and feared that the environment was being given too high a priority compared with development; far from suffering the impacts of pollution caused by industrialization, poor countries suffered the "pollution of poverty." Faced with this concern, the phrase "sustainable development" was used explicitly to smooth over the apparent dichotomy between economic growth based on industrialization, and associated adverse environmental impacts. The phrase captured perfectly the idea (or the belief) that it was possible to have development without adverse environmental side-effects (Clarke and Timberlake 1982). The essence of this idea, and the power of its political appeal, changed very little in the two decades that elapsed before the Stockholm Conference's successor at Rio de Janeiro in 1992.

The Stockholm Conference was held in June 1972, and was attended by 113 nations. Agreement was eventually reached on 26 Principles. These argued that development need not be impaired by environmental protection (Principle 11), and indeed that development was needed to improve the environment (8). Environmentally benign development was to be achieved by integrated development planning (13), rational planning to resolve conflicts between environment and development (14), aid, and particularly money, to pay for environmental safeguards (9 and 12), and reasonable prices for exports (10). However, of the 109 Recommendations for Action, only 8 addressed the question of development and the environment, and they were "extraordinarily negative" (Clarke and Timberlake 1982, p. 12), concerned chiefly with minimizing possible costs of environmental protection. There was little apparent awareness of the links between environment and poverty, and few ideas about how "sustainable development" was to be brought about. More was achieved at the meeting held at Cocoyoc in Mexico in October 1974. This did what Stockholm had not, and looked at environmental problems from the perspective of the Third World, and particularly the Third World poor. The resulting "Cocoyoc Declaration" pointed to the problem of the maldistribution of resources and to the inner limits of human needs as well as the outer limits of resource depletion. It pointed to basic needs, and called for a redefinition of development goals and global lifestyles.

The most conspicuous result of the Stockholm Conference was probably the creation of the United Nations Environment Program (UNEP), though it was neither as large nor as powerful an organization as some had hoped, and lacked the independent budget that a new UN agency would have commanded. It was located in Nairobi, an important symbolic location in the Third World but remote from the centers of power within the UN system in Europe and North America. Notwithstanding these constraints, UNEP has introduced a series of important initiatives, including the Regional Seas Programme, the Global Environment Monitoring System, and organization of the UN Conference on Desertification, in Nairobi in 1977 (United Nations 1977). UNEP also played a catalytic role in the preparation of *The World Conservation Strategy*, published in 1980.

Work on a strategy for nature conservation began in 1975, and from this narrow root the notion of a "world conservation strategy" grew (Boardman 1981; McCormick 1989). In 1975 IUCN joined UNEP, UNESCO, and FAO to form the Ecosystem

Conservation Group. In 1977 UNEP commissioned IUCN to draft a document to provide "a global perspective on the myriad conservation problems that beset the world and a means of identifying the most effective solutions to the priority problems" (Munro 1978).

Sustainable Development: the Mainstream

Sustainable development was codified for the first time in *The World Conservation Strategy* (WCS), prepared by IUCN with finance provided by the UN Environment Program and the World Wildlife Fund (IUCN 1980). Since then it has been further developed through the report of the World Commission on Environment and Development, *Our Common Future* (Brundtland 1987), and the follow-up to the WCS, *Caring for the Earth* (IUCN 1991), before its appearance in *Agenda 21* at the Rio Conference in 1992. These documents differ, but have a remarkably consistent core of ideas, a "mainstream" that has persisted through the two decades between Stockholm and Rio (Adams 1990, 1993). At the heart of these documents is a vision of sustainable development strongly influenced by science, by wildlife conservation, by concerns for multilateral global economic relations, and by an emphasis on the rational management of resources to maximize human welfare.

The World Conservation Strategy identifies three objectives for conservation. First, the maintenance of "essential ecological processes." These are "governed, supported or strongly moderated by ecosystems and are essential for food production, health, and other aspects of human survival and sustainable development" (para 2.1). These "life-support systems" include agricultural land and soil, forests, and coastal and freshwater ecosystems. Threats include soil erosion, pesticide resistance in insect pests, deforestation and associated sedimentation, and aquatic and littoral pollution. The second objective is the preservation of genetic diversity, the genetic material both in different varieties of locally-adapted crop plants or livestock and in wild species. This genetic diversity is both an "insurance" (for example against crop diseases) and an investment for the future (e.g. crop breeding or pharmaceuticals). The WCS's third objective is "the sustainable development of species and ecosystems," particularly fisheries, wild species which are cropped, forests and timber resources, and grazing land. These objectives are then broken down into a list of priority requirements, drawn up on the basis of criteria of significance (how important is it?), urgency (how fast is it getting worse?), and irreversibility. The first two objectives, to conserve ecological processes and genetic diversity, demand the rational planning and allocation of land use: giving crops priority on the best land (but not on marginal land), and setting aside and controlling areas such as watersheds and littoral zones for appropriate management only. The conservation of genetic diversity demands site-based protection of ecosystems and the timely creation of banks of genetic material. The WCS then discusses the priorities for national action (Sections 8–14). These are based on the preparation of separate national strategies (by governments or by non-government organizations) which review development objectives in the light of the conservation objectives. They establish priority requirements, identify obstacles, and propose cost-effective ways of overcoming them, determining priority ecosystems and species for conservation and establishing a practical plan of action.

The elements of the sustainable development ideas in *Our Common Future* extend the ideas in *The World Conservation Strategy* considerably. They blend environmental concerns (e.g. the need to achieve a sustainable level of population, and the desirability of conserving the resource base and reorientating technology) with development concerns (e.g. the fundamental goal of meeting basic needs and the need to build environmental factors into economic decision-making). Economic growth is seen as the only way to tackle poverty, and hence to achieve environment-development objectives. It must, however, be a new form of growth, sustainable, environmentally aware, egalitarian, integrating economic and social development: "material- and energy-intensive and more equitable in its impact" (Brundtland 1987, p. 52). The Brundtland Report's vision of sustainable development rests firmly on the need to maintain and revitalize the world economy. In short, "more rapid economic growth in both industrial and developing countries, freer market access for the products of developing countries, lower interest rates, greater technology transfer, and significantly larger capital flows, both concessional and commercial (Brundtland 1987, p. 89).

Caring for the Earth picks up themes from both the preceding documents. It opens with a chapter setting out "principles to guide the way towards sustainable societies," and these nine principles offer a structure for the rest of the report. They blend both the ethical ("respect and care for the community of life"), the humanitarian ("improve the quality of human life"), the scientific/environmentalist ("keep within the earth's carrying capacity"), and the pragmatic ("provide a national framework for integrating development and conservation"). Its central argument is much the same as those of its predecessors, although more carefully and fully expressed: "we need development that is both people-centred, concentrating on improving the human condition, and conservation-based, maintaining the variety and productivity of nature. We have to stop talking about conservation and development as if they were in opposition, and recognise that they are essential parts of one indispensable process" (IUCN 1991, p. 8).

Sustainable Development at Rio, 1992

The documents debated and agreed at the United Nations Conference in Rio de Janeiro in 1992 build very directly onto the evolving mainstream of ideas dominating public debate about the environment and development during the 1980s. The same themes appear in both the *Rio Declaration* (table 21.1) and the much larger text of *Agenda 21*. However, the creation of these texts was far from straightforward and harmonious. The "Rio process" was in practice a mutual bludgeoning between teams of diplomats to produce texts that gave least away to perceived national interests. In particular, the distinction between the views of countries in the industrialized "North" and the underdeveloped "South" became steadily more glaring in the run-up to, and during, the conference. There was difference over the key problems (for the industrialized countries, global atmospheric change and tropical deforestation; for unindustrialized countries, poverty and

Table 21.1 The 27 Principles of the Rio Declaration on Environment and Development.

1. Human beings are at the center of concerns for sustainable development. They are entitled to a healthy and productive life in harmony with nature.
2. States have, in accordance with the Charter of the United Nations and the principles of international law, the sovereign right to exploit their own resources pursuant to their own environmental and developmental policies, and the responsibility to ensure that activities within their jurisdiction or control do not cause damage to the environment of other States or of areas beyond the limits of national jurisdiction.
3. The right to development must be fulfilled so as to equitably meet developmental and environmental needs of present and future generations.
4. In order to achieve sustainable development, environmental protection shall constitute an integral part of the development process and cannot be considered in isolation from it.
5. All States and all people shall cooperate in the essential task of eradicating poverty as an indispensable requirement for sustainable development, in order to decrease the disparities in standards of living and better meet the needs of the majority of the people in the world.
6. The special situation and needs of developing countries, particularly the least developed, and those most environmentally vulnerable, shall be given special priority. International actions in the field of environment and development should also address the interests and needs of all countries.
7. States shall cooperate in a spirit of global partnership to conserve, protect, and restore the health and integrity of the Earth's ecosystem. In view of the different contributions to global environmental degradation, States have common but differentiated responsibilities. The developed countries acknowledge the responsibility that they bear in the international pursuit of sustainable development in view of the pressures their societies place on the global environment and of the technologies and financial resources they command.
8. To achieve sustainable development and a higher quality of life for all people, States should reduce and eliminate unsustainable patterns of production and consumption and promote appropriate demographic policies.
9. States should cooperate to strengthen endogenous capacity-building for sustainable development by improving scientific understanding through exchanges of scientific and technical knowledge, and by enhancing the development, adaptation, diffusion and transfer of technologies, including new and innovative technologies.
10. Environmental issues are best handled with the participation of all concerned citizens, at the relevant level. At the national level, each individual shall have appropriate access to information concerning the environment that is held by public authorities, including information on hazardous materials and activities in their communities, and the opportunity to

participate in decision-making processes. States shall facilitate and encourage public awareness and participation by making information widely available. Effective access to judicial and administrative proceedings, including redress and remedy, shall be provided.

11. States shall enact effective environmental legislation. Environmental standards, management objectives and priorities should reflect the environmental and developmental context to which they apply. Standards applied by some countries may be inappropriate and of unwarranted economic and social cost to other countries, in particular developing countries.
12. States should cooperate to promote a supportive and open international economic system that would lead to economic growth and sustainable development in all countries, to better address the problems of environmental degradation. Trade policy measures for environmental purposes should not constitute a means of arbitrary or unjustifiable discrimination or a disguised restriction on international trade. Unilateral actions to deal with environmental challenges outside the jurisdiction of the importing country should be avoided. Environmental measures addressing transboundary or global environmental problems should, as far as possible, be based on an international consensus.
13. States shall develop national law regarding liability and compensation for the victims of pollution and other environmental damage. States shall also cooperate in an expeditious and more determined manner to develop further international law regarding liability and compensation for adverse effects of environmental damage caused by activities within their jurisdiction or control to areas beyond their jurisdiction.
14. States should effectively cooperate to discourage or prevent the relocation and transfer to other States of any activities and substances that cause severe environmental degradation or are found to be harmful to human health.
15. In order to protect the environment, the precautionary approach shall be widely applied by States according to their capabilities. Where there are threats of serious or irreversible damage, lack of full scientific certainty shall not be used as a reason for postponing cost-effective measures to prevent environmental degradation.
16. National authorities should endeavor to promote the internationalization of environmental costs and the use of economic instruments, taking into account the approach that the polluter should, in principle, bear the cost of pollution, with due regard to the public interest and without distorting international trade and investment.
17. Environmental impact assessment, as a national instrument, shall be undertaken for proposed activities that are likely to have a significant adverse impact on the environment and are subject to a decision of a competent national authority.
18. States shall immediately notify other States of any natural disasters or other emergencies that are likely to produce sudden harmful effects on

the environment of those States. Every effort shall be made by the international community to help States so afflicted.

19. States shall provide prior and timely notification and relevant information to potentially affected States on activities that may have a significant adverse transboundary environmental effect and shall consult with those States at an early stage and in good faith.
20. Women have a vital role in environmental management and development. Their full participation is therefore essential to achieve sustainable development.
21. The creativity, ideals, and courage of the youth of the world should be mobilized to forge a global partnership in order to achieve sustainable development and ensure a better future for all.
22. Indigenous people and their communities, and other local communities, have a vital role in environmental management and development because of their knowledge and traditional practices. States should recognize and duly support their identity, culture, and interests and enable their effective participation in the achievement of sustainable development.
23. The environment and natural resources of people under oppression, domination, and occupation shall be protected.
24. Warfare is inherently destructive of sustainable development. States shall therefore respect international law providing protection for the environment in times of armed conflict and cooperate in its further development, as necessary.
25. Peace, development, and environmental protection are independent and indivisible.
26. States shall resolve all their environmental disputes peacefully and by appropriate means in accordance with the Charter of the United Nations.
27. States and people shall cooperate in good faith and in a spirit of partnership in the fulfilment of the principles embodied in this Declaration and in the further development of international law in the field of sustainable development.

---

the problems that flow from it), and over responsibility for finding solutions. As at Stockholm in 1972, there was fear on the part of Third World countries that their attempts to industrialize would be stifled by restrictive international agreements on atmospheric emissions. They also feared that their freedom to use natural resources within their boundaries would be constrained by agreements imposed by industrialized countries that had themselves become wealthy precisely by polluting freely and while removing their own forests.

These tensions between governments are clear in the UNCED documents. The *Rio Declaration* is not the strong and sharp "Earth

Charter" originally conceived by the Conference Chairman, Maurice Strong, but contains a long list of 27 Principles, "a bland declaration that provides something for everybody" (Holmberg et al. 1993, p. 7). It opens with the statement that "human beings are at the centre of concerns for sustainable development. They are entitled to a healthy and productive harmony with nature" (table 21.1). Many of the principles are uncontentious (e.g. 4, the need to integrate conservation and development; or 5, on the eradication of poverty). Others were more closely fought over at Rio, particularly those that address the central issue of the conference, international action and international responsibility. Thus Principle 2 notes the sovereign right of countries to develop, while Principle 7 establishes the notion of "common but differentiated responsibilities" for the global environment. Hidden behind a bland comment that "states shall cooperate in a spirit of global partnership to conserve, protect and restore the health and integrity of the world's ecosystem," responsibility here basically forces the burden of greatest action on developed countries (Holmberg et al. 1993). Even the text of the *Rio Declaration* (compact by the standards of *Agenda 21*) is self-contradictory, and the US delegation released an "interpretative statement" that effectively dissociated them from a number of the principles agreed. These included the notion of a right to development, in Principle 3 (they argued that "development is not a right . . . on the contrary development is a goal we all hold"; Holmberg et al. 1993, p. 30), and also rejected any interpretation of Principle 7 that suggested any form of international liability.

By no means all the debate at Rio was between rich and poor countries, for there was a major input from non-governmental organizations (NGOs). These were represented at the parallel Global Forum, physically and psychologically distant from the main conference. NGOs fought hard to win a place at the various preparatory meetings that prepared the conference documents, and some remained close to government delegations through the conference itself. Some 1,400 lobbyists were accredited to the main conference, although almost all were excluded from the official negotiating sessions (Holmberg et al. 1993). However, there were a far larger number distanced from the formal proceedings of the conference, on the separate Global Forum site, working with each other and through the specially convened "Earth Parliament." Chapter 27 of *Agenda 21* concerns NGOs, and amidst some bland and empty statements recognizes their important role in achieving sustainable development. They had a significant influence on the way issues like "empowerment" were approached in *Agenda 21*. Whilst NGOs were mostly locked out of formal negotiations at Rio, their growing political significance is revealed both by the debates in the preparation for and at the conference itself and by the size and moral authority of the Global Forum. Grassroots groups, and particularly womens' groups, are clearly to be key elements in future debates about the environment and development (e.g. Ekins 1992), although the conference also began to emphasize the distance between the powerful, wealthy, and influential NGOs of industrialized countries and the "grassroots" in the sense of groups formed among the poor of the urban and rural Third World.

The documents produced at Rio, and the debates behind them, are important, although in fact they add relatively little to previous thinking about sustainable development. They record the evolving debate between developers and environmentalists, between those who wish to exploit natural resources and those who wish to conserve them. They also highlight the increasingly important debate about the environment between the poor and the rich countries. Here Rio simply continues the war of words begun at Stockholm two decades earlier, and carried forward through the Brandt process and at numerous other meetings.

This debate has seemed relatively new to First World environmentalists used to thinking about problems of greenhouse warming or biodiversity loss as generalized global problems, rather than as problems with specific political, economic, and geographic contexts. It is far more familiar to those aware of the debates in recent decades about development, and particularly the failure of calls for a New International Economic Order (NIEO). Despite the doubling of the size of the Global Environmental Facility, the UN Conference on Environment and Development failed to stimulate the scale of financial support necessary to implement *Agenda 21*. Approximately $2.5 billion was pledged, compared with some $125 billion needed (Holmberg et al. 1993). This is the main source of disappointment in UNCED. It did not set a new agenda for development, let alone for a new international economic order. However, there is some prospect that the sustainable development ideas of Rio will not simply sink as the NIEO did, because the conference did generate a series of specific agreements in the form of the Convention on Climate Change, the Biodiversity Treaty (and the Desertification Treaty now being negotiated), and it created the Commission on Sustainable Development. Imperfect though these are, the Rio Conference thus started to fit the fine words of sustainable development into the structures and mechanisms of international development action. The extent to which that process bears fruit remains to be seen.

Geography and Sustainable Development

Sustainable development has already proved a spectacularly popular phrase in both general and academic writing about environment and development. It has roots and a large following in many disciplines, but probably none is so relevant to it as geography (Kates 1987). Where else can the science of the environment be married with an understanding of the economic, political, and cultural change that we call development? What other discipline offers insights into both environmental change and environmental management, and who but geographers can cope with the diversity of environments and countries, and the sheer range of spatial scales, at which it is necessary to work to understand processes of human use of nature and the dynamics of the environment?

Kates (1987) identifies the human environment as one of the four key traditions of academic geography. George Perkins Marsh's *Man and Nature*, written in 1864, was described as "the most important and original American geographical work of the nineteenth century" because of the revolution it brought about in understanding of the ways that people transformed their surroundings (Lowenthal 1958, p. 246). Marsh's ideas were recalled, and claimed for geography, at the symposium held by the Wenner-Gren Foundation at Princeton, New Jersey, in 1955, *Man's Role in Changing the Face of the Earth* (Thomas 1956), chaired by

Carl Sauer, Marston Bates, and Lewis Mumford. That meeting, in its turn, was recalled forty years later in a volume that both renewed the commitment of geographers to the study of human impacts upon the earth, and noted the scale and extent of that impact (*The Earth as Transformed by Human Action*, Turner et al. 1990).

Sustainable development draws on both the technocentrist and ecocentrist axes of environmentalism (O'Riordan 1981). On the one hand, many people writing about sustainable development propose rational, technical solutions to environmental problems (better environmental planning, and clean technologies, for example). Other proponents believe that sustainable development must involve more radical changes to economy and society (for example, zero growth or local self-sufficiency), or new attitudes to nature (for example, a concern for the rights of other species). Geographers have tended to emphasize the technocentric end of this spectrum. Resource management has long been an important theme in geography (e.g. Burton and Kates 1965; O'Riordan 1971; Mitchell 1979; c.f. Johnston 1983), as has the application of these ideas to the Third World, both in general terms (e.g. Omara-Ojungu 1992) and in more specific studies of arid lands (e.g. Heathcote 1983) or water resources (e.g. Agnew and Anderson 1992). These ideas have been particularly strong in North America, and have been extensively influenced by ideas about "rational utilization" or "wise use" of natural resources in the utilitarian conservation philosophy developed in the USA in the first decades of the twentieth century (Hays 1959). The notion of efficiency in resource use played an important role in the agencies created to implement Roosevelt's "New Deal" in the 1930s (Wallach 1991), when organizations like the Bureau of Reclamation and the Tennessee Valley Authority inscribed the principles of rational resource management on American landscapes.

Technocentric ideas about environmental management were strengthened in European geography by the growth of empires, particularly in Africa. In England, the Royal Geographical Society moved from exploration to a concern for geography in universities in the nineteenth century (Stoddart 1986), while in France geographers were closely involved in French imperial ambitions

(Heffernan 1990). In the twentieth century geographers mapped the new colonial territories, sanctioning by their skills the partitioning and expropriation of land (Stone 1980). Geography was well fitted to colonial service: climate, hydrology, soils, land use, settlements, markets, and transportation were all grist to the geographer's mill as the discipline became established in European universities, and developed its own institutional life.

Nonetheless, despite its apparent natural advantages, geography failed to provide the intellectual leadership for the environmental revolution of the 1960s and 1970s (Kates 1987). Notwithstanding experience in areas like natural hazards, neither the quantitative revolution nor the explosion of radical thought provided a strong foundation for understanding the key issues of environment and development that began to emerge. Here the utilitarian ideas about sustainable development that emerged in *The World Conservation Strategy* in 1980 are important, for they reflect the idea established in geography that natural resources should be scientifically identified and rationally assessed, and their utilization planned to maximize the efficiency of their use and the equity of their allocation.

However, if geographers are interested in following the "road still beckoning" (Kates 1987) to play a major role in debates about people and environment, there is now a far richer range of debates within the discipline that are relevant. Developments in physical geography, and techniques such as modeling, offer greater sophistication in scientific understanding of environmental change (e.g. Roberts 1994; Huggett 1993); radical scholarship has placed the rather static and complacent debates of some resources geography into political economic context (e.g. Watts 1983; Hewitt 1983; Blaikie 1985); and cultural geographers have looked anew at nature and at the discourses of science about it, to challenge the conventional ideas that underpin the "environment – development" debate (e.g. Fitzsimmons 1989; Katz and Kirby 1991; Evernden 1992; Simmons 1993a). In the 1990s, geographers have at hand the empirical and theoretical tools with which to both enter and lead pragmatic debates about sustainable development, and with which to explore and understand their implications. Both tasks are important, for the discipline and for their potential contribution to the growing international debate.

Dialogue and Debate over Sustainable Development

The documents of the sustainable-development mainstream have been produced by a process of consensus-creation. *The World Conservation Strategy* and *Caring for the Earth* were both debated at IUCN General Assemblies, and extensively with the various partners producing them. *Our Common Future*, and the various products of the Rio Conference, were even more obviously the fruit of political negotiation. It would be much too easy, however, to conclude from the carefully constructed rhetoric that has resulted, and from the extent to which at least the phrase "sustainable development" has been adopted, that these ideas are unproblematic. The superficial conformity of writing and thinking about sustainable development hides very real and fierce battles behind the scenes over meanings. First, the utilitarian mainstream of sustainable development is matched by counter-currents drawn from very different views of nature. Secondly, mainstream sustainable-development thinking is remarkably light on coherent theory, and in particular it fails to tackle the fundamental questions of the political economy of development.

First let us briefly consider some of the less utilitarian streams in thinking about environment and development. O'Riordan (1988) draws an important distinction between sustainable utilization and sustainability. Sustainable utilization is rational and utilitarian, and (as has been argued above) has dominated thinking about sustainable development since *The World Conservation Strategy*, although its roots (in geography and elsewhere) are much older. O'Riordan identifies a "triad" of concerns (basic needs, eco-development, and sustainable utilization) that have placed sustainable utilization "at the very heart of the development process" (p. 38). At base it is a scientific concept, drawn into the WCS from notions of sustained yield and the harvesting of renewable resources. Sustainability, on the other hand, is a more complex and less manageable concept embracing additional ethical features, concerned with the "right" management of nature. There is no doubt that a large part of the environmentalist interest in the concept of sustainable development has very little to do with the rational management of resources, or even with basic needs or inter- or intra-generational equity. It has to do with the moral and ethical treatment of nature. Such ideas are driven both by traditional concerns about extinction (and their corollary of preservation) and by new ideas about bioethics and the intrinsic rights of other species, or even of inanimate nature (e.g. "deep ecology," Devall and Sessions 1985). Within these radical alternatives to utilitarian environmentalism there is great diversity, and little agreement. Lewis (1992) identifies five distinct streams of thought. Two fall within the broad area of deep ecology, "harsh deep ecology" (which he labels "primitivism") and "soft deep ecology" (or "antihumanist anarchism"). Two lie broadly in the tradition of socialist thought ("humanist eco-anarchism" and "eco-marxism"). The last, "radical eco-feminism," he believes cuts across the rest. It is itself complex and fractured. Despite the appealing synthesis that appears to be offered by sustainable development, below the surface are hidden the seeds of fierce debates about ideologies of nature and highly fractured and contested visions of environmentalism.

The second problem for sustainable development is the way in which it skates over political and economic debates (Adams 1990). Mainstream sustainable-development thinking is built upon the conventional vision of a managed Keynesian world economy, mutual trading to mutual advantage, and the environmentalism of both the 1960s and 1970s. Ian Stewart commented on the Brundtland Report: "Given the Commission's target growth rates, and despite the central role that equity plays in the argument, the gaps between rich and poor will remain of destabilising proportion" (Stewart 1988, p. 119). The documents that set the agenda of sustainable development identify the problems of poverty and inequality, but they have little to say in the way of explanation of their causes, and hence perhaps hardly begin to formulate ways to overcome them. *Caring for the Earth*, for example, has two aims, "to keep human activities within the earth's carrying capacity and to redress the imbalances of security and opportunity between richer and poorer parts of the world" (IUCN 1991, p. 3): fine words, but beyond a simplistic appropriation of standard neo-populist concerns (e.g. basic needs, health, literacy, the particularly significant needs of children and women), there are no clear agendas for action.

Sustainable development is synthetic, and constructed within the confines of current convention in northern-dominated debate about development. The sustainable-development mainstream takes no position on contemporary debates in development (for example, the role of the state, or of aid, or the structure of the world economy), let alone developing new theoretical ideas of its own. Radical thoughts about the world economy or about the nature of this "development" (be they Marxist, anarcho-communist, or eco-feminist) are simply excluded, and the sustainable-development debate has remarkably little to contribute to the "impasse" of development theory in the 1990s (c.f. Schuurman 1993).

The lack of theoretical rigor and radical bite in sustainable development is not surprising. The mainstream texts are all about negotiation, and it is this political process of negotiation that explains the widespread support they command. They not only avoid shaking the status quo, but have become so rounded as to be all things to all readers. All sides of debates about environment and development are able to find substantial numbers of things on which they agree in these documents, precisely because they are not analytically very precise (Redclift 1987; O'Riordan 1988; Lélé 1991).

It is in the area not of theory but of praxis that sustainable development is having its primary effect. The tension between radicalism and conformism within sustainable-development thinking (c.f. Adams 1990) is slowly but surely being resolved in favor of a pragmatic reform of existing systems of decision-making, particularly through the incorporation of environmental assessment and environmental economics into the design of development projects and programs (e.g. Munasinghe 1993). Sustainable development might seem simply a wholly new approach to development, built on new ideas and a new consensus between environmentalists and development agencies, and the First and Third Worlds. In practice, behind the green rhetoric, change is both less dramatic and less extensive. Despite its radical promise, it is very much business as usual in the world of development.

PART VI

Conclusion

Sea-Land
5

22

Remapping the World: What Sort of Map? What Sort of World?

Peter J. Taylor, Michael J. Watts and R. J. Johnston

Introduction

We have tried to organize our contributors around five overarching themes each containing four chapters. The authors, on the other hand, have had other ideas; they have broken free of the strictures imposed by our neat organization, and that is how it should be. As we explained in chapter 1, the division into parts with "geo" prefixes was simply a pedagogic stance since global change is intrinsically multidisciplinary in nature. Hence we can find discussion on postmodernism and the environmental crisis in Part I, "Geoeconomic Change," on cultural relativities and social movements in Part II, "Geopolitical Changc," on political breakdowns, economic restructuring, and environmental problems in Part III, "Geosocial Change," on new divisions of labor, and the economic and political consequences of communication in Part III, "Geocultural Change," and on economic growth and the limitations of politics in Part V, "Geoenvironmental Change." All of this is very worthy of geography in its best holistic tradition, as the liberal arts degree that is most relevant for the world in which students are going to spend the rest of their lives. This interweaving of a rich variety of global themes cannot, should not, be too neatly summarized therefore, but should be left to stand in its own connected diversity.

In this concluding chapter we will highlight one basic similarity and one key difference that provide a dual focus on the inherent complexities of contemporary global change. Both draw upon our discussion of the space–time concepts introduced in the opening chapter but here we emphasize the various ways in which the authors have informed our original discussion. First, all the chapters show an acute awareness of the dangers of globalization as a superficial slogan. Global change does not in any sense make other geographical scales disappear, quite the reverse in fact: the rise of "globalization" coincides with a simultaneous affirmation of "localization" as places both of control (e.g. world cities) and of resistance (e.g. new nationalisms). There is much debate about the form that the global–local nexus takes at the end of the twentieth century but there is no doubt as to its importance. Secondly, the chapters reveal a diversity of terms to describe the rapid changes they analyze: "restructuring" is the customary way of describing geoeconomic change, geopolitical change is about "new order," in geosocial change "transitions" are commonplace, geocultural change is about "new identities," and it is in discussions of geoenvironmental change that the harbingers of "crisis" and "catastrophe" are most likely to be met. All of these terms have different meanings for the nature of change not just in terms of substantive content but also in terms of degree of change. We investigate the latter below.

Global–Local Nexus

The treatment of geographical scale has ranged across the whole gamut of possible positions in the chapters above. Some contributors have argued very strongly that what is global is local and *vice versa*; others, that scale is a crucial dimension for understanding social practices as, for instance, with representation in democratic theory and as the necessary ordering framework for understanding society–nature relations. Can such obverse positions be equally true beyond their specific subject matters? This is a difficult question and our answer hinges on how we deal with the perennial problem of the relation between theory and practice in social science. Clearly, in conceptualizing our world we must not fall into the trap of thinking that geographical scales exist separately from the social practices that create and continue to modify them. Logically, it is not possible to carry out a social activity at one particular scale but not at other scales. But this does not mean scale is neutral or inert in social activities. It is equally clear that some practices are facilitated by focusing at particular scales because effects of practices are typically concentrated at particular scales (Taylor 1993a, pp. 40–7). But since scales are not simply given but are themselves products of social activities, interpretations must be measured, reflecting the messy world we live in. Slogans such as "act locally, think globally" have important propaganda value in some circumstances but they are inadequate for a reasoned theoretical, or indeed political, understanding of global change. For instance, the implication of by-passing the state, in the above slogan, is especially problematic for many social practices, as is recognized above, and not just by the authors writing about geopolitical change.

Recent concern for geographical scale has concentrated on the global and the local as opposite ends of the range of social-space possibilities. But even these two seemingly straightforward concepts are by no means unproblematic (Smith 1993). The global implies a worldwide universalism, whereas the reality is that the processes of globalization are quite uneven. In the communications revolution, for example, the majority of humanity is "out of the loop" and there is little prospect of large swaths of the "South" being hooked into the system in the foreseeable future (Castells 1993). Similarly we can ask how local is local? If local implies community then only small neighborhoods and villages can be "true" communities based upon face-to-face interactions. Most studies treat localities as the local scale which is defined in terms of a town or city and its dependent region. This involves many definitional problems, but there is an underlying notion of local economy and society in a symbiotic relationship facing the outside world (Cooke 1989). However, despite these problems the chapters have been able to uncover a global–local nexus bridging a wide range of social practices. By nexus we mean that a complex connectivity exists between the two limiting geographical-scale possibilities. Five such examples can be easily identified from the chapters above.

Localities are often portrayed as "economic victims" of global forces, where investment decisions made thousands of miles away can make or break communities. This is sometimes known as the regional dilemma in a new market-led world. But life is never that simple. Localities are not inert population aggregates, they are constituted of people and their social networks that can, and do, devise practices to attract, retain, boost, and otherwise ameliorate forces that seem to be beyond control. This global–local nexus is, of course, enormously complex both economically and politically. This complexity is often related to a second nexus concerning the recent rise of multiple ethnic rebellions, religious revivals, and nationalisms. Each can be interpreted in part as local resistances to the homogenizing global political forces that favor larger and larger political spaces to counter economic globalization. These movements attempt to generate a politics sensitive to local needs, in reaction to the destructive and destabilizing impact of economic restructuring. A third nexus is a broader formulation to what lies behind ethnic revivals. The postmodern celebration of diversity in all its forms – gender, race, sexuality, physical ability, as well as religion and ethnicity – derives from a critique of the meta-narratives of modernity as a sort of intellectual globalization. The global implied by modernized space is countered by the local identified through diversity in places. But diversity is not universally accepted and intolerance has created a fourth "new world disorder" nexus wherein global changes are translated into numerous local conflicts. Sometimes surrogate wars for outside powers, the contemporary world is ablaze with political "flash points," civil wars that destroy localities, creating millions of refugees. These are the places we gasp at in horror on our television screens in the comfortable world. The fifth nexus treats the broader notion of destruction, the overloading of ecosystems locally to the point of producing uninhabitable localities, which can culminate in the destruction of the earth as a living system.

The key point about these interlocking global–local nexuses is that they each represent real tensions between activities and consequences that are separated by geographical scale. The conundrum is that there is no easy way to overcome such problems of remoteness. No sooner is a solution found to one particularly onerous and dangerous situation than the world has moved on, spawning another series of related problems. For instance, many people looked forward to the end of the Cold War as a means of solving some very crucial problems, notably the threat of nuclear war. The Cold War is over but the problems have not disappeared; worries about nuclear proliferation, and many more problems, have arisen. All we can say is that the world is now different, not necessarily better. And so we must return to the one unchanging fact of our world – that it is forever changing.

Remapping

In the opening chapter we listed a selection of contemporary news stories to illustrate the global nature of the world we are living in. This concluding chapter has been written less than six months later and our restless world has moved on. Africa has come to the fore with stories of massive contrasting fortunes. In South Africa a successful all-race election has taken place with little violence and a fair ballot in most provinces. But celebrations of the election of Nelson Mandela coincide with an unprecedented slaughter of people in Rwanda, where a civil war has produced hundreds of thousands of corpses in just a few weeks. In neighboring Uganda, Lake Victoria has been declared a disaster zone due to the pollution of rotting bodies flowing down the rivers from Rwanda. Six months ago few people had even heard of the little Central African country of Rwanda. Now it will become a watchword for brutality and descent into an earthly hell. This mixture of hope and despair extends far beyond one continent. In Gaza and Jericho Palestinians are tentatively taking over a small part of their lands as a first step toward peace in this part of the Middle East. In contrast, in Northern Ireland hopes for peace were temporarily squashed as the paramilitaries from both communities launched a tit-for-tat killing spree. Meanwhile civil war has broken out in Yemen. . . . By the time these words are being read the world will have moved on again, creating a different geography of hope and despair.

Geopolitical changes tend to hog the newspaper headlines but they are not necessarily the most important changes that are happening. For instance, the 1994 signing of the new GATT treaty will have profound effects on the livelihoods of millions of people in the coming years. We do not know yet the impact upon race relations in Europe of the legitimation of neo-Fascists by their inclusion in government (as a result of the Italian general election) for the first time since the defeat of Nazism. In contrast, one of the most conservative of churches, the Anglican community, has accepted the ordination of women in its heartland, England, providing an important ideological boost for the cause of equality between the sexes. This is all part of what we termed "all change" in our opening chapter. The question is, how can we interpret it? Our contributors have identified a range of ideas about the nature of this change. Social change may be speeding up, it may exhibit regular rhythms, or perhaps these arguments for continuity have got it wrong and we are experiencing a discontinuity, a rupture with what has gone before. We have at our disposal quite a large lexicon for describing this change – restructuring, transitions, cycles, trends, and numerous concepts that have the prefix "post" – but as yet no clear articulation of terminology to encompass the wide range of global changes covered in this book.

For the title of both this chapter and the book we have described change using the cartographic metaphor of remapping. Given its spatial connotations this may be deemed particularly appropriate for a geography text but our motives go beyond such niceties. We treat the notion of remapping as a means for thinking about open-ended processes of social change in order to dispel some pernicious myths about what is happening. We identify six remappings which we believe to be fundamental for a basic understanding of the current trajectory of our world. First, there is a reconfiguration of capitalism occurring with new divisions of labor, an enhanced role for finance, and new possibilities of control through communication and computing-technology innovations. This package of changes is disrupting traditional capital–labor relations within countries, but there is no simple trend to economic globalization. Related to this is the second remapping, which involves the promotion of markets as resource allocators in an increasingly deregulated world. But markets do not exist in a social vacuum. Evangelical marketers in both the old communist world and the Third World have confused market principles with market practices. All social activity is premised on the existence of rules: deregulation presupposes an alternative reregulation.

Which brings us to the third remapping, in the challenge to state sovereignty. Clearly the functions of the state have changed appreciably, but this does not mean the demise of the state. The legitimacy it gives to markets and other social activities continues to be a necessary prop to cope with social change. Without the state, capitalism's creative destruction, which is continually reproducing the system, degenerates into private mafias such as the market warlords in the former USSR, and short-term economic destruction. What is required is the fourth remapping, the construction of new civil societies in which peoples can engage with social change through their states. Through much of the world, social institutions have been eliminated in massive reform programs but without, as yet, new democratic institutions to take their place. This challenges the multi ethnic nature of most states, since democracy presupposes a single "demos," or people, through which to operate. The fifth remapping genuinely transcends the state in the form of environmental change. The environment is no respecter of political boundaries, and, when a "world leader" such as President George Bush attends the Environmental Summit at Rio promising to look after America's national interests, we can appreciate the necessity for a new politics that environmentalists are struggling to create. The final remapping encompasses much that is implied in the previous remappings. Modernity is being debated, which means that our centuries-old faith in automatic social progress is being re-evaluated. Does this latest critique of modernity represent something quite different from the long history of challenges to the notion of Progress, or Science, or Enlightenment? The answer to this question turns on whether one views postmodernism as taken from the same cloth as those earlier movements that bemoaned the loss of custom, tradition, and some mystic past.

All of these remappings imply profound social change, although whether this is in the form of continuity – more of the same – or discontinuity – a turning point to something different – remains open for debate. As editors we disagree amongst ourselves on this. But our purpose here is not to polish the crystal ball and predict the unpredictable; rather, we conclude by considering the remappings as geographies of global change. And at one level that geography is very simple. Contemporary trends are producing a more polarized world: according to the United Nations Development Program, between 1960 and 1990 differentials between the richest and poorest countries increased by 20 percent (UNDP 1993). At the personal level this is good news to most readers of this book since, as we noted at the beginning of the book, you are part of the comfortable world of relative affluence. With average luck, each of you should be able to find a pleasant niche in this world – which is, of course, one of the reasons you are reading this book, as part of reading for a degree. This is not a callous view, it is a realistic one defined by three authors who are also fully integrated into the comfortable world. But what of the rest of humanity, in their world of struggle?

While the comfortable world debates who will win the head-to-head economic competition between America, Europe, and Japan (Thurow 1992), those who are not combatants in this great contest already know who the real losers are. While the "Big Three" have hammered out a trade agreement through GATT and have got the rest of the world to go along with them, it should come as no surprise that sub-Saharan Africa is predicted to be the region to suffer most from the treaty's trade provisions. And this is to follow years of net outflows of billions of dollars from the continent, to a degree that it almost disappeared from the world economic map altogether. No wonder the Mexican Zapatist guerrilla leader, Subcomandante Marcos, has described NAFTA and the general opening up of the international economy as a "death sentence" for the poor. For many millions, that is what it will be. And these will include people in the former "Second World." So-called "shock therapy" in Russia and Eastern Europe is not only taking a massive economic toll in unemployment; precipitous collapses in entitlements and welfare are causing a rise in death rates among vulnerable groups – in Russia the death rate as a whole went up 22 percent in 1992 alone (*New York Times*, 1 March 1994, p. 1). Overall, the prospects for vast tracts of the former Second World and Third World are increasingly bleak. And this relates, of course, to the political turmoil in these regions. For all the democratic hype at the end of the Cold War, the political victory of liberal democratic states over communist states has no direct bearing on the transferability of liberal democracy across the world, beyond an initial "honeymoon period" of winner's euphoria. As we have emphasized before, social institutions have to be constructed, and that requires a suitable context. Increasing economic polarization both between and within countries is a most unsuitable context for creating institutions that define a political equality. Of course the idea of diffusing liberal-democratic practices is not new. It was tried after decolonization and failed almost everywhere. As before, instead of creating new liberal-democratic states, the likely outcome is liberal-democratic interludes where the military or authoritarian movements take advantage of the political opportunities that will inevitably arise from the continuing economic difficulties of the poorer countries (Taylor 1993a, pp. 270–6). Kaplan's (1994, pp. 48–60) image of a bifurcated world emerges to haunt us: "Part of the globe is inhabited by Hegel's . . . Last Man, healthy and well-fed . . . [T]he other part by Hobbes' First Man, condemned to a life that is 'poor, nasty, brutish and short'."

Whatever happened to "one world" in all of this? However unevenly, our world is interconnected through the vortex of globalization. Polarization could work as a sustainable system if the world were populated by rather dim economic men and women. But it is not; it is full of human beings with hopes and dreams and expectations. These will have to be accommodated across the world to prevent a politics of polarization destroying all. An interconnected world is an easily sabotaged world. Terrorism, the politics of the weak, has followed a definite upward trend. It used to be confused with Cold War conflict but, with the latter out of the way and terrorism showing no signs of abating, we can expect more and greater upheaval of our comfortable lives. Globally this throws up some nice ironies: in 1993 US tourists were avoiding Europe in case they got bombed, while European tourists were cancelling bookings in the USA in case they got mugged. This last example is important. The world of struggle interpenetrates the world of comfort, and increasingly so with the growth of polarization within countries. When will the protection costs of comfort outweigh the economic benefits of living in a comfortable world? Let us hope our world leaders realize that increasing polarization is not a sustainable condition for humanity, before the lessons of terrorism become clear for all to see.

Bibliography

Academy of Sciences (1993) *Population Summit of the World's Scientific Academies*. Washington: National Academy Press.

Adams W. M. (1990) *Green Development: Environment and Sustainability in the Third World*. London: Routledge.

Adams, W. M. (1993) Sustainable development and the greening of development theory. In F. Schuurman (ed.), *Beyond the Impasse: new directions in development theory*. London: Zed Books.

Aglietta, M. (1987) *A Theory of Capitalist Regulation: the US experience*. London: Verso.

Agnew, C. and E. Anderson (1992) *Water Resources in the Arid Realm*. London: Routledge, pp. 207–27.

Agnew, J. (1987) *Place and Politics: the Geographical Mediation of State and Society*. London: Allen Unwin.

Agnew, J. and P. L. Knox (1989) *The Geography of the World Economy*. London: Edward Arnold.

Agulhon, M. (1981) *Marianne in Battle: Republican Imagery and Symbolism in France, 1798–1880*. Cambridge: Cambridge University Press.

Aksoy, A. and K. Robins (1992) Hollywood for the 21st century: global competition for critical mass in image markets. *Cambridge Journal of Economics*, 16, 1, 1–22.

Alber, J. (1988) Is there a crisis of the welfare state? Cross-national evidence from Europe, North America and Japan. *European Sociological Review*, 4, 3, 181–207.

Alvares, C. (1992) *Science, Development and Violence*. Delhi: Oxford University Press.

Amin, A. and N. Thrift (1992) Neo-Marshallian nodes in global networks. *International Journal of Urban and Regional Research*, 16, 571–87.

Amin, S. (1976) *Unequal Development: An Essay on the Social Formations of Peripheral Capitalism*. New York: Monthly Review Press.

Ampel, N. M. (1991) Plagues – what's past is present: thoughts on the origin and history of new infectious diseases. *Reviews of Infectious Diseases*, 13, 658–65.

Anderson, B. (1983) *Imagined Communities*. London: Verso.

Anderson, B. (1992) The new world disorder. *New Left Review*, 193, 3–14.

Anderson, K. (1992) *Vancouver's Chinatown: Racial Discourse in Canada, 1875–1980*. Montreal: McGill-Queen's University Press.

Anzaldua, G. (1987) *Borderlands/La Frontera: The New Mestiza*. San Francisco: Spinster/Aunt Lute.

Appadurai, A. (1990) Disjuncture and difference in the global cultural economy. In M. Featherstone (ed.), *Global Culture. Nationalism, Globalization and Modernity*. Newbury Park, CA: Sage, pp. 295–310.

Appadurai, A. (1991) Global ethnoscapes: notes and queries for a transnational anthropology. In R. G. Fox (ed.), *Recapturing Anthropology: Working in the Present*. Santa Fe: School of American Research Press, pp. 191–210.

Appleyard, R. T. (1992) Migration and development: a global agenda for the future. *International Migration*, 30, 1/2, 17–54.

Arnold, G. (1993) *The End of the Third World*. London: Macmillan.

Ashley, R. (1987) The geopolitics of geopolitical space: towards a critical social theory of international politics. *Alternatives*, 12, 403–34.

Auletta, K. (1993a) Raiding the global village. *The New Yorker*, 2 August, pp. 25–30.

Auletta, K. (1993b) The last studio in play. *The New Yorker*, 4 October, pp. 77–81.

Auty, R. (1993) *Sustaining Development of Mineral Economies*. London and New York: Routledge.

Ayers, R. U. (1992) Toxic heavy metals: materials cycle optimization. *Proceedings National Academy of Sciences*, 89, 815–20.

Ball (1990) The Process of International Contract Labour Migration from the Philippines: The Case of Filipino Nurses. Unpublished Ph.D. dissertation, University of Sydney, Australia.

Bandyopadhyay, J. and V. Shiva (1988) Political economy of ecology movements. *Economic and Political Weekly*, June, pp. 1223–32.

Barff, R. and J. Austen (1993) "It's Gotta be da Shoes": domestic manufacturing, international subcontracting, and the production of athletic footwear. *Environment and Planning A*, 25, 8, 1103–14.

Barnes, T. (1992) Reading the texts of theoretical economic geography. In T. Barnes and J. Duncan (eds), *Writing Worlds: Discourse, Text and Metaphor in the Representation of Landscape*. London: Routledge, 118–35.

Barrett, M. (1991) *The Politics of Truth: From Marx to Foucault*. Cambridge: Polity Press.

Bauer, P. T. (1966) *Economic Analysis and Policy in Underdeveloped Countries*. London: Routledge and Kegan Paul.

Bell, D. (1994) Erotic topographies. *Antipode*, 26, 96–100.

Bell, D., J. Binney, J. Cream, and G. Valentine (1994) *Landscapes of Desire*. London: Routledge.

Bello, W. (1994) *Dark Victory*. Oakland: Food First.

Benedick, R. E. (1991) *Ozone Diplomacy*. Cambridge, MA: Harvard University Press.

Benenson, A. S. (1990) *Control of Communicable Diseases in Man*, 15th edn. Washington, DC: American Public Health Association.

Beneria, L. (1992) Accounting for women's work: the progress of two decades. *World Development*, 20, 11, 1547–60.

Beneria, L. and S. Feldman (1992) *Unequal Burden, Economic Crises, Persistent Poverty, and Women's Work*. Boulder, CO: Westview Press.

Berger, M. T. (1993) Civilising the South: the US Rise to Hegemony in the Americas and the Roots of "Latin American Studies" 1898–1945. *Bulletin of Latin American Research*, 12, 1, 1–48.

Bhabha, H. K. (ed.) (1990) *Nation and Narration*. London: Routledge.

Bhashkar, R. (1989) *Reclaiming Reality*. Oxford: Verso.

Biswas, M. R. and A. K. Biswas (1984) Complementarity between environment and development processes. *Environmental Conservation*, 11, 35–43.

Blaikie, P. (1985) *The Political Economy of Soil Erosion*. London: Longman.

Blaikie, P. and H. Brookfield (1987) *Land Degradation and Society*. London: Routledge.

Blaut, J. (1994) *The Colonizer's Model of the World: Geographical Diffusionism and Eurocentric History*. New York: Guilford.

Blonsky, M. (1992) *American Mythologies*. New York: Oxford University Press.

Bluestone, B. (1990) The impact of schooling and industrial restructuring on recent trends in wage inequality in the United States. *AEA Papers and Proceedings*, 80, 2, 303–7.

Bluestone B. and B. Harrison (1982) *The Deindustrialization of America*. New York: Basic Books.

Boardman, R. (1981) *International Organisations and the Conservation of Nature*. Bloomington, IN: Indiana University Press.

Bongaarts, J. (1985) The fertility inhibiting effects of the intermediate fertility variables. In F. Shorter and H. Zurayk (eds), *Population Factors in Development Planning in the Middle East*. Cairo: Population Council, pp. 152–69.

Bongaarts, J. (1992) Population growth and global warming. *Population and Development Review*, 18, 299–319.

Bonifice, P. and J. Fowler (1993) *Heritage and Tourism in "The Global Village."* London: Routledge.

Bookchin, M. (1986) *The Modern Crisis*. Philadelphia: New Society Publishers.

Booz·Allen & Hamilton (1993) *The Changing Environment for UK Broadcasters and its Economic Implications*. London: ITV Network Association.

Bordo, S. (1990) Feminism, postmodernism and gender-scepticism. In L. Nicholson (ed.), *Feminism/Postmodernism*. London: Routledge, pp. 133–56.

Boserup, E. (1980) *Population and Technology*. Oxford: Blackwell.

Bouvier, M., D. Pittet, L. Loutan, and M. Starobinski (1990) Airport malaria: a mini-epidemic in Switzerland. *Schweizerin Medizin Wochenschreiber*, 120, 1217–22.

Boyer, R. (1990) *The Regulation School: a Critical Introduction*. New York: Columbia University Press.

Bradley, D. J. (1988) The scope of travel medicine. In R. Steffen et al., *Travel Medicine: Proceedings of the First Conference on International Travel Medicine. Zurich, Switzerland, April 1988*. Berlin: Springer Verlag, pp. 1–9.

Bramwell, A. (1989) *Ecology in the 20th Century. A History*. New Haven, CT: Yale University Press.

Braudel, F. (1980) *On History*. London: Weidenfeld and Nicolson.

Braudel, F. (1984) *Civilization and Capitalism, 15th–18th century*. Vol. 3: *The Perspective of the World*. New York: HarperCollins.

Brecher, J., J. Brown Childs, and J. Cutler (eds) (1993) *Global Visions: Beyond the New World Order*. Boston: South End Press.

Brimblecombe, P. (1987) *The Big Smoke*. London: Routledge.

Brookfield, H. C. (1992) "Environmental colonialism," tropical deforestation, and concerns other than global warming. *Global Environmental Change*, 2, 93–6.

Brundtland, H. (1987) *Our Common Future*. Oxford: Oxford University Press (for the World Commission on Environment and Development).

Brydon, L. and S. Chant (1984) *Women in the Third World: Gender Issues in Rural and Urban Areas*. London: Edward Elgar.

Bulatao, R., E. Bos, P. Stephens, and M. Vu (1990) *World Population Projections 1992–3 Edition*. Baltimore: World Bank/Johns Hopkins.

Bunker, S. (1985) *Underdeveloping the Amazon: Extraction, Unequal Exchange, and the Failure of the Modern State*. Urbana and Chicago: University of Illinois Press.

Bunker, S. (1991) The Political Economy and Ecology of Raw Materials Extraction and Trade. University of Wisconsin-Madison, unpublished manuscript.

Burger, J. (1990) *Gaia Atlas of First Peoples: a Future for the Indigenous World*. New York: Anchor Books.

Burton, I. and R. W. Kates (eds) (1965) *Readings in Resource Management and Conservation*. Chicago: University of Chicago Press.

Busch, L., A. Bonnano and W. Lacy (1989) Science, technology and the restructuring of agriculture. *Sociologia Ruralis*, 29, 2, 118–30.

Bushrui, S., I. Ayman and E. Lazlo (eds) (1993) *Transition to a Global Society*. Oxford: One World Publications.

Caldwell, L. K. (1984) Political aspects of ecologically sustainable development. *Environmental Conservation*, 11, 299–308.

Cardiff, D. and P. Scannell (1987) Broadcasting and national unity. In J. Curran, A. Smith, and P. Wingate (eds), *Impacts and Influences: Essays on Media and Power in the Twentieth Century*. London: Methuen, pp. 157–73.

Cardoso, F. H. (1972) Dependency and development in Latin America. *New Left Review*, 74 (July–August), pp. 83–95.

Cardoso, F. H. and E. Faletto (1979) *Dependency and Development*. Berkeley: University of California Press.

Carmichael, A. G. and A. M. Silverstein (1987) Smallpox in Europe before the seventeenth century: virulent killer or benign disease? *Journal of the History of Medicine and Allied Sciences*, 42, 147–68.

Carpenter, R. A. (1989) Do we know what we are talking about? *Land Degradation and Rehabilitation*, 1, 1–3.

Carter, F. W. (1993) Ethnicity as a cause of migration in Eastern Europe. *Geojournal*, 30, 241–8.

Castells, M. (1983) *The City and the Grassroots*. Berkeley: University of California Press.

Castells, M. (1989) *The Informational City. Information Technology, Economic Restructuring and the Urban-Regional Process*. Oxford: Blackwell.

Castells, M. (1993) The informational economy and the new international division of labour. In M. Carnoy et al., *The New Global Economy in the Information Age*. University Park, PA: Pennsylvania State University Press.

Chandra, R. (1992) *Industrialization and Development in the Third World*. London: Routledge.

Chase-Dunn, C. (1985) The systems of world cities 800 A.D.–1975. In M. Timberlake (ed.), *Urbanization in the World-Economy*. New York: Academic Press, 269–382.

Chase-Dunn, C. (1989) *Global Formation*. Oxford: Blackwell.

Chisholm, M. (1982) *Modern World Development*. Totowa NJ: Barnes and Noble.

Christopherson, S. (1991) *The Service Sector: a Labour Market for Women?* OECD/GD (91)212. Paris: Organisation for Economic Cooperation and Development.

Clarke, C., C. Peach, and S. Vertovec (eds) (1990) *South Asians Overseas*. Cambridge: Cambridge University Press.

Clark, G. L. (ed.) (1992) Special issue on "real" regulation. *Environment and Planning A*, 24, 615–27.

Clark, G. L. (1993) Global interdependence and regional development: business linkages and corporate government in a world of financial risk. *Transactions, Institute of British Geographers*, N.S. 18, 309–25.

Clarke, R. and L. Timberlake (1982) *Stockholm Plus Ten: Promises Promises? The Decade since the 1972 UN Environment Conference*. London: Earthscan.

Cleland, J. and J. Hobcraft (eds) (1985) *Reproductive Change in Developing Countries*. Oxford: Oxford University Press.

Cleveland, C. J., R. Constanza, C. A. S. Hall, and R. Kaufmann (1984) Energy and the US economy: a biophysical perspective. *Science*, 225, 890–97.

Cliff, A. D. and P. Haggett (1985) *The Spread of Measles in Fiji and the Pacific: Spatial Components in the Transmission of Epidemic Waves through Island Communities*. Canberra: Australian National University.

Cliff, A. D., P. Haggett, and J. K. Ord (1986) *Spatial Aspects of Influenza Epidemics*. London: Pion.

Cliff, A. D. and P. Haggett (1988) *Atlas of Disease Distributions: Analytical Approaches to Epidemiological Data*. Oxford: Blackwell.

Coale, A. (1982) A re-assessment of world population trends. *Population Bulletin of the United Nations*, 14, 1–16.

Coale, A. (1991) Excess female mortality and the balance of the sexes: an estimate of the number of missing females. *Population and Development Review*, 17, 3, 517–23.

Cohen, J. L. (1987) Strategy and identity: new theoretical paradigms and contemporary social movements. *Social Research*, 52, 4, 663–716.

Commission of the European Community (1988) *The Future of Rural Areas*. Brussels.

Cooke, P. (1989) *Localities*. London: Unwin Hyman.

Corbridge, S. E. (1986) *Capitalist World Development: a Critique of Radical Development Geography*. London: Macmillan.
Corbridge, S. E. (1991) Third world development. *Progress in Human Geography*, 15, 3, 311–21.
Corbridge, S. (1993a) *Debt and Development*. IBG Studies in Geography. Oxford: Blackwell.
Corbridge, S. (1993b) Colonialism, post-colonialism and the political geography of the Third World. In P. J. Taylor (ed.), *Political Geography of the Twentieth Century*. London: Belhaven, pp. 171–205.
Cosgrove, D. and S. Daniels (eds) (1988) *The Iconography of Landscape*. Cambridge: Cambridge University Press.
Cutter, S. (1993) *Living with Risk*. London: Edward Arnold.
Daniels, S. (1993) *Fields of Vision: Landscape Imagery and National Identity in England and the United States*. Cambridge: Polity.
Daroesman, R. (1979) An economic survey of East Kalimantan. *Bulletin of Indonesian Studies*, 15, 3, 43–82.
Dasmann, R. F., J. P. Milton, and P. H. Freeman (1973) *Ecological Principles for Economic Development*. Chichester: Wiley.
Davidson, B. (1992) *The Black Man's Burden: Africa and the Curse of the Nation-State*. New York: Times Books.
Davis, K. (1963) The theory of change and response in demographic history. *Population Index*, 29, 345–66.
Davis, M. (1992) *Beyond Blade Runner: Urban Control. The Ecology of Fear*, Open Magazine Pamphlet 23. New Jersey: Westfield.
Denevan, W. M. (1992) The pristine myth: the landscape of the Americas in 1492. *Annals of the Association of American Geographers*, 82, 369–85.
Denitch, B. (1992) *After the Flood: World Politics and Democracy in the Wake of Communism*. Hanover, NH: Wesleyan University Press.
Descombes, V. (1993) *The Barometer of Modern Reason. On the Philosophies of Current Events*. Oxford: Oxford University Press.
Devall, B. and G. Sessions (1985) *Deep Ecology: Living as if Nature Mattered*. Salt Lake City: Peregrine Smith.
Diamond, I. and L. Quinby (eds) (1988) *Feminism and Foucault: Reflections on Resistance*. Boston, MA: Northeastern University Press.
Diamond, L. (ed.) (1992) *The Democratic Revolution: Struggles for Freedom and Pluralism in the Developing World*. New York: Freedom House.
Dicken, P. (1992) *Global Shift: the Internationalization of Economic Activity*, 2nd edn. New York: Guilford.
Dickson, M. (1993) Tremors on the television, *Financial Times*, 15 October.

DiMuccio, R. B. A. and J. N. Rosenau (1992) Turbulence and sovereignty in world politics. In Z. Mlinar (ed.), *Globalization and Territorial Identities*. Aldershot: Avebury, pp. 60–76.

Dollar, D. and E. Wolff (1993) *Competitiveness, Convergence, and International Specialization*. Cambridge, MA: MIT Press.

Domash, M. (1991) Towards a feminist historiography of geography. *Transactions, Institute of British Geographers*, N.S. 16, 95–104.

Donaghu, M. T. and R. Barff (1990) Nike just did it: international subcontracting, flexibility, and athletic footwear production. *Regional Studies*, 24, 537–52.

Dos Santos, T. (1970) The structure of dependence. *American Economic Review*, 60 (May), 231–6.

Douglas, M. (1966) *Purity and Danger: an Analysis of Concepts of Pollution and Taboo*. London: Penguin.

Douglas, M. (1992) A credible biosphere. In M. Douglas (ed.), *Risk and Blame: Essays in Cultural Theory*. London: Routledge, 255–70.

Douglas, M. and A. Wildavsky (1983) *Risk and Culture: an Essay on the Selection of Technological and Environmental Dangers*. Berkeley: University of California Press.

Downs, A. (1972) Up and down with ecology: the issue attention cycle. *The Public Interest*, 10, 38–50.

Driver, F. (1992) Geography's empire: histories of geographical knowledge. *Environment and Planning D: Society and Space*, 10, 23–40.

Du Bois, W. E. B. (1989) *The Souls of Black Folk*. New York: Bantam.

Dunbar, M. J. (1974) Arctic ecosystems and arctic resources. In E. Bylund (ed.), *Ecological Problems of the Circumpolar Area*. Lulea: Nordbottens Museum.

Duncan, J. S. (1989) The power of place in Kandy, Sri Lanka: 1780–1980. In J. A. Agnew and J. S. Duncan (eds), *The Power of Place: Bringing Together Geographical and Sociological Imaginations*. London: Unwin Hyman, 185–201.

Duncan, J. S. (1990) *The City as Text: the Politics of Landscape Interpretation in the Kandyan Kingdom*. Cambridge: Cambridge University Press.

Duncan, J. and N. Duncan (1988) (Re)-reading the landscape, *Environment and Planning D: Society and Space*, 6, pp. 17–26.

Dunning, J. H. (1981) *International Production and the Multinational Enterprise*. London: Allen and Unwin.

Dunning, J. H. (1983) Changes in the level and structure of international production: the last one hundred years. In M. Casson (ed.), *The Growth of International Business*. London: Allen and Unwin, 84–139.

Dunning, J. H., J. A. Cantwell, and T. A. B. Corley (1986) The theory of international production: some historical antecedents. In P. Hertner and G. Jones (eds), *Multinationals: Theory and History*. Brookfield, VT: Gower, 19–41.

Dunning, J. H. and J. A. Cantwell (1987) *IRM Directory of Statistics of International Investment and Production*. Basingstoke: Macmillan.

Durning, A. (1992) *How Much is Enough?* New York: W. W. Norton.

Dymski, G. and J. Veitch (1992) *Race and the Financial Dynamics of Urban Growth*. University of California, Riverside, Department of Economics, Working Paper 92–91.

Dzimbiri, Lewis B. (1993) Political and economic impacts of refugees: some observations on Mozambican refugees in Malawi. *Refuge*, 13, 6, 4–6.

Economist, The (1993) *Survey of the Food Industry*, 4 December 1993.

Ehrlich, P. (1968) *The Population Bomb*. London: Pan.

Ehrlich, P. (1971) Impact of population growth. *Science*, 171, 1212–17.

Ehrlich, P. and A. Ehrlich (1990) *The Population Explosion*. New York: Simon & Schuster.

Ehrlich, P., A. Ehrlich, and G. Daily (1993) Food security, population and environment. *Population and Development Review*, 19, 1–32.

Ekins, P. (1992) *A New World Order: Grassroots Movements for Global Change*. London: Routledge.

Elsom, D. (1992) *Atmospheric Pollution*. Oxford: Blackwell.

Emmanuel, A. (1972) *Unequal Exchange: a Study of the Imperialism of Trade*. New York: Monthly Review Press.

Environmental Project of Central America (1987) Militarization – the environmental impact in Central America. *Cultural Survival Quarterly*, 11, 3, 38–45.

Escobar, A. (1984) Discourse and power in development: Michel Foucault and the relevance of his work to the Third World. *Alternatives*, 10, 377–400.

Escobar, A. (1992a) Culture, economics, and politics in Latin American social movements theory and research. In A. Escobar and S. E. Alvarez (eds), *The Making of Social Movements in Latin America*. Boulder, CO: Westview Press, 62–85.

Escobar, A. (1992b) Imaging a post development era. *Social Text*, 31, 20–54.

Esping-Anderson, G. (1990) *The Three Worlds of Welfare Capitalism*. Cambridge: Polity.

Esteva, G. (1992) Development. In W. Sachs (ed.), *The Development Dictionary*. London: Zed Books, 6–25.

Eudey, A. A. (1988) Another defeat for the Nam Choan Dam, Thailand. *Cultural Survival Quarterly*, 12, 2, 13–16.

Evernden, N. (1992) *The Social Creation of Nature*. Baltimore: Johns Hopkins University Press.

Ewan, C., E. Bryant, and D. Calvert (1990) *Health Implications of Long-Term Climate Change*. Canberra, ACT: National Health and Medical Research Council of Australia.

Farvar, M. T. and J. P. Milton (eds) (1973) *The Careless Technology: Ecology and International Development*. London: Stacey.

Fernandes, W. and E. Thukral (eds) (1989) *Development, Displacement and Rehabilitation*. New Delhi: I.S.I.

Findji, M. T. (1992) From resistance to social movement: the indigenous authorities movement in Colombia. In A. Escobar and S. E. Alvarez (eds), *The Making of Social Movements in Latin America*. Boulder, CO: Westview Press, 112–33.

Fishman, J. (1972) *Language and Nationalism*. MA, Rowley: Newbury House.

FitzSimmons, M. (1989) The matter of nature. *Antipode*, 21, 106–20.

FitzSimmons, M. (1990) The social and environmental relations of the US agricultural regions. In P. Lowe, T. Marsden and S. Whatmore (eds), *Technological Change and the Rural Environment*. London: David Fulton, 8–32.

Flax, J. (1990) *Thinking Fragments: Psychoanalysis, Feminism and Postmodernism in the Contemporary West*. Berkeley and Los Angeles: University of California Press.

Flora, P. et al. (1983) *State, Economy and Society in Western Europe 1815–1975: a Data Handbook. Vol. I*. Frankfurt: Campus Verlag, and London: Macmillan.

Foster, H. (ed.) (1985) *Postmodern Culture*. London: Pluto Press.

Foucault, M. (1970) *The Order of Things: An Archeology of the Human Sciences*. London: Tavistock.

Foucault, M. (1977a) *Discipline and Punish: the Birth of the Prison*. London: Penguin Books.

Foucault, M. (1977b) *The Archeology of Knowledge*. London: Tavistock.

Foucault, M. (1978) *The History of Sexuality, Part I*. London: Penguin Books.

Foucault, M. (1979) *Discipline and Punish*. New York: Vintage Books.

Foucault, M. (1980) *Power/Knowledge*. Brighton: Harvester Press.

Frank, A. G. (1967) *Capitalism and Underdevelopment in Latin America*. New York: Monthly Review Press.

Franke, R. and B. Chasin (1992) Kerala: development without growth. *Earth Island Journal*, Spring, pp. 25–6.

Freeman, C. (1973) Malthus with a computer. In H. Cole et al. (eds), *Thinking about the Future: a Critique of the Limits to Growth*. London: Chatto and Windus, 5–55.

Fridenson, P. (1986) The growth of multinational activities in the French Motor Industry, 1890–1979. In P. Hertner and G. Jones (eds), *Multinationals: Theory and History*. Brookfield, VT: Gower, 157–68.

Friedland, W., A. Banton and R. Thomas (1981) *Manufacturing Green Gold*. Cambridge: Cambridge University Press.

Friedland, W. (1991) Introduction to W. Friedland, L. Busch, F. Buttel and A. Rudy (eds), *Towards a New Political Economy of Agriculture*. Boulder, CO: Westview Press, 1–36.

Friedmann, H. (1982) The political economy of food: rise and fall of the postwar international food order. *American Journal of Sociology*, 88, supplement 246–86.

Friedmann, H. and P. McMichael (1989) Agriculture and the state system. The rise and decline of national agricultures, 1870 to the present. *Sociologia Ruralis*, xxix, 93–117.

Friedmann, J. (1986) The world city hypothesis. *Development and Change*, 17, 69–83.

Friedmann, J. (1992) *Empowerment*. Oxford: Blackwell.

Friedmann, J. (1994) Where we stand: A decade of World City research. In P. Knox and P. J. Taylor (eds), *World Cities in a World-System*. Cambridge: Cambridge University Press.

Frobel, F., J. Heinrich, and O. Kreye (1980) *The New International Division of Labor*. Cambridge: Cambridge University Press.

Frognier, A., M. Quevit, and M. Stenbock (1982) Regional imbalances and centre–periphery relationships in Belgium. In S. Rokkan and D. Unwin (eds), *The Politics of Territorial Identity*. London: Sage.

Fujita, K. (1991) A world city and flexible specialization: restructuring the Tokyo metropolis. *International Journal of Urban and Regional Research*, 15, 269–84.

Fukuyama, F. (1992) *The End of History and the Last Man*. New York: Free Press.

Garitaonandia, C. (1993) Regional television in Europe. *European Journal of Communication*, 9, 3, 277–94.

Garson, J. P. (1992) Migration and interdependence: the migration system between France and Africa. In M. M. Kritz, Lin L. Lim and H. Zlotnik (eds), *International Migration Systems*. Oxford: Clarendon Press, 80–93.

Gellner, E. (1992) *Postmodernism, Reason and Religion*. London: Routledge.

Gibson, D. V., G. Kozmetsky, and R. W. Smilor (1992) *The Technopolis Phenomenon. Smart Cities, Fast Systems, Global Networks*. Lanham, MD: Rowman & Littlefield.

Giddens, A. (1985) *The Nation-state and Violence*. Cambridge: Polity.
Giddens, A. (1991) *Modernity and Self Identity. Self and Society in the Late Modern Age*. Cambridge: Polity Press.
Gillis, M. (1987) Multinational enterprises and environmental and resource management issues in the Indonesian tropical forest sector. In C. Pearson (ed.), *MNCS, Environment, and the Third World*. Durham, NC: Duke University Press, 64–89.
Gillis, M. (1988) Indonesia: public policies, resource management, and the tropical forest. In R. Repetto and M. Gillis (eds), *Public Policies and the Misuse of Forest Resources*. Cambridge: Cambridge University Press, 43–113.
Gilmore, J. S. (1975) *Boom Town Growth Management: a Case Study of Rock Springs – Green River, Wyoming*. Boulder, CO: Westview Press.
Gilroy, P. (1993) *The Black Atlantic. Modernity and Double Consciousness*. London: Verso.
Glacken, C. (1967) *Traces on the Rhodian Shore*. Berkeley: University of California Press.
Gleb, A. (1988) *Oil Windfalls: Blessing or Curse?* Washington, DC: Oxford University Press for the World Bank.
Goodin, R. E. and J. Le Grand (1987) *Not Only the Poor: the Middle Classes and the Welfare State*. London: Allen and Unwin.
Goodman, D., A. Sorj and J. Wilkinson (1987) *From Farming to Biotechnology*. Oxford: Basil Blackwell.
Goodman, D. and M. Redclift (1989) *The International Farm Crisis*. London: Macmillan.
Goodman, D. and M. Redclift (1991) *Refashioning Nature*. London: Routledge.
Gore, A. (1992) *Earth in the Balance. Forging a New Common Purpose*. London: Earthscan; Boston: Mifflin.
Goulborne, H. (1991) *Ethnicity and Nationalism in Post-imperial Britain*. Cambridge: Cambridge University Press.
Gould, P. (1990) *Fire in the Rain. The Democratic Consequences of Chernobyl*. Cambridge: Polity Press.
Gramsci, A. (1971) *Selections from the Prison Notebooks*, trans. Q. Haore and G. N. Smith. New York: International Publishers.
Gregory, D. (1994) *Geographical Imaginations*. Oxford: Blackwell.
Gregory, D. and J. Urry (eds) (1985) *Social Relations and Spatial Structures*. London: Macmillan.
Grubb, M. (1990) *Energy Policies and the "Greenhouse" Effect*. Aldershot: Dartmouth.
Guha, R. (1989) The Problem, *Seminar* (March), pp. 12–15.
Haber, W. (1993) Environmental attitudes in Germany: the transfer of

scientific information into political action. In R. J. Berry (ed.), *Environmental Dilemmas – Ethics and Decisions*. London: Chapman and Hall, 33–46.

Habermas, J. (1976) *Legitimation Crisis*. London: Heinmann.

Haggett, P. (1991) Some components of global environmental change. In B. A. Bannister et al., *Report of a Think Tank on the Potential Effects of Global Warming and Population Increase on the Epidemiology of Infectious Diseases*. Colindale: Public Health Laboratory Service, 5–14.

Haggett, P. (1992) Sauer's "Origins and dispersals": its implications for the geography of disease. *Transactions, Institute of British Geographers*, N.S. 17, 387–98.

Hall, S. (1984) The state in question. In G. McLennan, D. Held, and S. Hall (eds), *The Idea of the Modern State*. Milton Keynes: Open University Press.

Hall, S. (1991a) The local and the global. In A. King (ed.), *Culture, Globalization and the World System*, Department of Art and Art History, SUNY Binghamton, 19–39.

Hall, S. (1991b) Ethnicity: identity and difference. *Radical America*, 23, 9–20.

Hannertz, H. (1991) *Cultural Complexity*. New York: Columbia University Press.

Hannerz, U. (1992) The cultural role of world cities. In A. Cohen and K. Fukui (eds), *The Age of the City*. Edinburgh: Edinburgh University Press.

Hansen, A. (ed.) (1993) *The Mass Media and Environmental Issues*. Leicester: Leicester University Press.

Haraway, D. (1991) *Simians, Cyborgs and the Rediscovery of Nature*. London: Free Association Press.

Harvey, D. W. (1974) Population, resources and the ideology of science. *Economic Geography*, 50, 256–77.

Harvey, D. W. (1989) *The Condition of Postmodernity*. Oxford: Blackwell.

Harvey. D. W. (1990) Between space and time: reflections on the geographical imagination. *Annals of the Association of American Geographers*, 80, 418–34.

Harvey, D. W. (1993) From space to place and back again. In J. Bird et al. (eds), *Mapping the Futures*. London: Routledge.

Haubrich, J. G. and P. Wachtel (1993) Capital requirements and shifts in commercial bank portfolios. *Economic Review (Federal Reserve Bank of Cleveland)*, 29, 3, 2–15.

Hays, S. P. (1959) *Conservation and the Gospel of Efficiency: the*

Progressive Conservation Movement 1890–1920, Cambridge, MA: Harvard University Press.

Heathcote, R. L. (1983) *The Arid Lands: their Use and Abuse*. London: Longman.

Hecht, S. and A. Cockburn (1990) *The Fate of the Forest*. New York: Harper Collins.

Heffernan, M. J. (1990) Bringing the desert to bloom. French ambitions in the Sahara Desert during the late nineteenth century – the strange case of *la mer intérieure*. In D. Cosgrove and G. Petts (eds), *Water, Engineering and Landscape: Water Control and Landscape Transformation in the Modern Period*. London: Belhaven, 94–114.

Held, D. (1987) *Models of Democracy*. Cambridge: Polity Press.

Held, D. (1991) Democracy, the nation-state, and the global system. In D. Held (ed.), *Political Theory Today*. Stanford, CA: Stanford University Press.

Henderson-Sellars, A. and R. J. Blong (1989) *The Greenhouse Effect: Living in a Warmer Australia*. Kensington, Sydney: New South Wales University Press.

Herder, J. G. von (1968 [1783]) *Reflections on the Philosophy of the History of Mankind*, abridged and translated by T. Manuel. Chicago: University of Chicago Press.

Hershkovitz, L. (1993) Tiananmen Square and the politics of place. *Political Geography*, 12, 395–420.

Hertner, P. (1986) German multinational enterprise before 1914: some case studies. In P. Hertner and G. Jones (eds), *Multinationals: Theory and History*. Brookfield, VT: Gower, 113–34.

Hewison, R. (1987) *The Heritage Industry*. London: Methuen.

Hewitt, K. (1983) *Interpretations of Calamity: from the Viewpoint of Human Ecology*. Hemel Hempstead: Allen and Unwin.

Hillborn, R. (1990) Marine biota. In B. L. Turner II et al. (eds), *The Earth as Transformed by Human Action*. Cambridge: Cambridge University Press, 371–85.

Hirschmann, A. O. (1958) *The Strategy of Economic Development*. New Haven: Yale University Press.

Hjarvard, S. (1993) Pan-European television news: towards a European political public sphere? In P. Drummond, R. Paterson, and J. Willis (eds), *National Identity and Europe*. London: British Film Institute, 71–94.

Hobsbawm, E. (1990) *Nations and Nationalism since 1780*. Cambridge: Cambridge University Press.

Hobsbawm, E. and T. Ranger (eds) (1986) *The Invention of Tradition*. Cambridge: Cambridge University Press.

Holden, C. (1990) Climate experts say it again: greenhouse is real. *Science*, 248, 964–5.

Holdgate, M. W. (1979) *A Perspective of Environmental Pollution*. Cambridge: Cambridge University Press.

Holmberg, J., K. Thornson, and L. Timberlake (1993) *Facing the Future: Beyond the Earth Summit*. London: Earthscan/International Institute for Environment and Development.

Hönkopp, Elmar (1993) East–West migration: recent developments concerning Germany and some future prospects. In OECD, *The Changing Course of International Migration*. Paris: OECD, 97–104.

Hornung, M. (1993) *Critical Loads: Concept and Applications*. London: HMSO.

Horsman, R. (1981) *Race and Manifest Destiny: The Origins of American Racial Anglo-Saxonism*. Cambridge, MA: Harvard University Press.

Houghton, J., G. J. Jenkins, and J. J. Ephramus (1990) *Climate Change: the IPCC Scientific Assessment*. Cambridge: Cambridge University Press.

Howland, M. (1993) Technological change and the spatial restructuring of data entry and processing services. *Technological Forecasting and Social Change*, 43, 185–96.

Huggett, R. J. (1993) *Modelling the Human Impact on Nature: Systems Analysis of Environmental Problems*. Oxford: Oxford University Press.

Hughes, R. (1993) *Culture of Complaint: the Fraying of America*. New York: Oxford University Press.

Huminer, D., J. B. Rosenfeld, and S. D. Pitlik (1987) AIDS in the pre-AIDS era. *Reviews of Infectious Diseases*, 9, 1102–8.

Hung, W. (1991) Tiananmen Square: a political history of monuments. *Representations*, 35, 84–117.

Hymer, S. (1972) The multinational corporation and the law of uneven development. In J. Bhagwati (ed.), *Economics and World Order*. New York: Free Press, 113–40.

Hymer, S. (1976) *The International Operations of National Firms: a Study of Direct Foreign Investment*. Cambridge, MA: MIT Press.

Hymer, S. (1979) *The Multinational Corporation: a Radical Approach*. New York: Cambridge University Press.

Ignatieff, M. (1993) The Balkan tragedy. *New York Review of Books*, 13 May, p. 3.

Independent (1993) 27 October.

Independent on Sunday (1993) 10 October.

Innis, H. A. (1967) The importance of staple products. In W. T.

Easterbrook and M. H. Watkins (eds), *Approaches to Canadian Economic History*. Toronto: McClelland and Stewart.

Institute for Economics and Industrial Engineering (1993) Report delivered by the IEIE, Siberian Branch of the Russian Academy of Sciences, Novosibirsk, at Clark University, Worcester, MA, on 16 December 1993.

Intergovernmental Panel of Climatic Change (1990) *Scientific Assessment of Climate Change*. Report prepared for IPCC by Working Group 1. Geneva: World Meteorological Organization.

Itoh, M. (1992) The Japanese model of post-Fordism. In M. Storper and A. J. Scott (eds), *Pathways to Industrialisation and Regional Development*. London: Routledge, 116–34.

IUCN (1980) *The World Conservation Strategy*. Geneva: International Union for Conservation of Nature and Natural Resources, United Nations Environment Programme, World Wildlife Fund.

IUCN (1991) *Caring for the Earth: a Strategy for Sustainable Living*. Gland: IUCN.

Jackson, P. (1988) Street life: the politics of Carnival. *Society and Space*, 6, 213–27.

Jameson, F. (1991) *Postmodernism, or, the Cultural Logic of Late Capitalism*. London: Verso.

Jarman, N. (1992) Troubled images. *Critique of Anthropology*, 12, 179–91.

Jessop, B. (1989) Conservative regimes and the transition to post-Fordism: the cases of Great Britain and West Germany. In M. Gottdiener and N. Komninos (eds), *Capitalist Development and Crisis Theory: Accumulation, Regulation and Restructuring*. New York: St. Martin's Press.

Jessop, B. (1990) Regulation theories in retrospect and prospect. *Economy and Society*, 19, 2, 153–216.

Jessop, B. (1992) Fordism and post-Fordism: a critical reformulation. In M. Storper and A. J. Scott (eds), *Pathways to Industrialization and Regional Development*. London: Routledge, pp. 46–69.

Joekes, S. (1987) *Women in the World Economy, An INSTRAW Study*. New York: Oxford University Press.

Johnson, N. C. (1992) Nation-building, language and education: the geography of teacher recruitment in Ireland, 1925–55. *Political Geography Quarterly*, 11, 170–89.

Johnson, N. C. (1994) Sculpting heroic histories: the celebrating the centenary of the 1798 rebellion in Ireland. *Transactions of the Institute of British Geographers*, N.S. 19, 78–93.

Johnston, R. J. (1983) Resource analysis, resource management and the integration of physical and human geography. *Progress in Physical Geography*, 7, 127–46.

Johnston, R. J. (1989a) The individual and the world economy. In R. J. Johnston and P. J. Taylor (eds), The *World in Crisis*. Oxford: Blackwell, 200–30.

Johnston, R. J. (1989b) *Environmental Problems: Nature, Society and the State*. London: Belhaven.

Johnston, R. J. (1992) Laws, states and superstates. *Applied Geography*, 12, 211–20.

Johnston, R. J. (1993) The rise and decline of the corporate-welfare state: a comparative analysis in global context. In P. J. Taylor (ed.), *Political Geography of the Twentieth Century*. London: Belhaven, 115–70.

Johnston, R. J., P. J. Taylor, and J. O'Loughlin (1987) The geography of violence and premature death: a world-systems approach. In R. Vayrynen (ed.), *The Quest for Peace*. London: Sage, 241–59.

Johnston, R. J. and P. J. Taylor (eds) (1989) *A World in Crisis? Geographical Perspectives*, 2nd edn. Oxford: Blackwell.

Jones, H. (1990) *Population Geography*. London: Paul Chapman.

Jones, H. (1993) The small islands factor in modern fertility decline. In D. Lockhart et al. (eds), *The Development Process in Small Island States*. London: Routledge, 161–78.

Jowett, A. J. (1989) China: 40 years of demographic development. *Geography*, 74, 346–9.

Kahn, H. and C. L. Cooper (1993) *Stress in the Dealing Room. High Performers Under Pressure*. London: Routledge.

Kandiyoti, D. (1991) Identity and its discontents: women and nation. *Millennium: Journal of International Studies*, 20, 429–43.

Kane, E. J. (1990) Incentive conflict in the international risk-based capital agreement. *Economic Perspectives (Federal Reserve Bank of Chicago)*, 14, 3, 33–6.

Kaplan, R. (1994) The coming anarchy, *Atlantic Monthly*, February, 44–86.

Kapstein, E. B. (1992) Between power and purpose: central bankers and the politics of regulatory convergence. *International Organization*, 46, 1, 265–87.

Karl, T. R. (1993) Missing pieces of the puzzle. *Research and Exploration*, 9, 234–49.

Kasperson, J. X., Kasperson, R. E. and B. L. Turner II (eds) (1993) *Regions at Risk: International Comparisons of Threatened Environments*. Tokyo: United Nations University.

Kates, R. W. (1987) The human environment: the road not taken and the road still beckoning. *Annals of the Association of American Geographers*, 77, 4, 525–34.

Kates, R. W., B. L. Turner II, and W. C. Clark (1990) The great transformation. In B. L. Turner II et al. (eds), *The Earth as Transformed by Human Action*. Cambridge: Cambridge University Press, 1–18.

Katz, C. and A. Kirby (1991) In the nature of the thing: the environment and everyday life. *Transactions, Institute of British Geographers*, N.S. 16, 259–71.

Kay, C. (1989) *Latin American Theories of Development and Underdevelopment*. London: Routledge.

Kennedy, P. (1993) *Preparing for the Twenty-first Century*. New York: Random House.

Khlat, M. and M. Khoury (1991) Inbreeding and diseases: demographic, genetic and inbreeding perspectives. *Epidemiologic Reviews*, 13, 28–41.

Kidron, M. and R. Segal (1984) *The New State of the World Atlas*. New York: Simon & Schuster.

King, A. D. (ed.), *Culture, Globalization and the World-System*. Basingstoke: Macmillian.

King, A. D. (1990) *Global Cities. Post-Imperialism and the Internationalization of London*. London: Routledge.

Kotlyakov, V. M. (1991) The Aral Sea Basin: a critical environmental zone. *Environment*, 33, 1, 4–9, 36–8.

Laclau, E. (1985) New social movements and the plurality of the social. In D. Slater (ed.), *New Social Movements and the State in Latin America*. Amsterdam: CEDLA, 27–42.

Lam, Lawrence (1993) Focus on Southeast Asian refugees. *Refuge*, 13, 5, p. 1.

Lande, R. and J. Geller (1991) Paying for family planning. *Population Reports*, J39, Baltimore: Johns Hopkins University.

Landry, D. and G. Maclean (1993) *Materialist Feminisms*. Oxford: Basil Blackwell.

Laponce, J. A. (1984) The French language in Canada: tensions between geography and politics. *Political Geography Quarterly*, 3, 91–105.

Latouche, S. (1993) *In the Wake of the Affluent Society: an Exploration of Post-Development*. London: Zed Books.

Latour, B. (1993) *We Have Never Been Modern*. Brighton: Harvester Wheatsheaf.

Law, J. (1994) *Organizing Modernity*. Oxford: Blackwell.

Lawson, V. and Klak, T. (1993) An argument for critical and comparative research on the urban geography of the Americas. *Environment and Planning A*, 25, 8, 1071–84.

Lefebvre, H. (1991) *The Production of Space*. Oxford: Basil Blackwell.

Le Heron (1988) Food and fibre production under capitalism: a conceptual agenda. *Progress in Human Geography*, 12, 3, 409–30.

LéLé, S. M. (1991) Sustainable development: a critical review, *World Development*, 19, 607–21.

Levitt, T. (1983) *The Marketing Imagination*. New York: Free Press.

Lewis, M. (1992) *Green Delusions: an Environmentalist Critique of Radical Environmentalism*. Durham and London: Duke University Press.

Leyshon, A. (1993) Crawling from the wreckage: speculating on the future of the European Exchange Rate Mechanism. *Environment and Planning A*, 25, 1553–57.

Lipietz, A. (1986) New tendencies in the international division of labor: regimes of accumulation and modes of accumulation. In Allen J. Scott and Michael Storper (eds), *Production, Work, Territory: The Geographical Anatomy of Industrial Capitalism*. Boston: Allen and Unwin, 16–40.

Lipietz, A. (1987) *Mirages and Miracles: The Crises of Global Fordism*. London: Verso.

Lipietz, A. (1992a) The regulation approach and capitalist crisis: an alternative compromise for the 1990s. In M. Dunford and G. Kafkalas (eds), *Cities and Regions in the New Europe*. London: Belhaven, 309–34.

Lipietz, A. (1992b) *Towards a New Economic Order: Postfordism, Ecology and Democracy*. Cambridge: Polity.

Liverman, D. M. (1990) Vulnerability to global environmental change. In R. E. Kasperson, K. Dow, D. Golding and J. X. Kasperson (eds) *Understanding Global Environmental Change: the Contributions of Risk Analysis and Management*. Worcester, MA: ET Program, 27–44.

Livingstone, David N. (1992) *The Geographical Tradition*. Oxford: Blackwell.

Loftus, B. (1990) *Mirrors: William III and Mother Ireland*. Dundrum: Picture Press.

London Economics (1992) *The Potential Role of Market Mechanisms in the Control of Acid Rain*. Department of the Environment, Economic Research Series. London: HMSO.

Love, J. L. (1980) Raul Prebisch and the origins of the doctrine of unequal exchange. *Latin American Research Review*, 15, 3, 45–72.

Lovelock, J. E. (1979) *Gaia: a New Look at Life on Earth*. Oxford: Oxford University Press.

Lowenthal, D. (1958) *George Perkins Marsh: Versatile Vermonter*. New York: Columbia University Press.

Lowenthal, D. (1990) Awareness of human impacts: changing attitudes and emphases. In B. L. Turner II et al. (eds), *The Earth as Transformed by Human Action*. Cambridge: Cambridge University Press, 121–35.

Lowenthal, D. (1991) British national identity and the English landscape. *Rural History*, 2, 205–30.

Lyotard, J-F. (1986) *The Postmodern Condition*. Manchester: University of Manchester Press.

McCally, M. and C. K. Cassel (1990) Medical responsibility and global environmental change. *Annals of Internal Medicine*, 113, 467–73.

McCloskey, J. M. and H. Spalding (1989) A reconnaissance-level inventory of the amount of wilderness remaining in the world. *Ambio*, 18, 4, 221–7.

McCormick, J. S. (1989) *Reclaiming Paradise: the Global Environmental Movement*. Bloomington, IN: Indiana University Press.

McCrone, D. (1992) *Understanding Scotland: the Sociology of a Stateless Nation*. London: Routledge.

McDowell, L. (1991) The baby and the bathwater: deconstruction, diversity and feminist theory in geography. *Geoforum*, 22, 123–34.

McDowell, L. (1992a) Space, place and gender relations, part I: feminist empiricism and the geography of gender relations. *Progress in Human Geography*, 17, 157–79.

McDowell, L. (1992b) Space, place and gender relations, part II: identity, difference, feminist geographies and geometries, *Progress in Human Geography*, 17, 305–18.

McDowell, L. and D. Massey (1984) A woman's place? In D. Massey and J. Allen (eds), *Geography Matters*. Cambridge: Cambridge University Press.

Mace, R. (1976) *Trafalgar Square: Emblem of Empire*. London: Lawrence and Wishart.

McGrew, A. (1992) A global society? In S. Hall, D. Held, and T. McGrew (eds), *Modernity and Its Futures*. Oxford: Polity.

Machimura, T. (1992) The urban restructuring process in Tokyo in the 1980s: transforming Tokyo into a world city. *International Journal of Urban and Regional Research*, 16, 114–28.

Mack, P. (1990) *Viewing the Earth. The Social Construction of the Landsat Satellite*. Cambridge, MA: MIT Press.

McKeown, T. (1988) The *Origins of Human Disease*. Oxford: Blackwell.

Mackinder, H. J. (1904) The geographical pivot of history. *Geographical Journal*, 23, 421–42.

McMichael, P. and D. Myhre (1991) Global regulation versus the nation-state: agro-food systems and the new politics of capital, *Capital and Class*, 43, 83–105.

McNicoll, G. (1992) The United Nations' long-range population projections. *Population and Development Review*, 18, 333–40.

Malcolm, D. (1990) Hollywood is the enemy. *Guardian*, 29 November.

Malthus, R. (1976) *An Essay on the Principle of Population* (reprint). Harmondsworth: Pelican.

Manne, A. S. and R. G. Richels (1992) *Buying Greenhouse Insurance: the Economic Costs of CO_2 Emission Limits*. Cambridge, MA: MIT Press.

Mansell, R. (1993) *The New Telecommunications. A Political Economy of Network Evolution*. London: Sage.

Marglin, F. A. and S. A. Marglin (eds) (1990) *Dominating Knowledge: Development, Culture and Resistance*. Oxford: Clarendon Press.

Marsden T., R. Munton, S. Whatmore and J. Little (1986) Towards a political economy of capitalist agriculture. *International Journal of Urban and Regional Research*, 4, 498–521.

Marshall, B. (ed.) (1991) *The Real World*. Boston: Houghton Mifflin.

Martin, B. (1988) Feminism, criticism and Foucault. In I. Diamond and L. Quinby (eds), *Feminism and Foucault: Reflections on Resistance*. Boston, MA: Northeastern University Press.

Marx, K. (1976) *Capital* (reprint). London: Penguin.

Masquelier, A. (1992) Encounter with a Road Siren. *Visual Anthropology*, 8/1, 56–69.

Massey, D. (1984) *Spatial Divisions of Labour*. London: Macmillan; New York: Methuen.

Massey, D. (1991) A global sense of place. *Marxism Today*, June, pp. 24–9.

Mattelart, A. (1979) *Multinational Corporations and the Control of Culture*. Brighton: Harvester.

Meadows, D., D. Meadows, J. Randers and W. Behrens (1972) *Limits to Growth*. New York: Universe Books.

Meadows, D. H., D. L. Meadows and J. Randers (1992) *Beyond the Limits. Global Collapse or Sustainable Future*? London: Earthscan.

Melucci, A. (1989) *Nomads of the Present*. London: Radius.

Meyer, W. B. and B. L. Turner II (1992) Human population growth and global land-use/land-cover change. *Annual Review of Ecology and Systematics*, 23, 39–61.

Micklin, P. (1988) Desiccation of the Aral Sea: a water management disaster in the Soviet Union. *Science*, 241, 1170–76.

Mingione, E. (ed.) (1993) The new urban poverty and the underclass. Special issue of *International Journal of Urban and Regional Research*, 17, 3.

Mitchell, B. R. (1970) *European Historical Statistics 1750–1970*. London: Macmillan.

Mitchell, B. (1979) *Geography and Resource Analysis*. London: Longman.

Mitchell, B. R. (1983) *International Historical Statistics: the Americas and Australasia*. London: Macmillan.

Momsen, J. and V. Kinnaird (eds) (1992) *Different Places, Different Voices*. London: Routledge.

Moody, R. (ed) (1988) *The Indigenous Voice* (2 vols). London: Zed.

Morley, D. and K. Robins (1989) Spaces of identity: communication technologies and the reconfiguration of Europe. *Screen*, 30, 4, 10–34.

Morris, M. (1988) The Pirate's fiancee: feminists and philosophers, or maybe tonight it'll happen. In I. Diamond and L. Quinby (eds), *Feminism and Foucault: Reflections on Resistance*. Boston MA: Northeastern University Press, pp. 21–42.

Morris, M. (1992) The man in the mirror: David Harvey's "condition" of postmodernity. *Theory Culture and Society*, 10, 253–79.

Mosse, G. L. (1975) *The Nationalization of the Masses*. New York: Howard Fertig.

Mulgan, G. J. (1991) *Communication and Control. Networks and the New Economies of Communication*. Cambridge: Polity Press.

Munasinghe, M. (1993) *Environmental Economics and Sustainable Development*. World Bank Environment Paper No. 3. Washington: World Bank.

Munro, D. A. (1978) The thirty years of IUCN. *Nature and Resources*, 14, 2, 14–18.

Munton R. (1992) The uneven development of capitalist agriculture. The repositioning of agriculture within the food system. In K. Hoggart (ed.), *Agricultural Change, Environment and Economy. Essays in Honour of W. B. Morgan*. London: Mansell, pp. 25–48.

Murley, L. and M. Stevens (eds) (1991) *NSCA Pollution Glossary*. Brighton: National Society for Clean Air and Environmental Protection.

Naess, A. (1989) *Ecology, Community and Lifestyle*. Cambridge: Cambridge University Press.

Nairn, T. (1993) All Bosnians now? *Dissent*, Fall, 403–10.

Nandy, A. (1984) Culture, state and rediscovery of Indian Politics. *Economic and Political Weekly*, 19, 49, 2078–83.

Newson, M. D. (1992a) *Land, Water and Development. River Basin Systems and their Sustainable Management*. London: Routledge.

Newson, M. D. (1992b) Twenty years of systematic physical geography: prospects for a New Environmental Age. *Progress in Physical Geography*, 16, 2, 209–21.

Newson, M. D. (1994) *Scales and their Appropriateness for Integrating*

Land-Use Management: Science, Subsidiarity and Sustainability. Aberdeen: Macauley Land Use Research Institute.

Niess, F. (1990) *A Hemisphere to Itself: a History of US–Latin American Relations*. London: Zed Books.

Nohrstedt, A. (1993) Communication in the risk-society: public relations strategies, the media and nuclear power. In A. Hansen (ed.), *The Mass Media and Environmental Issues*. Leicester: Leicester University Press, 81–104.

Nordhaus, W. D. (1993) Reflections on the economics of climate change. *Journal of Economic Perspectives*, 7, 11–25.

Notestein, F. (1945) Population: the long view. In T. Schultz (ed.), *Food for the World*. Chicago: University of Chicago Press, 36–57.

Nriagu, J. O. and J. S. S. Lakshminarayana (1989) *Aquatic Toxicology and Water Quality Management*. New York: John Wiley and Sons.

OECD (1985) *The State of the Environment*. Paris: OECD.

OECD (1992a) *Trends in International Migration*. Paris: OECD.

OECD (1992b) *Development Cooperation: 1992 Report*. Paris: OECD.

OECD (1993a) *The Changing Course of International Migration*. Paris, OECD.

OECD (1993b) *World Energy Outlook to the Year 2010*. Paris: IEA/OECD.

Offe, C. (1985) New social movements: challenging the boundaries of institutional politics. *Social Research*, 52, 4, 817–68.

Ohmae, K. (1990) *The Borderless World: Power and Strategy in the Interlinked Economy*. London: Collins.

Olalquiaga, C. (1992) *Megalopolis. Contemporary Cultural Sensibilities*. Minneapolis: University of Minnesota Press.

Omara-Ojungu, P. H. (1992) *Resource Management in Developing Countries*. London: Longman.

Ong, A. (1987) *Spirits of Resistance and Capitalist Development: Factory Women in Malaysia*. Binghamton: State University of New York Press.

O'Riordan, T. (1971) *Perspectives on Resource Management*. London: Pion.

O'Riordan, T. (1981) *Environmentalism*, 2nd edn. London: Pion.

O'Riordan, T. (1988a) The politics of sustainability. In R. K. Turner (ed.), *Sustainable Environmental Management: Principles and Practice*. Boulder, CO: Westview Press.

O'Riordan, T. (1988b) The earth as transformed by human action: An international symposium. *Environment*, 30, 1, 25–8.

O'Riordan, T. and R. K. Turner (1983) *An Annotated Reader in Environmental Planning and Management*. Oxford: Pergamon Press.

O'Riordan, T. and S. Rayner (1991) Risk management for global environmental change. *Global Environmental Change*, 1, 2, 91–108.

Paehlke, R. (1989) *Environmentalism and the Future of Progressive Politics*. New Haven, CT: Yale University Press.

Pangle, T. L. (1992) *The Ennobling of Democracy: The Challenge of the Postmodern Age*. Baltimore: Johns Hopkins University Press.

Parnell, P. C. (1992) Time and irony in Manila squatter movements. In C. Nordstrom and J. Martin (eds), *The Paths to Domination, Resistance and Terror*. Berkeley: University of California Press, 154–76.

Pateman, C. and E. Gross (eds) (1986) *Feminist Challenges*. Boston, MA: Northeastern University Press.

Paz, O. (1992) *In Search of the Present*. New York: Harper.

Pearce, D., A. Markyanda and E. Barbier (1989) *Blueprint for a Green Economy*. London: Earthscan.

Peet, R. and M. Watts (1993a) Development theory and environment in an age of market triumphalism. *Economic Geography*, 69, 3, 227–53.

Peet, R. and M. Watts (eds) (1993b) Environment and Development, Parts I and II. *Economic Geography*, July and October, vol. 69, nos 3 and 4.

Pepper, D. (1984) *The Roots of Modern Environmentalism*. London: Croom Helm.

Pepper, D. (1991) *Communes and the Green Vision*. London: Green Print.

Pepper, D. (1993) *Eco-Socialism*. London: Routledge.

Peters, R. H. (1991) *A Critique for Ecology*. Cambridge: Cambridge University Press.

Petrella, R. (1991) World city-states of the future. *New Perspectives Quarterly*, 8, 59–64.

Pezzey, J. (1991) *Impacts of "Greenhouse" Gas Control Strategies on UK Competitiveness*. London: HMSO.

Philo, C. (1989) "Enough to drive one mad": the organisation of space in nineteenth century lunatic asylums. In J. Wolch and M. Dear (eds), *The Power of Geography*. London: Macmillan, 258–90.

Pierson, C. (1991) *Beyond the Welfare State? A New Political Economy of Welfare*. University Park, PA: Pennsylvania State University Press.

Popke, E. J. (1994) Recasting geopolitics: the discursive scripting of the International Monetary Fund. *Political Geography*, 13, 255–69.

Population Reference Bureau (1994) *World Population Data Sheet*. Washington: PRB.

Poole, T. (1993) Star in the East heralds TV revolution. *Independent on Sunday*, 18 April.

Porter, R. (1993) Baudrillard: history, hysteria and consumption. In C. G. Rojek and B. S. Turner (eds), *Forget Baudrillard?* London: Routledge, 1–21.

Portney, K. E. (1991) Public environmental decision making: citizen roles. In R. A. Chechile and S. Carlisle (eds), *Environmental Decision Making. A Multidisciplinary Perspective*. New York: Van Nostrand Reinhold, 195–216.

Powell, N. (1993) French redskins take on the cowboys. *Observer*, 19 September.

Pred, A. and M. J. Watts (1992) *Reworking Modernity: Capitalisms and Symbolic Discount*. Brunswick, NJ: Rutgers University Press.

Prendergast, C. (1992) *Paris in the Nineteenth Century*. Oxford: Blackwell.

Przeworski, A. (1991) *Democracy and the Market: Political and Economic Reforms in Eastern Europe and Latin America*. Cambridge: Cambridge University Press.

Putnam, R. D. (1993a) *Making Democracy Work: Civic Traditions in Modern Italy*. Princeton, NJ: Princeton University Press.

Putnam, R. D. (1993b) The prosperous community: social capital and public life. *The American Prospect*, 13, pp. 35–42.

Rabinow, P. (ed.) (1984) *The Foucault Reader: an Introduction to Foucault's thought*. London: Penguin.

Radcliffe, S. A. and S. Westwood (eds) (1993) *"Viva" Women and Popular Protest in Latin America*. London: Routledge.

Rees, J. (1990) *Natural Resources: Allocation, Economics, and Policy*. London and New York: Routledge.

Reich, R. B. (1991) *The Work of Nations: Preparing Ourselves for 21st-Century Capitalism*. New York: Knopf.

Reid, W. and R. R. Miller (1989) *Keeping options alive. The scientific basis for conserving biodiversity*. Washington DC: World Resources Institute.

Rimmer, P. (1993) Reshaping Western Pacific Rim cities. In K. Fujita and R. C. Hill (eds), *Japanese Cities in the World Economy*. Philadelphia: Temple University Press.

Roberts, N. (ed.) (1994) *The Changing Global Environment*. Oxford: Blackwell.

Robertson, R. (1990) Mapping the global condition. In M. Featherstone (ed.), *Global Culture. Nationalism, Globalization and Modernity*. Newbury Park, CA: Sage, pp. 15–30.

Robertson, R. (1991) Social theory, cultural relativity, and the problem of globality. In A. D. King (ed.), *Culture, Globalization and the World-System*. Basingstoke: Macmillan.

Robins, K. (1989) Reimagined communities? European image spaces, after Fordism. *Cultural Studies*, 3, 2, 145–65.

Robins, K. (1990) Global local times. In J. Anderson and M. Ricci (eds), *Society and Social Science: A Reader*. Milton Keynes, Open University, 196–205.

Robins, K. and F. Webster (1990) Broadcasting politics: communications and consumption. In M. Alvarado and J. O. Thompson (eds), *The Media Reader*. London: British Film Institute, 135–50.

Rodriguez, N. P. and J. R. Feagin (1986) Urban specialization in the world-system. *Urban Affairs Quarterly*, 22, 187–219.

Rogers, R. (1992) The politics of migration in the contemporary world. *International Migration*, 30, 1/2, 33–48.

Rolston, B. (1991) *Politics and Painting: Murals and Conflict in Northern Ireland*. Toronto: Fairleigh Dickinson University Press.

Rorty, R. (1989) *Contingency, Irony and Solidarity*. Cambridge: Cambridge University Press.

Rosenau, P. M. (1992) *Postmodernism and the Social Sciences: Insights, Inroads and Intrusions*. Princeton, NJ: Princeton University Press.

Rosenzweig, C. and D. Hillel (1993) Agriculture in a greenhouse world. *Research and Exploration*, 9, 208–21.

Rosenzweig, C. and M. L. Parry (1994) Potential impact of climate change on world food supply. *Nature*, 367, 133–8.

Rosqvist, R., M. Skurnik and H. Wolf-Watz (1987) AIDS in the pre-AIDS era. *Reviews of Infectious Diseases*, 9, 1102–1108.

Ross, S. (1990) Worldview Address, delivered at the Edinburgh International Television Festival, 26 August.

Rostow, W. (1960) *The Stages of Economic Growth: A Non-Communist Manifesto*. Cambridge: Cambridge University Press.

Routledge, P. (1993) *Terrains of Resistance: Nonviolent Social Movements and the Contestation of Place in India*. Westport: Praeger.

Royal Society of London and the American National Academy of Sciences (1992) *Population Growth, Resource Consumption and a Sustainable World*. London: Royal Society.

Royal Society of London (1994) *Population: the Complex Reality*. London: The Royal Society.

Sachs, W. (ed.) (1992a) *The Development Dictionary*. London: Zed Books.

Sachs, W. (1992b) Introduction. In W. Sachs (ed.), *The Development Dictionary*. London: Zed Books, 1–5.

Sahabat Alam Malaysia (1987) Sarawak – Orang Ulu fight logging. *Cultural Survival Quarterly*, 11, 4, 20–3.

Said, E. (1978) *Orientalism*. London: Routledge; New York: Pantheon.
Samuel, R. (ed.) (1989) *Patriotism: the Making and Unmaking of British National Identity*. London: Routledge.
Sandoval, C. (1991) US, Third World feminism: the theory and method of oppositional consciousness in the postmodern world. *Genders*, 10, 1–24.
Sassen, S. (1991) *The Global City*. Princeton: Princeton University Press.
Sayer, A. and R. Walker (1992) *The New Social Economy: Reworking the Division of Labor*. Cambridge, MA: Blackwell.
Scannell, P. (1989) Public service broadcasting and modern public life. *Media, Culture and Society*, 11, 2, 135–66.
Schiller, H. I. (1991) Not yet the post-imperialist era. *Critical Studies in Mass Communication*. 8, 13–28.
Schivelsbuch, W. (1986) *The Railway Journey. The Industrialisation of Time and Space in the Nineteenth Century*. Berkeley: University of California Press.
Schneider, S. H. and P. J. Boston (1991) *Scientists on Gaia*. Cambridge, MA: MIT Press.
Schoenberger, E. (1988) Multinational corporations and the new international division of labor: a critical appraisal. *International Regional Science Review*, 11, 2, 105–20.
Schumacher, E. F. (1973) *Small is Beautiful*. London: Sphere.
Schumpeter, J. (1952) *Socialism, Capitalism and Democracy*. London: Thames, p. 83.
Schuurman, F. J. (ed.) (1993) *Beyond the Impasse: New Directions in Development Theory*. London: Zed Books.
Scott, A. and Storper, M. (eds) (1993) *Pathways to Industrialization and Regional Development*. London: Routledge.
Sen, A. (1981) *Poverty and Famines*. Oxford: Clarendon.
Sen, A. (1990a) More than 100 million women are missing. *New York Review of Books*, 20 December.
Sen, A. (1990b) Food, economics and entitlements. In J. Dreze and A. Sen (eds), *The Political Economy of Hunger*, Oxford: Clarendon, pp. 35–50.
Senelle, R. (1989) Constitutional reform in Belgium: from unitarism towards federalism. In M. Forsyth (ed.), *Federalism and Nationalism*. Leicester: Leicester University Press.
Seton-Watson, H. (1977) *Nations and States: an Enquiry into the Origins of Nations and the Politics of Nationalism*. London: Methuen.
Seydlitz, R., S. Laska, D. Spain, E. W. Triche and K. L. Bishop (1993) Development and social problems: the impact of the offshore oil industry on suicide and homicide rates. *Rural Sociology*, 58, 1, 93–110.

Sheffer, Gabriel (1986) A new field of study: modern diasporas in international politics. In G. Sheffer (ed.), *Modern Diasporas in International Politics*. London: Croom Helm, 1–14.

Sheridan, A. (1980) *Michel Foucault: The Will to Truth*. New York: Tavistock.

Shiva, V. (1989) *Staying Alive: Women, Ecology and Survival in India*. London: Zed.

Shrestha, N. and J. Patterson (1990) Population and poverty in dependent states. *Antipode*, 22, 93–120.

Shrivastava, P. (1992) *Bhopal. Anatomy of a Crisis*, 2nd edn. London: Paul Chapman.

Simmons, I. G. (1989) *Changing the Face of the Earth*. Oxford: Blackwell.

Simmons, I. G. (1993a) *Interpreting Nature: Cultural Constructions of the Environment*. London: Routledge.

Simmons, I. G. (1993b) *Environmental History: A Concise Introduction*. Oxford: Blackwell.

Simon, J. (1986) Bright global future. In C. Southwick (ed.), *Global Ecology*. Sunderland: Sinauer, pp. 63–8.

Sklair, L. (1991) *Sociology of the Global System*. Baltimore: Johns Hopkins University Press.

Slater, D. (1992) On the borders of social theory: learning from other regions. *Society and Space*, 10, 307–27.

Slater, D. (1993a) The political meanings of development: in search of new horizons. In F. J. Schuurman (ed.), *Beyond the Impasse: New Directions in Development Theory*. London: Zed Books, pp. 93–112.

Slater, D. (1993b) The geopolitical imagination and the enframing of development theory. *Transactions: an International Journal of Geographic Research*, December, 18, pp. 419–37.

Smallman-Raynor, M., A. D. Cliff and P. Haggett (1992) *London International Atlas of AIDS*. Oxford: Basil Blackwell.

Smil, V. (1992) China's environment in the 1980's: some critical changes. *Ambio*, 21, 431–6.

Smith, A. D. (1986) *The Ethnic Origin of Nations*. Oxford: Basil Blackwell.

Smith, F. B. and M. J. Clark (1989) *Airborne Debris from the Chernobyl Incident*. Met. Sci. Paper 2, London: HMSO.

Smith, H. B. and D. Tirpak (eds) (1989) *Potential Effects of Global Climatic Change on the United States: Volume G. Health*. Washington, DC: US Government Printing Office (Environmental Protection Agency, EPA 230–05–89–057).

Smith, M. P. (1994) Can you imagine? Transnational migration and the globalization of grassroots politics. In P. Knox and P. J. Taylor (eds),

World Cities in a World-System. Cambridge: Cambridge University Press.

Smith, N. (1993) Homeless/global: scaling places. In J. Bird et al. (eds), *Mapping the Futures*, London: Routledge.

Smith, T. F., A. Srinivasan, G. Schochetman, M. Marcus and G. Myers (1988) The phylogenetic history of immunodeficiency viruses. *Nature*, 333, 573–5.

Snoddy, R. (1993) The film that can erase itself. *Financial Times*, supplement on Cable TV and Satellite Broadcasting, 6 October.

Soper, K. (1990) *Troubled Pleasures: Writings on Politics, Gender and Hedonism*. London: Verso.

Soper, K. (1993) Postmodernism, subjectivity and the question of value. In J. Squires (ed.), *Principled Positions: Postmodernism and the Rediscovery of Value*. London: Lawrence and Wishart.

Sorkin, M. (1992) *Variations on a Theme Park. The New American City and the End of Public Space*. New York: Hill and Wang.

Spivak, G. C. (1988) Can the subaltern speak? In C. Nelson and L. Grossberg (eds), *Marxism and the Interpretation of Culture*. Urbana: University of Illinois Press, pp. 271–313.

Spivak, G. C. (1991) Identity and alterity: an interview. *Arena*, 97, 65–76.

Springborg, P. (1992) *Western Republicanism and the Oriental Prince*. Cambridge: Polity Press.

Squires, J. (ed.) (1993) *Principled Positions: Postmodernism and the Rediscovery of Value*. London: Lawrence and Wishart.

Stewart, I. A. (1988) The Brundtland Commission: pathways to sustainable development. In A. Davidson and M. Dence (eds.), *The Brundtland Challenge and the Cost of Inaction*. Halifax, NS: Royal Society of Canada and Institute for Research on Public Policy, 117–22.

Stoddart, D. R. (1986) *On Geography and its History*. Oxford: Blackwell.

Stone, J. C. (1980) Imported technology for alien purposes: the case of the land surveyor in North-Western Rhodesia. In J. C. Stone (ed.), *Experts in Africa*. Aberdeen: Aberdeen University African Studies Group, 29–42.

Storper, M. and R. Walker (1989) *The Capitalist Imperative. Territory, Technology and Industrial Growth*. Oxford: Basil Blackwell.

Strange, S. (1982) Cave! Hic Dragones: a critique of regime analysis. *International Organization*, 36, 2, 479–96.

Strangeland, P. (1984) Getting rich slowly – the social impact of oil activities. *Acta Sociologica*, 27, 3, 215–37.

Strassoldo, R. (1992) Globalism and localism: theoretical reflections and some evidence. In Z. Mlinar (ed.), *Globalization and Territorial Identities*. Aldershot: Avebury, 35–59.

Strohmeyer, J. (1993) *Extreme Conditions: Big Oil and the Transformation of Alaska*. New York: Simon and Schuster.

Sturken, M. (1991) The wall, the screen, and the image: The Vietnam Veterans Memorial. *Representations*, 35, 118–42.

Summers, L. (1992) Let them eat pollution. *The Economist*, 8 February, p. 66.

Susman, P. (1989) Exporting the crisis: US agriculture and the third world. *Economic Geography*, 65, 4, 293–313.

Taussig, M. (1993) *Mimesis and Alterity*. London: Routledge.

Taylor, A. (1992) *Choosing our Future. A Practical Politics of the Environment*. London: Routledge.

Taylor, C. (1992) *Multiculturalism and the "Politics of Recognition."* An essay with commentary by A. Gutmann (ed.) et al. Princeton: Princeton University Press.

Taylor, P. J. (1993a) *Political Geography: World-Economy, Nation-State and Locality*, 3rd edn. Harlow: Longman.

Taylor, P. J. (1993b) Geopolitical world orders. In P. J. Taylor (ed.), *Political Geography of the Twentieth Century*. London: Belhaven.

Taylor, P. J. (1994) The state as container: territoriality in the modern world-system. *Progress in Human Geography*, 18, 151–62.

Taylor, P. W. (1986) *Respect for Nature*. Princeton: Princeton University Press.

Thomas, W. L., Jr (ed.) (1956) *Man's Role in Changing the Face of the Earth*. Chicago: University of Chicago Press.

Thompson, E. P. (1987) The rituals of enmity. In D. Smith and E. P. Thompson (eds), *Prospectus for a Habitable Planet*. London: Penguin.

Thrift, N. J. (1990) Transport and communication 1730–1914. In R. L. Dodgshon and R. Butlin (eds), *A New Historical Geography of England and Wales*, 2nd edn. London: Academic Press, 453–86.

Thrift, N. J. (1994) On the social and cultural determinants of international financial centres. In S. Corbridge, N. J. Thrift and R. L. Martin (eds), *Money, Power and Space*. Oxford: Blackwell.

Thrift, N. J. and A. Leyshon (1994) A phantom state? The detraditionalisation of money, the international financial system and international financial centres. *Political Geography*, 13, 299–327.

Thurow, L. (1992) *Head to Head: the Coming Economic Battle among Japan, Europe and America*. New York: Warner.

Tickell, A. and J. Peck (1992) Accumulation, regulation, and the geographies of post-Fordism. *Progress in Human Geography*, 16, 2, 190–218.

Tiffen, M., M. Mortimore and F. Gichuki (1994) *More People, Less Erosion*. Chichester: Wiley.

Todd, E. (1987) *The Causes of Progress; Culture, Authority and Change*. Oxford: Blackwell.

Touraine, A. (1985) An introduction to the study of social movements. *Social Research*, 52, 4, 749–87.

Toye, J. (1993) *Dilemmas of Development*, 2nd edn. Oxford: Blackwell.

Trustees for Alaska (1993) *Trustees Battles Lease Sale Adjacent to Arctic Refuge*. Summer.

Tucker, C. J., H. E. Dregne and W. W. Newcomb (1991) Expansion and contraction of the Sahara Desert from 1980 to 1990. *Science*, 253, 299–301.

Turner, B. L., II (1991) Comment on Ponting's view of environment and prehistory. *Environment*, 23, 4, 2–3, 45.

Turner, B. L. et al. (eds) (1990) *The Earth as Transformed by Human Action: Global and Regional Changes in the Biosphere over the Past 300 Years*. Cambridge: Cambridge University Press.

Turner, B. L., II, and K. W. Butzer (1992) The Columbian encounter and land-use change. *Environment*, 43, 16–20.

Turner, B. L., II, R. H. Moss, and D. L. Skole (eds) (1993) *Relating Land Use and Global Land-Cover Change: a Proposal for an IGBP-HDP Core Project*. IGBP Report No. 24 & HDP Report No. 5. Stockholm.

Turner, S., R. Hanham, and A. Portararo (1977) Population pressure and agricultural intensity. *Annals of the Association of American Geographers*, 67, 386–7.

United Nations (ed.) (1977) *Desertification: its Causes and Consequences*. Oxford: Pergamon Press.

United Nations (1993) *World Investment Report*. New York: United Nations.

UNCTC (1983/1988) *Transnational Corporations in World Development: Third Survey*. New York: United Nations.

United Nations Development Program (1992, 1993) *Human Development Report*. New York, Washington, and Oxford: Oxford University Press for the UNDP.

United Nations Environment Program (1991) *Environmental Data Report 1991*. Oxford: UNEP/Basil Blackwell.

United States Bureau of Land Management (1991) *Alaska Minerals: Facts, Figures and Trivia*. Anchorage: Bureau of Land Management, Alaska State Office, Division of Mineral Resources.

United States Committee for Refugees (1993) *World Refugee Survey 1993*. Washington DC: US Committee for Refugees.

United States Council on Environmental Quality (1982) Environmental Quality 1981: 12th Annual Report of the Council on Environmental Quality. Washington, DC: USCEQ.

Urry, J. (1990) *The Tourist Gaze*. London: Sage.

Valentine, G. (1993) Negotiating and managing multiple sexual identities: Lesbian time-space strategies. *Transactions, Institute of British Geographers*, N.S. 18, 237–48.

Vernon, R. (1992) Transnational corporations: where are they coming from, where are they headed? *Transnational Corporations*, 1, 2, 7–35.

Visvanathan, S. (1985) From the annals of the laboratory state. *Lokayan Bulletin*, 3, 4, 23–47.

Waddington, C. H. (1975) The origin. In E. B. Worthington (ed.), *The Evolution of the IBP*. Cambridge: Cambridge University Press, pp. 4–11.

Wagner-Pacifini, R. and B. Schwartz (1991) The Vietnam Veterans Memorial: commemorating a difficult past. *American Journal of Sociology*, 97, 376–420.

Walker, R. B. J. (1993) *Inside/Outside. International Relations as Political Theory*. Cambridge: Cambridge University Press.

Wallace, I. (1985) Towards a geography of agribusiness. *Progress in Human Geography*, 9, 491–514.

Wallach, B. (1991) *At Odds with Progress: Americans and Conservation*. Tucson: University of Arizona Press.

Wallerstein, I. (1974) *The Modern World-system: Capitalist Agriculture and the Origins of the European World-Economy in the Sixteenth Century*. New York: Academic Press.

Wallerstein, I. (1984) *Politics of the World-Economy*. Cambridge: Cambridge University Press.

Ward, M. (1983) *Unmanageable Revolutionaries: Women in Irish Nationalism*. Dingle: Brandon Press.

Waring, M. (1988) *If Women Counted*. New York: Harper & Row.

Warner, M. (1985) *Monuments and Maidens: the Allegory of the Female Form*. London: Picador.

Watts, M. J. (1983) *Silent Violence: Food, Famine and Peasantry in Northern Nigeria*. Berkeley, CA: University of California Press.

Watts, M. J. (1984) State, oil, and accumulation: from boom to crisis. *Environment and Planning D: Society and Space*, 2, 403–28.

Watts, M. J. (1991a) Visions of excess. *Transition*, 51, 124–41.

Watts, M. J. (1991b) Entitlements or empowerments? *Review of African Political Economy*, 51, 9–26.

Watts, M. J. (1992a) Capitalisms, crises, and cultures, I: notes toward a totality of fragments. In A. Pred and M. J. Watts, *Reworking Modernity*. New Brunswick, NJ: Rutgers University Press.

Watts, M. J. (1992b) Space for everything (a commentary). *Cultural Anthropology*, 7, 115–29.

Watts, M. J. and H. Bohle (1993) The space of vulnerability. *Progress in Human Geography*, 17, 43–68.

Weale, A. (1992) *The New Politics of Pollution*. Manchester: Manchester University Press.

Weihe, W. H. (1979) Climate, health and disease. *Proceedings of the First World Climate Conference*. Geneva: World Meteorological Organization, Paper No. 537.

Westwood, J. C. N. (1980) *The Hazard from Dangerous Exotic Diseases*. London: Macmillan.

Whatmore, S. (1991) *Farming Women: Gender, Work and Family Enterprise*. London: Macmillan.

Whitmore, T. M. and B. L. Turner II (1992) Landscapes of cultivation in Mesoamerica on the eve of the conquest. *Annals of the Association of American Geographers*, 82, 402–25.

Widgren, J. (1993) Movements of refugees and asylum seekers: recent trends in a comparative perspective. In OECD, *The Changing Course of International Migration*. Paris: OECD, pp. 87–96.

Williams, C. H. and A. D. Smith (1983) The national construction of social space. *Progress in Human Geography*, 7, 502–18.

Williams, F. (1989) *Social Policy: a Critical Introduction*. Cambridge: Polity.

Williams, M. (1990) Forests. In B. L. Turner et al., *The Earth as Transformed by Human Action*. Cambridge: University of Cambridge Press.

Williams, R. (1983) *Keywords*. London: Fontana.

Williams, R. (1985) *The Country and the City*. London: Chatto & Windus.

Williams, R. (1990) *Notes on the Underground*. Cambridge, MA: MIT Press.

Wilmoth, J. and P. Ball (1992) The population debate in American popular magazines, 1946–90. *Population and Development Review*, 18, 631–68.

Winram, S. (1984) The opportunity for world brands. *International Journal of Advertising*, 3, 7, 17–26.

Wolin, S. S. (1960) *Politics and Vision: Continuity and Innovation in Western Political Thought*. Boston: Little Brown.

Wolin, S. S. (1989) *The Presence of the Past: Essays on the State and the Constitution*. Baltimore: Johns Hopkins University Press.

Woodehouse, T. (1990) *Replenishing the Earth, The Right Livelihood Awards, 1986–1989*. Bideford: Green Books.

Woods, R. (1982) *Theoretical Population Geography*. London: Longman.

Woods, R. (1989) Malthus, Marx and population crises. In R. J. Johnston and P. J. Taylor (eds), *A World in Crisis?* Oxford: Blackwell, 151–74.

World Bank (1986/1992/1993) *World Development Report*. Oxford: Oxford University Press.

World Bank (1992) *World Development Report 1992: Development and the Environment*. New York: Oxford University Press for the World Bank.

World Health Organisation Commission on Health and the Environment (1992) *Report of the Panel on Food and Agriculture*. Geneva: WHO.

World Resources Institute (1991, annual) *World Resources*. Oxford: Oxford University Press.

Worster, D. (1985) *Nature's Economy, a History of Ecological Ideas*. Cambridge: Cambridge University Press.

Worthington, E. B. (1983) *The Ecological Century: a Personal Appraisal*. Cambridge: Cambridge University Press.

Wright, P. (1985) *On Living In an Old Country*. London: Verso.

Wynne, B. (1990) Sheepfarming after Chernobyl: a case study in communicating scientific information. In H. Bradby (ed.), *Dirty Words. Writings on the History and Culture of Pollution*. London: Earthscan, 139–60.

Wynne, B. (1992) Uncertainty and environmental learning. Reconceiving science and policy in the preventive paradigm. *Global Environmental Change*, 2, 2, 111–27.

Young, I. M. (1990) *Justice and the Politics of Difference*. Princeton; N.J.: Princeton University Press.

Young, J. E. (1992) *Mining the Earth*. Worldwatch Paper 109. Washington, DC: Worldwatch Institute.

Zimmermann, E. W. (1951) *World Resources and Industries*. New York: Harper.

Contributors

JOHN AGNEW is Professor of Geography in the Maxwell School of Citizenship and Public Affairs, Syracuse University. His major research interests are in international political economy, political geography, and the urban geography of Italy. His books include *Place and Politics* (Allen and Unwin, 1986) and *The United States in the World-Economy* (Cambridge University Press, 1987), and he is co-author of *The Geography of the World Economy* (Edward Arnold, 1994).

W. M. ADAMS is a Lecturer in Geography at the University of Cambridge. His research concerns environment, conservation, and development in the UK and the Third World. His books include *Green Development: Environment and Sustainability in the Third World* (Routledge, 1990) and *Wasting the Rain: Rivers, People and Planning in Africa* (Earthscan, 1992).

RICHARD BARFF is Associate Professor and Chair in the Geography Department at Dartmouth College. His main research interests center on analyzing the linkages between international and internal systems of labor exchange and immigration, cycles and instability in New England's economy, and the geography of low-technology industry.

GAVIN BRIDGE is a graduate student at Clark University. He is engaged on a project to investigate the social and environmental impacts of economic and political transition in the former Soviet Union.

SUSAN CHRISTOPHERSON is an Associate Professor in the Department of City and Regional Planning, Cornell University. Her primary research field is the organization and regulation of labor markets, where she has published widely. She is actively engaged in international policy research for the United Nations Conference on Trade and Development, and the Development Working Party on Women's Role in the Economy.

ANDREW CLIFF is Reader in Theoretical Geography at the University of Cambridge. His research focuses on spatial analysis and medical geography. He is co-author of *Atlas of Disease Distributions* (Blackwell, 1988) and *Atlas of AIDS* (Blackwell, 1992). He directs a major Wellcome Trust grant on disease change with Peter Haggett.

JODY EMEL is an Associate Professor in the Geography Department and is Director of the Center for Land Water and Society in the George Perkins Marsh Institute at Clark University. Most of her research is within two broad areas: the cultural and economic history of resource production regions and the social and legal theory of resource development and allocation institutions.

ALLAN FINDLAY is Professor of Geography at the University of Dundee and has worked as Research Associate to the UN International Labor Office, Geneva. His research interests are international labor migration, population and development, and applied population research. He is editor of the *Scottish Geographical Magazine*.

PETER HAGGETT is Professor of Urban and Regional Geography in the University of Bristol. His research focuses on spatial analysis and medical geography. He is co-author of *Atlas of Disease Distributions* (Blackwell, 1988) and *Atlas of AIDS* (Blackwell, 1992). He directs a major Wellcome Trust grant on disease change with Andrew Cliff.

NUALA L. JOHNSON is a Lecturer in Geography at University College, London. Her research focuses on nation-building, the iconography of public monuments, and the heritage industry in Ireland.

R. J. JOHNSTON is Vice-Chancellor of the University of Essex. From 1974 to 1992 he was Professor of Geography at the University of Sheffield. His recent works include *Environmental Problems* (Belhaven, 1989), *A Question of Place* (Blackwell, 1991), and, as co-editor, *A World in Crisis?* (Blackwell, 1991) and *The Dictionary of Human Geography* (Blackwell, 1994).

NURIT KLIOT is an Associate Professor in the Department of Geography, University of Haifa. Her research focuses upon the politics of water resources, refugees and migration, and the environment of Cyprus. She has recently published *Water Resources and Conflict in the Middle East* (Routledge, 1994).

PAUL KNOX is Associate Dean of the College of Architecture and Urban Studies and Professor of Urban Affairs at Virginia Polytechnic Institute and State University. His research focuses on urbanization and global change. Recent publications include *The Restless Urban Landscape* (Prentice Hall, 1993), *Urbanization* (Prentice Hall, 1994), and, as co-author, *The Geography of the World Economy* (Edward Arnold, 1994).

LINDA MCDOWELL is a Lecturer in Geography at the University of Cambridge and a Fellow of Newnham College. Her main interests are in the social and economic restructuring of contemporary Britain, especially the feminization of the labor force, and in feminist theory. She is co-author of *Landlords and Property* (Cambridge University Press, 1989) and co-editor of *Defining Women* (Polity, 1993).

WILLIAM B. MEYER is Research Assistant Professor in the George Perkins Marsh Institute at Clark University. His interests include the human dimensions of global change and environmental history. He is co-editor of *Changes in Land Use and Land Cover* (Cambridge University Press, 1994).

MALCOLM NEWSON is Professor of Physical Geography at Newcastle University. His research interests are hydrology and environmental management, with emphasis on the integrated management of river basins. His two most recent books are *Land, Water and Development* (Routledge, 1992) and *Hydrology and the River Environment* (Oxford University Press, 1994).

JOE PAINTER is a Lecturer in Geography at the University of Durham. His research interests are in geographical aspects of state and local governance, labor relations, and the culture of political institutions. He has published on the restructuring of the local state in Britain and the use of regulation theory in analyzing political change.

SUSAN M. ROBERTS is an Assistant Professor in the Department of Geography, University of Kentucky. Her main field of research is contemporary global political economy, particularly the international finance system. This includes work on offshore financial centers in the Caribbean.

KEVIN ROBINS is a Lecturer in Geography affiliated to the Centre for Urban and Regional Development Studies, Newcastle University. He has been part of the Economic and Social Research Council's Programme on Information and Communications Technology and is author of *Geografia dei Media* (Baskerville, 1993) and co-editor of *The Regions, the Nations and the BBC* (British Film Institute, 1993).

PAUL ROUTLEDGE is Leverhulme Fellow in the Geography Department, University of Bristol. His research interests lie in development, culture, and resistance. He is author of *Terrains of Resistance: Nonviolent Social Movements and the Contestation of Place in India* (Praeger, 1993).

DAVID SLATER is Professor of Social and Political Geography at Loughborough University of Technology. He researches broadly on the geography and politics of development. He is author of *Territory and State Power: the Peruvian Case* (Macmillan, 1989) and guest editor of *Social Movements and Political Change in Latin America* (Latin American Perspectives, Sage, 1994).

PETER J. TAYLOR is Professor of Political Geography at Newcastle University. His research interests focus on the changing nature of the state and comparative hegemonies. His books include *Britain and the Cold War: 1945 as Geopolitical Transition* (Pinter, 1990), *Political Geography: World-Economy, Nation-State and Locality* (Longman, 1993), and he has edited *Political Geography of the Twentieth Century* (Belhaven, 1993) and co-edited *World in Crisis?*(Blackwell, 1989).

NIGEL THRIFT is Professor of Geography at the University of Bristol. His research interests include the social and cultural determinants of the international financial system. geographies of financial exclusion, social and cultural theory and countries of the Pacific Basin. He has co-edited *The Socialist Third World* (Blackwell, 1987), *Class and Space: the Making of Urban Society* (Routledge, 1987), *New Models in Geography* (Unwin-Hyman, 1989) and *Money, Power and Space* (Blackwell, 1994).

B. L. TURNER II is Professor of Geography and Director of the George Perkins Marsh Institute at Clark University. He has interests in nature-society relationships ranging from ancient Maya agriculture and environment to contemporary agricultural change in the tropics and global land-use change and has written *The Earth Transformed by Human Action* (Cambridge University Press, 1990) and is co-editor of *Changes in Land Use and Land Cover* (Cambridge University Press, 1994). He is co-chair of the joint International Geosphere-Biosphere Programme and Human Dimensions Programme planning a core research project on global land-use/cover changes.

MICHAEL J. WATTS is Director of the Institute of International Studies and Professor of Geography at the University of California, Berkeley. His research interests are in African agrarian change and on the role of local, national and international power and politics in Third World development. He is author of *Silent Violence: Food, Famine and Peasantry in Northern Nigeria*

(University of California Press, 1983) and co-author of *Reworking Modernity* (Rutgers University Press, 1992).

SARAH WHATMORE is Reader in Human Geography at the University of Bristol. Her research interests include agriculture, property and nature. She is currently engaged in an Economic and Social Research Council research fellowship in their Global Environmental Change Programme. She has published *Farming Women: Gender, Work and Family Enterprise* (Macmillan, 1991).

Further Reading

Further reading to chapter 2: A Hyperactive World

Castells, M. (1989) *The Informational City*. Oxford: Blackwell.
Clark, G. L. (ed.) (1992) Special issue on "real" regulation, *Environment and Planning A*, 24, 5.
Harvey, D. W. (1989) *The Condition of Postmodernity*. Oxford: Blackwell.
Jameson, F. (1991) *Postmodernism*. London: Verso.
Law, J. (1994) *Organizing Modernity*. Oxford: Blackwell.

Further reading to chapter 3: From Farming to Agribusiness: the Global Agro-food System

Bernstein, H., B. Crow, M. Mackintosh, and C. Martin (eds) (1990) *The Food Question: Profits versus People*.
Friedland, W., L. Busch, F. Buttel, and A. Rudy (eds) (1991) *Towards a New Political Economy of Agriculture*. Boulder, CO: Westview.
Goodman, D. and M. Redclift (1991) *Refashioning Nature*. London: Routledge.
Munton, R. (1992) The uneven development of capitalist agriculture. In K. Hoggart (ed.), *Agricultural Change, Environment and Economy*. London: Mansell.
Political Geography (1993) The globalization of agriculture, vol. 12, part 3 (special issue).
Sociologia Ruralis (1989) Agricultural change, vol. 23, part 2 (special issue).

Further reading to chapter 4: Multinational Corporations and the New International Division of Labor

Dicken, P. (1992) *Global Shift: the Internationalization of Economic Activity*. New York: Guilford.

Dunning, J. H. (1992) *Multinational Enterprises and the Global Economy*. Reading, MA: Addison-Wesley.

Kenney, M. and R. Florida (1993) *Beyond Mass Production: the Japanese System and its Transfer to the US*. New York: Oxford University Press.

Noponen, H., J. Graham, and A. R. Markusen (eds) (1993) *Trading Industries, Trading Regions*. New York: Guildford.

Further reading to chapter 5: Trajectories of Development Theory: Capitalism, Socialism, and Beyond

Corbridge, S. (1993) Colonialism, post-colonialism and the political geography of the Third World. In P. J. Taylor (ed.), *Political Geography of the Twentieth Century*. London: Belhaven.

Hettne, B. (1990) *Development Theory and the Three Worlds*. London: Longman.

Sachs, W. (ed.) (1992) *The Development Dictionary*. London: Zed.

Schuurman, F. (ed.) (1993) *Beyond the Impasse: New Directions in Development Theory*. London: Zed.

Slater, D. (1993) The geopolitical imagination and the enframing of development theory, *Transactions: an International Journal of Geographic Research*, N.S. 18, 419–37.

Further reading to chapter 6: Democracy and Human Rights after the Cold War

Attfield, R. and B. Wilkins (eds) (1992) *International Justice and the Third World*. London: Routledge.

Christie, D. (1984) Recent calls for economic democracy, *Ethics*, 95, 112–28.

Dunn, J. (1993) *Western Political Theory in the Face of the Future*, 2nd edn. Cambridge: Cambridge University Press.

Lukes, S. (1993) Five fables about human rights, what it would be like if . . . , *Dissent*, Fall, 427–37.

Mansbridge, J. (1990) Feminism and democracy, *The American Prospect*, 1, 126–39.

Further reading to chapter 7: The Renaissance of Nationalism

Anderson, B. (1983) *Imagined Communities*. London: Verso.
Bhabha, H. K. (ed.) (1990) *Nation and Narration*. London: Routledge.
Daniels, S. (1993) *Fields of Vision: Landscape Imagery and National Identity in England and the United States*. Cambridge: Polity.
Hobsbawm, E. (1990) *Nations and Nationalism since 1780*. Cambridge: Cambridge University Press.
Warner, M. (1985) *Monuments and Maidens: the Allegory of the Female Form*. London: Picador.

Further reading to chapter 8: Global Regulation and Trans-state Organization

Europa World Yearbook (annual). London: Europa.
Diehl, P. F. (ed.) (1989) *The Politics of International Organizations: Patterns and Insights*. Chicago: Dorsey.
Kapstein, E. B. (1992) Between power and purpose: central bankers and the politics of regulatory convergence, *International Organization*, 46, 265–87.
Statesman's Year Book (annual). New York: St. Martin's.

Further reading to chapter 9: The Regulatory State: the Corporate Welfare State and Beyond

Esping-Anderson, G. (1990) *The Three Worlds of Welfare Capitalism*. Oxford: Polity.
Jessop, B. et al. (eds) (1991) *The Politics of Flexibility*. London: Elgar.
McLennan, G., D. Held, and S. Hall (eds) (1984) *The Idea of the Modern State*. Milton Keynes: Open University Press.
Offe, C. (1984) *Contradictions of the Welfare State*. London: Hutchinson.
Pierson, C. (1991) *Beyond the Welfare State*. University Park, PA: Pennsylvania State University Press.
Williams, F. (1989) *Social Policy: a Critical Introduction*. Oxford: Polity.

Further reading to chapter 10: Population Crises: the Malthusian Specter?

Findlay, A. (1991) Population and environment: reproduction and production. In P. Sarre (ed.), *Environment, Population and Development*, 3. (London: Hodder and Stoughton, p. 38.)
Jones, H. (1990) *Population Geography*, 2nd edn. London: Paul Chapman, ch. 7.
Royal Society (1994) *Population: the Complex Reality*, Section 3: Demographic transition in a gender perspective. London: Royal Society.
Woods, R. (1989) Malthus, Marx and population crises. In R. Johnston and P. Taylor (eds), *A World in Crisis?* Oxford: Blackwell, pp. 151–74.

Further reading to chapter 11: Global Migration and Ethnicity: Contemporary Case-Studies

Kubat, Daniel (ed.) (1993) *The Politics of Migration Policies*. New York: Center for Migration Studies.
Loesher, Gil and Leila Monahan (1990) *Refugees and International Relations*. Oxford: Clarendon House.
Refugee Participation Network (1993) *Refugees in Europe*, RPN no. 14. Oxford: Refugee Studies Programme.
Weiner, Myron (1993) *International Migration and Security*. Boulder, San Francisco, and Oxford: Westview Press.
de Zayas, Alfred Maurice (1988) A historical survey of Twentieth Century expulsions. In Anna C. Bramwell (ed.), *Refugees in the Age of Total War*. London: Unwin Hyman, pp. 15–33.

Further reading to chapter 12: Changing Women's Status in a Global Economy

Beneria, Lourdes and Shelley Feldman (eds) (1992) *Unequal Burden: Economic Crisis, Persistent Poverty, and Women's Work*. Boulder, CO: Westview Press.
Cockburn, Cynthia and Susan Ormrod (1993) *Gender and Technology in the Making*. London: Sage.
Enloe, Cynthia (1989) *Bananas, Beaches and Bases: Making Feminist Sense of International Politics*. Berkeley: University of California Press.
McDowell, Linda (1991) Life without father and Ford: the new gender

order of Post-Fordism, *Transactions Institute of British Geographers*, N.S. 16, 4, 400–419.

Momsen, Janet (1991) *Women and Development in the Third World.* New York: Routledge.

Rubery, Jill (ed.) (1988) *Women and Recession.* New York: Routledge.

Further reading to chapter 13: Disease Implications of Global Change

Cliff, A. and P. Haggett (1988) *Atlas of Disease Distributions: Analytical Approaches to Epidemiological Data.* Oxford: Blackwell.

Institute of Medicine (1992) *Emerging Infections: Microbial Threats to Health in the United States.* Washington, DC: National Academy Press.

Smallman-Raynor, M., A. Cliff, and P. Haggett (1992) *London International Atlas of AIDS.* Oxford: Blackwell.

Steffen, R. et al. (1988) *Travel Medicine.* Berlin: Springer Verlag.

Further reading to chapter 14: World Cities and the Organization of Global Space

Featherstone, M. (ed.) (1990) *Global Culture.* Newbury Park, CA: Sage.

Friedmann, J. (1986) The world city hypothesis, *Development and Change*, 17, 69–83.

King, A. D. (1990) *Global Cities.* London: Routledge.

Knox, P. L. and P. J. Taylor (eds) (1985) *World Cities in a World-System.* Cambridge: Cambridge University Press.

Sassen, S. (1991) *Global City.* Princeton, NJ: Princeton University Press.

Sklair, L. (1991) *The Sociology of the Global System.* Baltimore: Johns Hopkins University Press.

Sudjic, D. (1992) *The 100-Mile City.* London: André Deutsch.

Further reading to chapter 15: The New Spaces of Global Media

Featherstone, Mike (ed.) (1990) *Global Culture.* London: Sage.

Nordenstreng, Kaarle and Herbert I. Schiller (eds) (1993) *Beyond National Sovereignty: International Communication in the 1990s.* Norwood, NJ: Ablex.

Schiller, Herbert I. (1993) *Mass Communications and American Empire*, 2nd edn. Boulder, CO: Westview Press.

Schneider, Cynthia and Brian Wallis (eds) (1988) *Global Television.* New York: Wedge.

Further reading to chapter 16: Resisting and Reshaping the Modern: Social Movements and the Development Process

Elkins, P. (1992) *A New World Order*. New York: Routledge.
Hecht, S. and A. Cockburn (1990) *The Fate of the Forest*. New York: Harper Collins.
Sachs, W. (ed.) (1992) *The Development Dictionary*. London: Zed.
Said, E. (1979) *Orientalism*. New York: Vintage.
Shiva, V. (1989) *Staying Alive: Women, Ecology and Survival in India*. London: Zed.

Further reading to chapter 17: Understanding Diversity: the Problem of/for "Theory"

Gregory, D. (1994) *Geographical Imaginations*. Oxford: Blackwell.
Grossberg, L., C. Nelson, and P. Treichler (eds) (1992) *Cultural Studies*. London: Routledge.
Harvey, I. (1989) *The Condition of Postmodernity*. Oxford: Blackwell.
Johnston, R. J. et al. (1994) *The Dictionary of Human Geography*. Oxford: Blackwell.
McDowell, L. (1992) Space, place and gender relations, parts I and II, *Progress in Human Geography*, 17, 157–79 and 305–18.
Rose, G. (1993) *Feminism and Geography*. Oxford: Polity.
Squires, J. (ed.) *Principled Positions*. London: Lawrence and Wishart.

Further reading to chapter 18: The Earth Transformed: Trends, Trajectories, and Patterns

Clark, W. C. and R. E. Munn (eds) (1986) *Sustainable Development of the Biosphere*. Cambridge: Cambridge University Press.
Goudie, A. (1993) *The Human Impact on the Natural Environment*, 4th edn. Oxford: Blackwell; Cambridge, MA: MIT Press.
Kasperson, J. X., R. E. Kasperson, and B. L. Turner, II (eds) (1994) *Regions at Risk: Comparisons of Threatened Environments*. Tokyo: United Nations University Press.
Stern, P. C., O. R. Young, and D. Druckman (1992) *Global Environmental Change: Exploring the Human Dimensions*. Washington, DC: National Academy Press.
Turner, B. L., II, W. C. Clark, R. W. Kates, J. F. Richards, J. Y. Mathews, and W. B. Meyer (eds) (1990) *The Earth as Transformed by Human Action*. Cambridge: Cambridge University Press.

Further reading to chapter 19: The Earth as Input: Resources

Gedicks, A. (1992) *The New Resource Wars: Native and Environmental Struggles against Multinational Corporations*. Boston: South End Press.

Proceedings of the National Academy of Sciences (1992) Industrial ecology: papers from a colloquium, pp. 793–1148.

Rees, J. (1990) *Natural Resources: Allocation, Economics and Policy*. London: Routledge.

Wescoat, J. (1991) Resource management: the long-term global trend, *Progress in Human Geography*, 15, 81–93.

Further reading to chapter 21: Sustainable Development?

Adams, W. M. (1990) *Green Development: Environment and Sustainability in the Third World*. London: Routledge.

Holmberg, J., K. Thornson, and L. Timberlake (1993) *Facing the Future: Beyond the Earth Summit*. London: Earthscan.

McCormick, J. (1989) *Reclaiming Paradise: the Global Environmental Movement*. London: Belhaven.

Redclift, M. (1987) *Sustainable Development: Exploring the Contradictions*. London: Methuen.

Index

Note: Figures and Tables are in *italic*.